GEOMETRY IN
CONDENSED MATTER PHYSICS

GEOMETRY IN CONDENSED MATTER PHYSICS

Edited by **J. F. Sadoc**

Laboratoire de Physique des Solides
Université de Paris-Sud

World Scientific
Singapore • New Jersey • London • Hong Kong

Published by

World Scientific Publishing Co. Pte. Ltd.

P O Box 128, Farrer Road, Singapore 9128

USA office: 687 Hartwell Street, Teaneck, NJ 07666

UK office: 73 Lynton Mead, Totteridge, London N20 8DH

Library of Congress Cataloging-in-Publication Data

Geometry in condensed matter physics / J.F. Sadoc, editor ;
   contributions by Y. Bouligand . . . [et al.].
       p.   cm. –– (Directions in condensed matter physics; vol. 9)
   ISBN 9810200897
    1. Condensed matter.   2. Geometry.   I. Sadoc, J. F.   II. Series.
   QC173.4.C65G46   1990
   530.4'1––dc20   90–45518
                                                        CIP

Printed in Singapore by Utopia Press.

FOREWORD

The study of condensed matter was at first dominated by the "crystal", because its atomic structure was known very precisely through X-ray diffraction. The power of that experimental method relies upon the rigorous periodicity of the crystalline lattice.

When this lattice is absent, or not well defined, as in amorphous or ill-crystallised materials, one is unable to obtain directly an image of the atomic structure. The only information which can be drawn from the diffraction patterns is the "Radial Distribution Function", which is insufficient to describe the atomic structure.

The disordered states of solids become more and more important in physics and technology. Besides the amorphous materials, various cases of intermediary states of order have been found: liquid crystals, high polymers, quasicrystals, etc.. A good knowledge of the atomic structures is essential for the understanding of the physical properties.

The more generally used method is to imagine a structure and to compare the theoretical predictions with the observations. But the models are often only semi-quantitative and they are based upon rather arbitrary assumptions. Over the years, the progress has been slow and the results not entirely satisfactory. For instance for glassy metals, the fit with the observed Radial Distribution Function is acceptable, but the calculated density is definitely too low.

Since about fifteen years ago, new theoretical ideas have profoundly changed the situation. The central subject of this book edited by J. F. Sadoc is a thorough description of the new concepts, followed by their applications to different types of disordered solid materials.

The basic difficulty which the building of amorphous models encounters is the following. One starts with a group of a few atoms having a structure justified by the interactions between neighbours. But it is not possible to enlarge the model regularly. For instance, 4 atoms are closely packed if they are sitting on the vertices of a tetrahedron: it is impossible to fill the space with a set of tetrahedra in contact. That is forbidden by the geometry, by the "mathematical reality". The new idea is that, if it is not possible in our euclidean 3D space, it is possible in other spaces with different geometries (curved 3D space, 6D spaces, etc. ...) Let us recall that the tiling of an Euclidian plane with regular pentagons is impossible, but it is possible on the surface of a sphere.

Of course, one must come back from the abstract space to our real space. That is the second point of the new theory: the regular figure is transformed into a model in real space by mathematical operations (projections) or systematic introduction of defects (disclinations) which describe the necessary irregularities of the disordered solid. One "flattens" a curved space as one flattens an orange peel by making some cuts in it.

The big advantage is that the disorder in the models is no longer due to arbitrary choices. One starts from a perfectly defined object and the disorder in our space is in some sense explained by the passage out of the initial abstract space. The disorder may be quantitatively defined. Moreover, it appears that some peculiarities of the structure of disordered materials are not due to special properties of that matter but they are simply consequences of geometrical or topological laws.

All the authors of this book have tried to introduce the characteristics of the special geometries without using a too abstract mathematical style. But the readers must be warned: it is not at all easy to become familiar with properties of strange spaces because they are in contradiction with our experience in the real world. The necessary intellectual effort is the price to pay for a better understanding of the disorder, which is still mysterious in comparison to the simple and beautiful order in matter.

The different chapters of the book deal with various types of disordered structures. They are written by specialists who are known for their original contribution in the field. The unifying theme is the use of special geometries; the reader will find from one chapter to another slightly different approaches which may facilitate his understanding.

It is a fact that many original papers are very difficult to read; the authors concentrate upon what is new and, to save room, neglect to recall the elements, which are supposed to be known. The result is that such papers are in practice reserved for a handful of well trained specialists.

Now the scientific community needs another kind of book or review paper; especially written to attract newcomers in the field, and to help them to assimilate the new developments. It is the duty of the specialists not to abandon the beginners alone along the roads newly opened, far away from the front of the progressing science.

That is just what J. F. Sadoc and his coauthors have successfully done in their book. We must congratulate them in having given a good example of the type of work which could be called "initiation of advanced physicists to new methods".

A. Guinier
January 1990

PREFACE

In Physics, as in any scientific research, we find several tendencies.

Derivation of physical properties from first principles could be opposed to phenomenological studies; on one side descriptive observations and deductive approaches are also examples of these possible tendencies.

What is the role played by geometry in these approaches of the condensed matter sciences?

At the beginning of this century, it could be thought that geometry, and mainly geometry in its visual and intuitive aspects, was a good tool for description but was not involved in first principles of physics.

For instance, Crystallography was developed in a very elaborate mathematical theory of symmetry, but this science was unable to predict structures and phase transitions.

Quantum mechanics, statistical mechanics were used very efficiently, but geometry was only used as a structural parameter introduced in a phenomenological way.

The theory of defects has been a revolution showing that geometry and topology could offer basic concepts in condensed matter physics.

During the last twenty years geometry has put a new insight in that field. With its two aspects: intuitive or more formal it is an ingredient in the understanding of physical properties.

It is all the different aspects of the fascinating power of geometry in condensed matter sciences that I have tried to gather in this book.

I want to acknowledge all authors for their contributions, in this direction, and Prof. A. Guinier who has read the manuscript, given interesting comments and agreed to write the foreword.

The Editor

CONTENTS

Chapter III. WHICH UNIVERSE FOR BLUE PHASE?
Elizabeth DUBOIS-VIOLETTE and Brigitte PANSU

Chapter I
GAUGE THEORY AND GEOMETRY OF CONDENSED MATTER

Nicolas RIVIER

GAUGE THEORY AND GEOMETRY OF CONDENSED MATTER

Nicolas Rivier

MSD 223
Argonne National Laboratory
Argonne, IL 60439, USA*
Institut de Physique Expérimentale
Université de Lausanne
1015 Dorigny-Lausanne, Switzerland**

> *L'intelligence de l'exactitude
> n'empêche pas le plaisir de
> l'inexactitude. (R. Magritte,
> Leçons de Choses, 1962)*

1. INTRODUCTION

1.1 The Symmetry of Disorder

A glass is extremely homogeneous, as early X-ray crystallographers have discovered to their surprise: "...one of the most interesting discoveries made in the comparatively early history of X-ray analysis was the fact that silk or even paper are more crystalline than glass." (Dame Kathleen Lonsdale, quoted in ref.[1], p. 26). In fact, the scale of any regular arrangement is orders of magnitude smaller than the smallest microcrystallites (10 versus 100 to 1000 interatomic distances).

Can this homogeneity be expressed as a mathematical symmetry or invariance ? At first sight, the situation does not look

---

* Work supported by the U.S. Department of Energy, BES-Materials Sciences, under contract #W-31-109-ENG-38.

** Work supported by the Herbette Fundation.

promising: Glasses do not have the generative translational and rotational symmetry of crystals, neither microscopic ruler nor compass. In fact, in glasses this symmetry is completely broken and the space group is trivial. Nevertheless, glasses have overall homogeneity and isotropy, a symmetry experienced by getting lost in a forest, without external beacons like fixed stars, sun, or compass. Every atom, every tree is as good a reference point as any other, even though disorder implies that they have different local environments and characteristics. By contrast, in a regular orchard or plantation, this local environment is the same everywhere, and provides a (local) meter and compass, even in a foggy night. The random forest does not offer these geodesic bearings, and one gets lost. Its local reference frame has arbitrariness, which is expressed as a gauge invariance (invariance of the physical properties of the glass with respect to a local, or gauge transformation).[2,3]

We are familiar in classical and quantum mechanics with the relationship between symmetry, degeneracy, invariance of the physical properties of the system under transformations probing the symmetry, and conservation of the infinitesimal generator of these transformations. This relationship provides a conceptual framework of great generality, and a tool (like Bloch's theorem) for obtaining an explicit solution to a particular problem. When the system is disordered, its invariance is only statistical: It is only on average that the system remains the same. One might fear that symmetry would only be apparent at the very end of the calculation, when an average is made over the various members of the statistical ensemble, and that its conceptual guidance and simplifying power would be as intellectually illuminating as a *deus ex machina*. Not so: Gauge theory[2,3] is a method using this invariance at the onset of the calculation.

It is true that the average needed to exhibit the invariance is only local ("...using, like all macroscopic theories, physical quantities which are averages taken over "physically infinitesimal" volume elements, and neglecting the microscopic fluctuations of these quantities, due to the molecular structure of the material"[4]), but this local average affects profoundly the geometrical framework, because it connects different points in space in a way which is essentially frustrated, i.e. path-dependant. Again, we refer to the mixing up of geodesic paths experienced upon getting lost in a forest. The geometrical concept which embodies frustration is a (non-trivial) fiber bundle.[5] It is the geometrical framework of gauge theories, where the local average is replaced by projection, albeit in a complicated and high-dimensional configuration space.

Local invariance restricts very much the possible forms of the free energy or Hamiltonian one can write for glasses, and imposes remarkable geometry and physical behaviour. "Every circumstance left unspecified* in the statement of a problem (here, the reference frame) defines an invariance property which the solution must have if there is to be any definite solution at all. The transformation group (here, the gauge group of local transformations), which expresses these invariances mathematically, imposes definite restrictions on the form of the solution, and in many cases fully determines it".[6]

Disordered materials are homogeneous overall (one can get lost - without external bearings - in a forest, which then looks like an intrinsically homogeneous and isotropic geometrical environment, but one in which one is unable to keep walking in a "straight" line and a "constant" directions, as if the geodesics were all entangled**), but this homogeneity is very different from the usual symmetry of crystals or empty space, of a plantation or of the ocean (again, without stars, moon, sun or compass). The traditional symmetry is probed by displacements, which are finite in the case of the crystal or the plantation, and infinitesimal in empty space or at sea, that form a group of transformations (automorphisms) leaving the structure exactly unchanged.

---

*   Here, by necessity (disorder) rather than by ignorance.

**   The reader is encouraged (with caution) to experience this feeling, by walking or skiing in a thick fog. One quickly loses sense of the direction of steepest descent. Local roughness of the ground makes it worse.

By contrast, after any displacement in our forest or disordered material, one faces an environment which is different from the original (a few fir trees instead of birches forming a slightly different pattern), but this difference will not help find one's way out. The difference is subjective - *this* tree *here*, which *I* am facing *now* is a birch - even if the existence of these fluctuations is objective, a physical definition of disorder. The forest or the soap froth as a whole is an objective, homogeneous, space-filling collection of fluctuating elements. The automorphisms probing this overall homogeneity are not displacements, but local transformations of the elements - a few fir trees here into birches - which leave unchanged the overall structural and physical properties of the material. These automorphisms are <u>gauge transformations</u>, and the overall homogeneity is gauge invariance.

The state of the system is characterized locally by physical parameters. Local change of these parameters may not leave the system strictly invariant, but it does so statistically, insofar as the relevant physical quantities of the system, and the physical laws, involve some local average which is itself invariant. This statistical invariance does not look very useful as a practical tool, but it can be reformulated in a fiber bundle representation of the gauged system, where it is only the projection of the system on a smaller configuration space which remains strictly invariant. The parameters are the quantities to be gauged. The main problem is to replace average by projection. This will be done explicitly in §5, where we shall see that parameter fluctuations and local average are not only a convenience, but a necessity imposed by conflicting geometrical requirements.

Our metaphysical foggy forest should justify the opening statement of this chapter, that gauge invariance is the symmetry of disorder. In the rest of the chapter, we shall prove this statement, and exploit it in several examples.

Let us begin with two examples of gauge invariance without disorder, one simple but involving quantum mechanics, the other classical and familiar, but more subtle.

The simplest example of a gauge transformation is an arbitrary change of the phase of an electron wavefunction at any point in space. The probability of finding an electron at a point, proportional to the modulus square of the wave function, remains unchanged. Only the relative phase shift of two partial wavefunctions, split, taken along different paths in space (and thus phase-shifted differently) and recombined at a point, will produce observable interference patterns. This example indicates that the phase shift at any point is not quite arbitrary. Phases at nearby points must be connected smoothly, and this may induce a path-dependant phase shift, similar to the two brothers starting at the same point on the same floor of the castle which they are exploring, and ending up on a different floor after being split by the staircase.

Swimming (in a viscous environment, where there is no kinetic energy, and the swimmer is automatically at the bottom of its local potential well, a situation experienced by a paramecium in water or a human in syrup) is a gauge theory.[7] The shape of the swimmer is exactly the same at the end and at the beginning of each stroke, yet, it moves forward. The swimmer's center of mass is the gauged quantity, as was the phase of the electron.

The natural geometric environment of gauge theories is the fiber bundle. It is a simple, though abstract object, which enables us to introduce clearly the various elements of a gauge theory, independently of the - often much more sophisticated - physical environment.

1.2 The Geometry of Fiber Bundles

A gauge field, geometrically, is a connection in a fiber bundle. A fiber bundle consists of a <u>base space</u> (the ordinary space in which the electron moves, the space of possible shapes of the swimmer, or the angle of the turntable of Fig. 1), a total space (position and phase of the electron, shape and position of the swimmer, angle of the turntable and orientation of the ribbon), and a map that projects every point of the total space onto a point in the base space. The set of all points in the total space that are mapped onto the same point

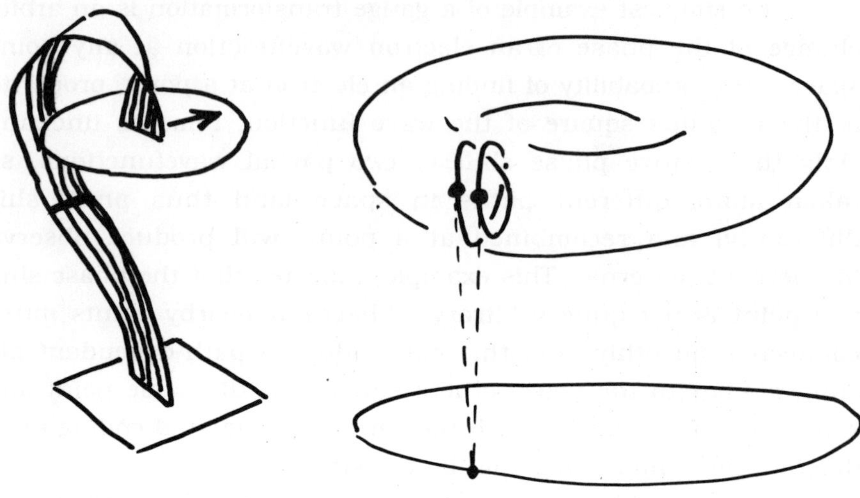

Fig.1.   Schematic model of D.A.Adams's antitwistor[8] and its fiber bundle representation. The orientation of the turntable makes up the base space (a circle), that of the ribbon constitutes the fiber (a double circle). The total space is a double torus. If one restricts the orientation of the ribbon to its equilibrium configuration, the fiber becomes two points and the total space, a double circle. The connection is effected mechanically. Rotation by 4π is the identity operation. Useful for a (rotating) space station needing supply of gas, water and electricity by cable from Earth.

in the base is called the _fiber_. The fiber of every point in the base must be topologically the same, and is represented by a single, ideal fiber (which is the electronic phase, the length of the swimming-pool, or the orientation of the ribbon in Fig.1). In most condensed matter applications, the base space is the ordinary space occupied by the material, and the fiber represents "the parameter to be gauged". For an Ising magnet, the fiber has two points, corresponding to spin up and spin down. In a castle, the fiber has as many points as there are storeys, while the base space is the plan of the castle. A superconductor is described by a complex order parameter, the Cooper pair wave function, whose phase is its fiber. A hollow cylinder and a Moebius strip can be viewed as fiber bundles, both with a circle as base and a segment of straight line, their height, as fiber. The two bundles are locally identical, but differ globally: The cylinder is a trivial bundle, while the Moebius strip is twisted.

A trivial fiber bundle is one for which the total space is the direct product of the base space by the fiber. Examples are an Ising ferromagnet, a superconductor without vortices, a castle without staircases, or the cylinder.

On the other hand, a bundle is not trivial if it ends up <u>twisted</u>, like the Moebius strip: The orientation of the fiber is modified as one moves along a close contour in base space. Note that the fiber at the end of the circumnavigation must be indistinguishable from that at the beginning, since there is one unique, representative fiber, hence only finite rotations by multiple of π are permitted. These are called <u>large gauge transformations</u>. They are topologically stable, and cannot be undone by infinitesimal corrections along the contour. Other example of twisted bundles are a real castle with staircases, (trace a closed circuit on the plan around the staircase; where do you end?), a superconductor with vortices, and an Ising spin glass with frustrated plaquettes.

After the gauged parameter, the next essential concept is that of a <u>connection</u>, or <u>gauge field</u>. It describes how the orientation of the fiber changes as one goes along a path in base space, or how the

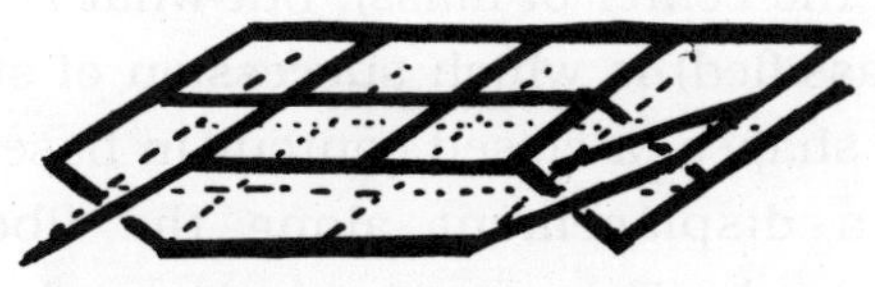

Fig.2. Fiber bundle for an Ising spin glass (§4.1). x = frustrated plaquette, —— = ferromagnetic, ……. = antiferromagnetic interaction in the base space. The two points of the fiber represent spin up or down configurations, in arbitrary order.

path in base space is <u>lifted</u> into a path of the fiber bundle. Usually, in condensed matter and in most physical examples, this connection is a natural one: In an Ising magnet, it is carried through by the exchange interaction J_{ij} between two neighbouring spins at i and j along the path, that is, by the energy in the bond linking the two spins. Spin up (down) at i is connected to spin down (up) at j if $J_{ij}<0$ is antiferromagnetic (Fig.2). See Bernstein and Phillips[5] for other examples, and for further reassurances as to the naturalness of the connection process. In D.A. Adams antitwistor[8] (Fig.1), the connection is mechanically obvious. We shall encounter in §4 an example in covalent glasses which is slightly more elaborate,[9] but a careful analysis of the energy carried by a bond leads easily to the correct representation.

The only difficult example which I have encountered is that of a gauge description of swimming in fluid with high Reynolds number[7] (e.g. paramecium in water). Here the ordinary space along which the paramecium moves is the fiber, while the (complicated) space of all possible shapes is the base. A successful stroke is a large gauge transformation. A trivial bundle is useless. The problem for our primordial Mark Spitz is to find an efficient connection. Connection is established rather naturally (for example by pseudopodal deformation and flow of the internal fluid to create a displacement of the center of mass), but what is not obvious (and has not yet been classified) is which succession of strokes (with identical initial and final shapes, a closed contour in base space) is non-trivial, and leads to a displacement along the fiber, i.e. propels the paramecium forward. The same analysis can be applied to divers (of Olympic standard - the final body shape must be identical to the initial one).

In most cases when the base is continuous, the connection is carried through parallel transport, which is the prescription that the orientation of the fiber (or of any transported object) remains constant along a geodesic in base space. (See eq. (6.11) for a mathematical definition). It follows that if two geodesics intersect at more than one point, and unless the fiber is orthogonal to the base space,

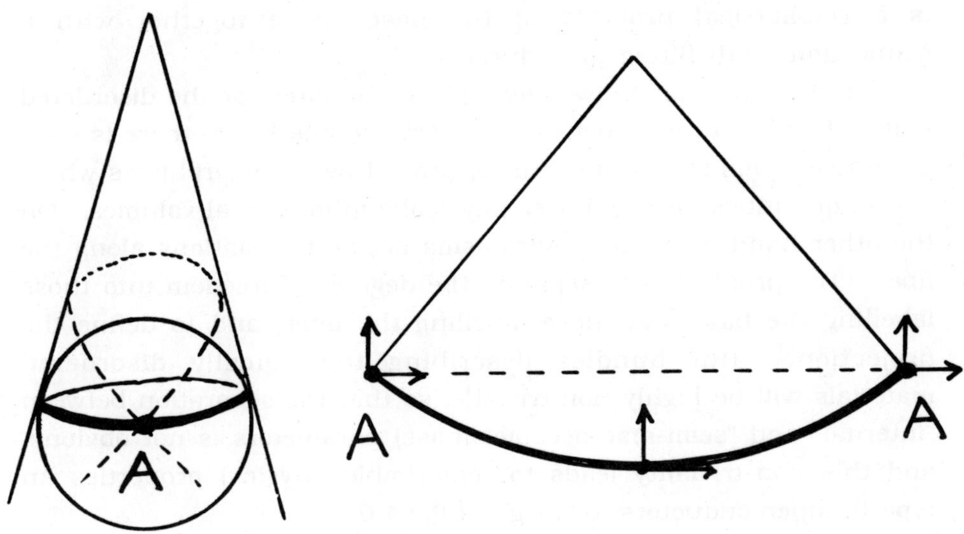

Fig.3.   Example of parallel transport and non-unicity of a local reference frame in the presence of curvature. The frame at A has been rotated after the (parallel) round-the-world trip. The cone, tangent to the trajectory, can be flattened on a plane to define parallel transport and exhibit the rotation caused by curvature. Alternatively, one sees that some geodesics on the cone (broken line) self-intersect.

the bundle is non-trivial (Fig.3), as with a sphere as base space.[*]

The total <u>curvature</u> of the region in the base surrounded by a closed contour is given by the distance made along any one of the fiber by the lifted path. Even if there is no parallel transport (as in discrete bundles like spin glasses or covalent glasses), the natural path lifting operation generalizes the notion of curvature. Thus, a non-frustrated plaquette in a spin glass is flat, and a frustrated plaquette is curved or warped.[10)] Note that this notion of curvature is independent of the precise nature of the fiber, that is of the type of spins (Ising, XY, Heisenberg, Potts, etc.) or objects considered. It

---

[*]   This is a sufficient, but not a necessary condition for non-triviality: A bundle with a negatively curved based space is also non-trivial.

is a geometrical property of the base space together with a connection (path lifting procedure).

The reader may be wondering how the forest or the disordered material will fit into a fiber bundle. On the one hand, there is non-generative symmetry or invariance, probed by automorphisms which act on quantities averaged over physically infinitesimal volumes. On the other hand, these automorphisms act as translations along the fiber. The problem is to separate the degrees of freedom into those labelling the base and those labelling the fiber, and to define the projection. But bundles describing topologically disordered materials will be highly non-trivial* , so that the separation between "internal" and "semi-macroscopic" (base) parameters is not obvious, and this non-triviality leads to remarkable physical properties (in type II superconductors, or in glass (§3,4,6)).

1.3   What Can Be Gauged ?

What is the quantity to be gauged, constituting the fiber? In the original application of Weyl,[2] it is the measure of the length of a segment at a given point in space, the local scale of the metric tensor $g'_{ij}(\mathbf{x}) = \lambda(\mathbf{x}) \, g_{ij}(\mathbf{x})$. (See §6).** In classical applications, it is an internal, local parameter, like the phase of an electron or of the superconducting order parameter, a spin in spin glass or the isospin of an elementary particle (changing a proton into a neutron). Its particular value influences that of the matter nearby, through a specific interaction called gauge field or connection, which is the vector potential in electromagnetism or the exchange interaction in magnetism. The geometrical influence of the gauged parameter is indirect and felt through conflict:[10] One point receives conflicting

---

*   There is a lot of frustration in glasses (a high density of odd lines, see §2), so that the condensed matter physicist does not have the problems of his cosmologist colleague who needs "le sérieux coup de pouce de l'imagination et de l'intuition pour oser élargir *à tout l'espace-temps* des observations qui, de fait, ne concernent qu'une minuscule partie de celui-ci".[11]

**   The German word is "Eichung". The translator[2] calls it "calibration" (in French, "étalonnage").

orders from its neighbours, or there is incompatibility between the initial and final values of the internal parameter obtained by going from neighbour to neighbour around a closed loop. So the gauged quantity may be projected, but should not be projected out.

In the geometrical context of this book, the gauged quantity defines the fiber. It is the local configuration at some point in space within the condensed material (the base space), the species and shape of a tree in the forest in which one has got lost, the local strain in an elastic solid, the orientation of the bonds incident on a vertex of a continuous random network modelling an atom in a covalent glass, the position of a swimmer in a viscous liquid (a paramecium in water[7]), with a base space consisting of all its possible shapes), or the metric tensor in space-time.[2] It affects directly the geometry of its surroundings, and the gauge field carrying this influence is a geometrical connection which relates the stroke of a swimming paramecium to the boost it effects, measures the amount of rearrangement or non-uniformity required to accommodate the orientation of a silicon atom and its four bonding electronic orbitals, or expresses the fact that the vertical in Rio is not parallel to that in Paris. (Fig.3). The gauge field appears as an essential element of the geometry, and it is also minimally coupled* to the local parameter which is being "gauged". Railways gauges may vary from country to country, with a gauge field to transmit this fact continuously across the border (as in Port Bou**). The alternative (not entertained in this chapter) is to change trains (as in le Châtelard).

A disordered material is therefore characterized, at first sight, by spatial fluctuations of the internal parameter(s) (the species of the tree, the size and shape of a soap bubble or the orientation of a silicon atom with its four orbitals, a local frame called Vierbein, or

---

*   Through the covariant derivative of the matter field, exclusively. See eq.3.3.

**   A similar mismatch occurs at the Russian border, apparently because the Tsar confused the item that was to be calibrated when he took the decision.

14

tetrapod) of its constituting elements (matter field). Furthermore, the material is organized in space (it fills space, at the very least, often with perfect short-range order, - silica glass consists of corner-sharing regular tetrahedra), so that a local transformation of an element must necessarily affect its immediate neighbors as well, through the connection or gauge field.

It is this connection which defines the geometry of the fiber bundle, besides the geometry of base space. Notably, the lift of a closed contour in base space is a path in the bundle defined by the connection, on which any local parameter must be returned to its original value (if the bundle is flat within the contour), or to a value equivalent the original, namely to another point on the original fiber (if it is twisted). Moreover, this consistency requirement, which is built in the bundle, must be retained under any gauge transformation. In short, the bundle is <u>gauge-invariant</u>. Geometry appears therefore as a topological organization of ordinary space (the base space of the bundle), a partition into simplices, a graph[9], together with the loci of frustration (of non-triviality in the bundle), objects of co-dimension 2 (i.e. of dimension D-2 in D-dimensional space, points in 2D, lines in 3D): Vertices carry the material elements (trees, atoms) with their internal parameter. Edges carry the connection (gauge field), and faces or plaquettes, the consistency requirement, but it will turn up that frustrated plaquettes are threaded through like necklaces, by closed lines.[12]

Some lack of consistency or incompatibility is physically permissible (vortices as excitations), or geometrically necessary (frustration). The phase of the superconducting order parameter or the direction of a planar spin can rotate by any multiple of 2π, the direction of the liquid crystal molecule by π, or the four orbitals of a silicon atom may have been permuted after following a closed path. The parameter is then returned to the same point in the manifold M of internal states in a non-trivial fashion. A plaquette whose boundary has this property is deemed frustrated. Frustration[10] is a geometrical property which is invariant under gauge transformation, as will be seen in several examples.

1.4　Defects and Disorder

Incompatibility in gauge theory is a generalization of the theory of defects in condensed matter.[13] If an internal parameter (matter field) $\phi(\mathbf{r})$ is given everywhere in a continuum (the Euclidean space R3, for example), this defines a mapping between R3 (base space) and the manifold M of states of the internal parameter (the fiber of the bundle, with equivalent values identified[*])

$$\phi(\mathbf{r}): \ R3 \to M \ . \tag{1}$$

Defects are then detected and classified by the elements of the homotopy groups of M, that is by the classes of non-contractible closed contours (surfaces, etc.) in M which are the images of closed contours within the material. Any contour in R3 can be shrunk in a simply-connected (bulk, non-punctured) solid. If its image in M remains finite, it represents a defect.

Gauge transformations act on the internal parameter, and define explicitly the manifold M, which may have non-trivial homotopy groups. This generalizes defect theory to disordered materials, by eliminating reference to $\phi(\mathbf{r})$ as an order parameter, and to M as a coset G/H of the group of general transformations by the symmetry group (point - or space group) of the ordered material.

More importantly, the fact that gauge transformations define a connection enables us to deal with disordered networks (like covalent glasses and spin glasses) and not only with a continuum. The emphasis is shifted from the properties of the internal parameter, to incompatibility in connections around plaquettes. Defects have become geometrical objects, instead of the focus of some damage of the order parameter.

The reader needs no convincing of the essential part that defects play in governing the physical properties of ordered condensed matter, from dislocations in crystals to disclinations in

---

[*]　For example, the fiber of the antitwistor of Fig.1 consists of a circle covered twice. The internal manifold M, which identifies rotations differing by 2π, is one single circle.

liquid crystals (despite Kittel relegating dislocations as the runt of Solid State Physics). It has been argued that the same is true for glasses, which owe their anomalous but universal properties to geometrical frustration, a line "defect" better described as odd line or 2π-disclination.[12]

But, what is a the meaning of a "defect" in a disordered material, or of an individual spaghetti in a plateful? It is a geometrical entity, topologically stable (it cannot be annealed out by small transformations), which emerges naturally from gauge theory.

This chapter uses glasses and other disordered materials as an illustration of the geometry of gauge symmetry. If the list of references is thin for a review, it is because this view of glass has not found universal acceptance as yet. It originated, in spin glasses, with Toulouse[10] and with Dzyaloshinskii and Volovik.[14] There is an excellent recent review by Venkataraman and Sahoo.[15] The author and coworkers have shown that representations of glass (as an elastic continuum,[16] as a projected,[17] or decorated crystal in curved space[18]) are gauge invariant. Gauge is the local geometry of disorder.

1.5 Summary

§2 is a description of the geometry of the base space of disordered materials. Disordered materials are essentially frustrated, they contain topological defects, which form closed lines. §3 introduces gauge invariance in continuous space. Gauge invariance imposes the introduction of a gauge field, which appears almost automatically. We will meet the main physical consequences of gauge invariance (with minimal coupling), illustrated in superconductors and in glasses. The physical nature of gauge and matter fields will also be established. §4 is on gauge invariance in discrete space, with applications on spin glasses and on covalent glasses. Here, the aptest geometrical framework is that of a (discrete and twisted) fiber bundle. Again, we will see a surprising physical consequence of the twisted structure of the bundle (the tunneling modes in glass, with the - astonishing in system without generative symmetry - degeneracy ($<10^4$ eV) that they imply). §5 is on projection and disorder, and will show (in a simple model), how disorder can be reformulated as an invariance and a projection, i.e., as a fiber bundle. The long §6 is an application to the continuum theory of elasticity with defects (dislocations and disclinations) for crystalline and glassy media. Relationship between dislocation theory and non-Euclidean geometry have been recognized since the late forties,[19,20] but the gauge invariance of the former has never been fully exploited. This can be done in glasses, which are thereby able to accommodate defects that 3D crystals cannot do. A short conclusion (§7) follows.

2. GEOMETRY OF DISORDED MATERIALS

2.1 A Random Network

We have seen in the first section that the geometrical structure of a gauge theory is that of a fiber bundle. This section describes, in the typical case of disordered condensed matter, the geometry of the base space of the fiber bundle, that is, the geometrical scaffolding on which live the various physical quantities or fields discussed in this chapter. The only essential result for the rest of the chapter is the existence of topological line "defects" in disordered solids, which are characterized by their existence rather than by their intensity ("to be or not to be", rather than "how much"), and can be described as 2π-disclinations in the continuum, or as odd lines in network glasses.[12] These line defects are also the loci in the base space where the fiber bundle is non-trivial.

Disordered solids, from foams to glasses, belong structurally to one of two main classes, networks and packings. The structure of both classes is described on a microscopic scale by a <u>regular graph</u>[9].

In covalent or network glasses, vertices and edges of the graph are atoms and bonds, respectively, faces are (shortest) rings (usually puckered), and cells do not have a direct geometrical interpretation. Atoms have fixed valence (short-range order) determined by chemistry, and vertices have the same coordination z (z=4 for amorphous silicon, z=3 for arsenic) everywhere in the graph, which is then called regular graph, or <u>continuous random network</u>.

In random packings (metallic glasses, sandstone, etc.), only atoms or grains have a direct physical essence, and neighborhood must be ascribed through some partition of space, the simplest of which is the Voronoi partition, in which a point in space is assigned to the territory of the nearest atom. Electoral commissions, Mayan communes, predators, real estate developers, or mathematicians like Laguerre and Gaultier, have devised their own variations.[21] Every atom is thus ascribed a (convex) polyhedron or cell, and the packing is a space-filling assembly of such cells. The assembly of Voronoi polyhedra, with their (planar) faces, (straight) edges and vertices, form a graph, called <u>Voronoi froth</u>, which is, again, regular ($z=4$ in 3D and $z=3$ in 2D): Any vertex with higher coordination can be split into several regular vertices by an infinitesimal deformation of the packing (or by one more aggressive fox); it is neither topologically, nor structurally stable and occurs with negligible probability in a random arrangement.

Both types of graph are regular, but for different reasons: Chemistry for covalent glasses, "random avoidance of the niceties of adjustment"[22], i.e. entropy, for random packings. Cellular or Voronoi froths also have a dual (not regular) graph, with atoms, cells, etc. as its vertices, joined by an edge if they are adjacent. The cells of the dual graph are tetrahedra, because the original froth is regular. The dual graph is therefore a simplicial partition of space. Random simplicial decompositions have been used to describe the structure of metallic glasses[23,24] (obtained by decurving the regular polytope {3,3,5} with 600 tetrahedra in curved space), but also by Regge as a geometrical mean of imparting curvature in general relativity[25, ch.42], or as an attempt[26] to recover infinitesimal translational invariance in lattice gauge theory (unsuccessful, since the homogeneity of glass is not a generative symmetry, but a gauge invariance (see §1)).

An infinite random network can be constructed systematically by applying repeatedly, at points chosen at random but uniformly so, the same local elementary transformation, which could be a bond switch[21,27], or the local commutator between two successive decurving operations.[18,28] The network can also be built by hand.

The first two of the three main structural features of glasses or of any large, space filling, random structures, can be distinguished at a glance:

1) Non-collinearity of local frames.

2) Overall, but non-generative homogeneity.

3) Topological "defects" (odd lines[12]) or non-planar inclusions[29]*).

The third is less obvious, but essential. If 3) is missing, one may still have a random structure, but without topological disorder, as with a disordered but crystalline alloy, or with a Mattis random magnet.[10] These three features are characteristic of the amorphous** state of matter, and can be taken as its structural definition. We have already discussed objective homogeneity, despite different and non-collinear local frames, in the Introduction.

There are two types of automorphisms, leaving the network strictly, or statistically, invariant. The former are permutations in the labelling of the bonds which carry the energy. Labelling the edges has no physical significance, but it is, as we shall see in §4, essential in describing the configuration(s) of the system.

The latter are bond switches,[21,27,29] or other elementary topological transformations (bubble disappearance, cell division) governing the evolution of a soap froth, a biological tissue or a mosaic of

---

* Kuratowski's $K_{3,3}$. This is the topological defect induced by a parity-conserving switch between bonds in the same ring. A subgraph containing the ring is no longer planar. This switch is used to generate random networks containing even rings only, as structural models for a-GaAs (without any wrong bond).[29] Such a network can therefore have topological disorder even without odd lines. It is not known whether a-GaAs contains these non-planar inclusions, and, if it does, what are their physical effects.

** Do not confuse amorphous state of matter or amorphous structure, with the amorphous continuum of infinitesimal geometry.[2] The latter is taken in the sense of the Genesis (without any form, like the earth in the beginning), as a continuum of points without connection or metric. The former is a metric space with an affine connection between each point and its neighbourhood, where parallel transport of vectors (direction) and congruent transport of segments (length) are defined by the structure (frame at each point).

metallurgical grains. They act like atomic collisions in gases to keep invariant the local average properties and correlations. Detailed balance gives the values of the local properties and correlations that are fixed under elementary topological transformations. Aboav's relation,[21] obeyed by all natural 2D networks and froths, and its, yet unverified counterpart in 3D,[30] give the invariant correlations of the topological properties of the network under these bond switches. Bond switches have also been used to construct random networks on a computer, starting from a crystalline structure like diamond. A very large number of switches, and simulated annealing are required to reach randomness, but invariance of observable physical properties like the diffraction spectrum under further switches (stationarity) and independence from the initial configuration or the annealing procedure are manifestly achieved.[27]

2.2 Topological Defects

A topologically stable constituent (="defect") of a material (counting as a configuration in the partition function, and in the topological entropy of a liquid or a glass) retains its identity and its existence under small, continuous deformations of the system, as well as under structural automorphisms, strict or statistical. These include flow or motion of the system. A dislocation in a crystal is topologically stable, an arbitrary elastic distortion, usually not. Topological stability is expressed as a conservation law. This type of conservation law is known geometrically as Poincaré's identity, expressing the fact that the boundary (∂) of a boundary vanishes, thus

$$\partial.\partial \equiv 0 . \tag{1}$$

This identity has a dual,[*] a differential equation over the fields, called Bianchi's identity, d.d $\equiv$ 0. For example, see the Maxwell equations

---

[*]   Duality is taken as in Stokes's theorem, with the integral of a n-form **A** over the boundary of a (n+1)-volume (e.g. the circulation of vector potential: **A** over the contour enclosing surface S, n=1) equal to the integral of the exterior derivative d**A** of the n-form - itself an

div $\mathbf{B} = 0$ and $\partial_t\mathbf{B} + \text{curl } \mathbf{E} = 0$ in electromagnetism. We shall meet other examples of Bianchi identities in §7. Suffices to say now that they are topological conservation laws.

What are then the "defects" in a glass, random network or elastic continuum with trivial space group (no generative symmetry)? Networks have rings, some of them odd. Suppose that odd rings are non-bounding (because zero is manifestly bounding, is even, and the number of sides is not a conserved parameter in the absence of rotation symmetry*). Poincaré's identity (1) requires that *odd rings cannot be found in isolation, but are threaded through by uninterrupted lines, which form closed loops or terminate at the surface of the material.*[12] Thus, odd rings are not isolated elements, but part of linear necklaces which are topologically stable. They are characterized by their essence (oddness) rather than by their identity. Nevertheless, because they are lines, they find it difficult to annihilate each other, despite their algebra modulo 2. However, they can cross each other without topological obstruction.[13]

Proof of this conservation law is elementary.[12,31] Here is one. Any uninterrupted line must puncture an arbitrary, closed surface S inside the material an <u>even</u> number of times (there is always an exit for the odd line entering S). Consider such a surface, intersecting the network at its vertices (this can always be arranged by small deformation of S). S contains therefore some vertices, edges and rings of the network, which triangulate it (ditto). So, for any S, there must be an even number of odd rings on S.

Ascribe a weight $J_e = -1$ to every edge e on S, and an index $\Phi_f = \Pi_{e \in f} J_e = \pm 1$ (even/odd ring) to every ring f on S. Then $\Pi_{f \in S} \Phi_f = $

$(n+1)$-form - over the $(n+1)$-volume (e.g. the magnetic flux: curl $\mathbf{A}$ over S).

* A disclination (rotation dislocation) is required to transform a 6-sided ring into a 5-sided ring. But rotation (by $2\pi/6$) is not a symmetry available to be broken in glasses, and cannot be singled out metrically. Topologically, of course, it makes sense, but only because it changes the parity of the ring.

$\Pi_{e\in S}\, J_e^2 = 1$. (J_e enters twice in the product because every edge separates two faces on S). q.e.d.[31]

For some amorphous materials, like polymer glasses, it may be too difficult or complicated to construct a random network describing the structure. The simplest, least specific and semi-macroscopic description of any glass is as an elastic continuum, with a trivial space group (i.e. without any translational or rotational symmetry). By contrast, a crystal can also be represented as an elastic continuum, but with either full (Volterra continuum, the toy model of dislocation theory) or discrete (one of the 230 space groups) symmetry. Then, odd lines are the only structural constituents surviving the transition from discrete network to a continuum. They take the form of 2π-disclinations.

This can be shown very simply, by using the Burgers contour technique of dislocation textbooks, which is a particular application of homotopy theory of defects.[31] Take a local configuration around a closed contour in space, while maintaining its orientation relative to the reference frames at all points on the contour (parallel transport, eq.(6.11)). The configuration, upon returning to the starting point, either has the same orientation, or any difference is physically irrelevant. The latter case applies to crystals, where possible misfits are elements of the space group, which label in turn dislocations and disclinations. In glasses, the space group is trivial, and the original orientation must be recovered. Nevertheless, there are still two possible transformations of the configuration: A rotation by 4π is homotopic (continuously deformable) to the identity, whereas rotation by 2π entangles the links between local configuration and the rest of the system (like the ribbon of Fig.1).*

Random networks and disordered elastic continua sustain therefore, algebraically, the same topological defect, an uninterrupted odd line. It is also the only one: translations are homotopically trivial, so that dislocations of any kind are not topologically stable in glasses. Neither are point defects. It may be

---

* SO(3) is not simply-connected. $\pi_1(SO(3)) = Z_2$. (Space group: $T3\wedge SO(3)$. $\pi_1(T3)=1$, $\pi_2(SO(3))=1$).

argued that, while odd lines in networks are geometrical objects in the base space, 2π-disclinations are constructed, through parallel transport, in total space. But their projection leaves a core singularity, which must be excluded from the base space. Otherwise, one would have an excitation, not a structural defect.

Finally, odd lines also appear in the curved space representation of amorphous materials[23]: Glasses are perfect crystals (ideal state) in curved space, which are projected or iteratively decurved[24,18], with disorder, into Euclidean space (natural state). In going from the ideal state (crystal in curved space) to the relaxed natural state, two types of topological defects have been introduced, both disclinations:

1) Decurving disclinations (whose action is labelled and measured in the curved, ideal state). They are essentially negative and are characterized by their intensity (since the space they decurve is crystalline). Decurving disclinations are found in metallic crystalline (Frank-Kasper or tetrahedrally close-packed[32]) phases, where they are called main skeleton or Frank-Kasper lines, as well as in metallic glasses.[18] The crystalline relatives of covalent glass are the many phases of crystalline silicon, germanium or ice (diamond, wurzite, BC8, etc.)[33], but systematic decurving and identification of the disclination network remains to be done.

2) Disordering disclinations. They have strength 2π only, and their action is measured in the natural state of the system. In metallic glasses, they form small closed loops, which consist microscopically, and in reference to the ideal, curved space, of a neutral wool-ball (a "pelote")[18,28] of negative and positive disclinations.*

---

* It is not easy to see how a wool-ball of lines in an ideal curved space should become a single Z_2 loop in the flat continuum. This correspondence comes from the topological properties of the manifold of internal states of the gauge transformation $T(x)$ (eq.3.2), whose general form is known explicitly in decurving procedures.[18,59] The only delicate step is to extrapolate to the continuum from an infinite sequence of discrete decurving operations. But we know how to proceed with our experience in macroscopic electrodynamics[4] and in spin glasses,[9,34] where the

All descriptions concur: the base space, geometrical substrate of amorphous materials has topological defects, which are uninterrupted lines characterized by their existence (oddness) only.

3.   GAUGE INVARIANCE IN CONTINUOUS SPACE

3.1   General Formalism

This key section exhibits the inescapable effects of gauge invariance.

The thermodynamic and ground state configuration of condensed materials can be described semi-microscopically - over length scales larger than the interatomic spacing a - in terms of a physical field or order parameter, the matter field $\phi(\mathbf{x})$. This is the starting point of the Ginzburg-Landau-Wilson theory of phase transitions, but its relevance does not depend on the occurrence of a phase transition, or on an "order" parameter for that matter. The free energy of the material is a functional of $\phi(\mathbf{x})$, which can represent strain in elasticity, magnetization in magnets or a complex order parameter in superconductors. In glasses, $\phi(\mathbf{x})$ describes the orientation of the local frame, or, in a more abstract but algebraically simple formulation[18], the infinite sequence of decurving operations responsible for mapping the polytope {3,3,5} (the ideal archetype of close-packed materials in curved space) into the relaxed, local configuration of the glass in Euclidean space.

One defines a local transformation $\mathbf{T}(\mathbf{x})$ from $\phi(\mathbf{x})$ to another, physically equivalent local configuration $\phi'(\mathbf{x})$,

$$\phi'(\mathbf{x}) = \mathbf{T}(\mathbf{x}) \, \phi(\mathbf{x}) \, \mathbf{T}^{-1}(\mathbf{x}) \, , \qquad\qquad \mathbf{T}^{+}(\mathbf{x}) = \mathbf{T}^{-1}(\mathbf{x}) \qquad (1)$$

$\mathbf{T}(\mathbf{x})$ is the gauge transformation studied in this chapter. It refers here to a rotation of the local frame in the continuum. In general, it can be a bond-switching operation,[21,27] a permutation between the bonds[9] in a network, or the commutator between two sequences of

---

physical fields are averaged over small regions of physical size (plaquette size).

decurving operations.[18] Its essential property is that it must leave invariant all observable quantities, the energy of the material first of all.

$\mathbf{T}(\mathbf{x})$ can therefore be chosen as an element of the group of transformations that leaves the potential energy $V[\phi]$ (a functional of $\phi(\mathbf{x})$, not of its derivatives) invariant. $\phi(\mathbf{x})$, and the other fields $\mathbf{A}_i$, $\mathbf{F}_{ij}$ to be introduced below, have their own internal degrees of freedom, labelled by Greek indices, and their own manifold of internal states. The gauge transformation $\mathbf{T}(\mathbf{x})$ acts within this manifold, in which adjoint (+), trace (Tr) and inverse (-1) operations are also defined.

Disorder, namely non-uniformity of $\phi(\mathbf{x})$, causes strain and costs elastic energy. If $\phi(\mathbf{x})$ fluctuates smoothly, this energy is measured by a term proportional to $\mathrm{Tr}\{(\partial_i\phi)(\partial^i\phi)^+\}$ $(= \mathrm{Tr}\{(\partial_i\phi)(\partial^i\phi^+)\})$ (with summation convention over repeated indices and trace taken over internal (Greek) indices), where ∂_i is the ordinary gradient.[*] This is the familiar fluctuation contribution to the Ginzburg-Landau expansion of the free energy density. Unfortunately, this expression is not gauge invariant, and cannot therefore represent a physical (objective) quantity: The last two terms in

$$\partial_i\phi'(\mathbf{x}) = \partial_i[\mathbf{T}(\mathbf{x})\phi(\mathbf{x})\mathbf{T}^{-1}(\mathbf{x})] = \mathbf{T}(\partial_i\phi)\mathbf{T}^{-1} + (\partial_i\mathbf{T})\phi\mathbf{T}^{-1} + \mathbf{T}\phi(\partial_i\mathbf{T}^{-1}) \qquad (2)$$

spoil the covariance. Rather than the ordinary derivative, it is the <u>covariant derivative</u>,

$$D_i\phi(\mathbf{x}) = \partial_i\phi + g\,[\mathbf{A}_i,\phi] \qquad (3)$$

which is covariant:

$$D_i\phi' = (D_i\phi)' = \mathbf{T}\,(D_i\phi)\,\mathbf{T}^{-1}, \qquad (4)$$

and has a gauge-invariant contribution to the free energy.

---

[*]   Note the unusual $\partial_i = \partial_i^+$ here, because + operates in the (Greek) internal manifold, and ∂_i within (Latin) space.

Covariance has been achieved at the cost of introducing a new quantity, the gauge field $A_i(x)$ (the vector potential in electromagnetism, connection in differential geometry), which transforms as

$$A_i'(x) = T(x) A_i(x) T^{-1}(x) + (1/g) T(x) [\partial_i T^{-1}(x)] \qquad (5)$$

under the local gauge transformation $T(x)$. g is a coupling constant which is physically measurable (for example, through flux quantization in superconductors). $A_i^+ = - A_i$ is antihermitian. (Recall that the adjoint operation acts within the manifold of internal states; it does not affect the gradient ∂_i). Equation (3) shows that $A_i(x) \neq 0$ signals non-collinearity of the reference frames in the neighborhood of x.

So far, the sole purpose of the gauge field $A_i(x)$ has been to guarantee covariance. It has, however, the geometrical interpretation of a connection in differential geometry (§6) or in fiber bundles (§4),

$$g\, A^\alpha_{\beta i} = \Gamma^\alpha_{\beta i} \qquad (6)$$

as can be seen by comparing eq.(3) with the covariant derivative of a tensor of rank 2 (obtained by parallel transport of the tensor, similar to that given by eq.(6.11) for a vector).*

A non-uniform gauge field makes its own contribution to the free energy (if it did not, $A_i(x)$ would simply be a Lagrange multiplier enforcing an external, quenched constraint, rather than a dynamical variable), which has, to lowest order in the field $A_i(x)$, the gauge-invariant form $\mathrm{Tr}\{F_{ij}(x) F^{ij+}(x)\}$, where

$$F_{ij}(x) = \partial_i A_j(x) - \partial_j A_i(x) + g\, [A_i, A_j] \qquad (7)$$

---

* With all indices explicitly shown, this equation is

$$D_i \phi^\alpha_\beta = \partial_i \phi^\alpha_\beta + \Gamma^\alpha_{\delta i}\, \phi^\delta_\beta - \Gamma^\delta_{\beta i}\, \phi^\alpha_\delta$$

It is identical to (3).

is only gauge covariant in general

$$\mathbf{F_{ij}}' = \mathbf{F_{ij}}\,[\mathbf{A'}] = \mathbf{T(x)}\,\mathbf{F_{ij}(x)}\,\mathbf{T^{-1}(x)} \tag{8}$$

(see the curvature in differential geometry (5.9)). To be gauge-invariant (magnetic induction in electromagnetism), $\mathbf{F}$ must be linear in $\mathbf{A}$, or the gauge field must be abelian.

In the decurving model of glass,[18] $\mathbf{F_{ij}}$ turns out to be gauge invariant, abelian and therefore observable directly. Disorder is here a local commutation of the order of decurving operations. In general, one can always define a gauge invariant expression proportional to $\mathbf{F_{ij}}$, for a given structure of the glass, where the sources of non-collinearity (cores of disclinations) are very localized and do not overlap (see eq.(23) below). They can be excluded from the elastic medium Σ which they puncture, and act only as a surface contribution to the free energy. The structure is then characterized arcwise from some starting point, in a continuous version of the spanning tree used in electrical network theory,[35,9] and the gauge invariant $\mathbf{F_{ij}}$ measures the density of topological disorder.

Finally, it is convenient to exclude from the space Σ (the base space of the bundle, into which gauge and matter fields are put) the geometrical cores of the odd lines (loci of non-triviality of the bundle, sources of non-collinearity). Σ is then a flat Euclidean continuum, punctured by odd lines. This complicates slightly the geometry (Σ is multiply-connected), but simplifies enormously the algebra. Notably, the amplitude of the matter field (in the ground state configurations) is never singular within Σ. It is even constant within Σ in the extreme type II limit (§3.4). Boundary conditions on the punctures (the inner surfaces of Σ) are free, the only requirement being one of compatibility between configurations around a puncture. This allows the fields to take up configurations (e.g. rotation by 2π around a puncture) which would have been forbidden, or singular if the space had been simply-connected. This does not preclude the existence of (singular and mobile) vortex excitations at higher temperatures. Gauge invariance is unaffected, because we are dealing here with the geometry of base space, and not with the nature of the fiber and its transformations.

The gauge invariant free energy functional is thus,[3]*

$$F = \int_\Sigma d\mathbf{x}\, f(\mathbf{x}) \,, \qquad f(\mathbf{x}) = f[\phi(\mathbf{x}), \mathbf{A}_i(\mathbf{x})] \,, \qquad (9)$$

$$f(\mathbf{x}) = (1/2)\, \mathrm{Tr}\, \{\gamma\, (D_i \phi)(D^i \phi^+) + \mathbf{F}_{ij}\, \mathbf{F}^{ij+}\} + V(\sqrt{\mathrm{Tr}\phi\phi^+}).$$

γ measures the relative contributions to the free energy due to non-uniformity of matter and gauge field.** The potential V reduces to an additive constant in glasses and in extreme type II superconductors.[36]

Note that the coupling between matter (ϕ) and gauge ($\mathbf{A}_i$) fields is minimal, through the covariant derivative (3). A gauge field is necessary and sufficient to make the derivative covariant, and since this is the only reason for its introduction, one can assume that the covariant derivative is the only coupling between gauge and matter field. This simple reasoning leads to three essential physical consequences (§3.2 and Table I). The idea of a continuous, local gauge symmetry in continuous space is due to Weyl,[2] and has been generalized by Yang and Mills.[3] It becomes a purely geometrical theory in the "extreme type II limit" (in superconducting terminology[36]).***

---

* Σ is a (punctured) Euclidean continuum, with metric $g_{ij} = \delta_{ij}$, and there is no need to introduce a covariant volume element.

**   Only $g\mathbf{A}_i$ is defined by eq. (3). The magnitude and dimension of the gauge field $\mathbf{A}_i$ are chosen by convention so that its contribution to the free energy density, in the absence of matter field, is the invariant $\mathrm{Tr}\, \mathbf{F}_{ij}\, \mathbf{F}^{ij+}$, with a numerical constant (the free energy is the integral of this invariant over the covariant volume $\sqrt{(\det g_{ij})}\, d\mathbf{x}$). Here, following electromagnetism and superconductivity, we have chosen $1/2$ for the numerical constant. Weyl[2] uses $1/4$. In electromagnetism and superconductivity, the "dimensionless" constant depends of the system of units: In cgs, it is $1/8\pi$. In SI, it is $1/2\mu_0$.

***   The extreme type II limit is when the length scale (ξ) of fluctuations of the matter field is very much shorter than the gauge

3.2 Physical consequences of minimal coupling, Higgs mechanism

Consider the example of superconductivity: The matter field is a complex function, the superconducting order parameter or Cooper pair wave function $\phi(\mathbf{x}) = \rho(\mathbf{x}) \exp[i\theta(\mathbf{x})]$. The Ginzburg-Landau free energy density (9) reads, in this case (and in SI units-see second footnote, preceding page),

$$f(\mathbf{x}) = (1/2)\gamma|[\partial-(2\pi i e^*/h)\mathbf{A}(\mathbf{x})]\phi(\mathbf{x})|^2 + V(|\phi(\mathbf{x})|) + (\text{curl } \mathbf{A}(\mathbf{x}))^2/(2\mu_0)$$

$$= (1/2)\gamma\{(\text{grad}\rho)^2 +\rho^2 [\text{grad}\theta-(2\pi e^*/h)\mathbf{A}]^2\}+V(\rho)+(\text{curl}\mathbf{A})^2/(2\mu_0) \quad (10)$$

where $\mathbf{A}$ is the electromagnetic potential, $e^*=2e$ is the charge of the Cooper pair, and the potential $V(\rho)$ has a minimum at $\rho=\rho_0\neq0$ in the superconductor. The last term represents the energy of the electromagnetic field (in the absence of an external field). One recognizes the covariant derivative which ensures minimal coupling between gauge ($\mathbf{A}$) and matter (ϕ) fields. The Euler-Lagrange equations minimizing the free energy under variations of the fields are coupled equations in the amplitude ρ of the matter field (the celebrated Ginzburg-Landau equation) and in the gauge field together with the phase θ of the matter field (the London-Josephson

---

field's (λ) and than any other geometrical length in the problem (such as the average distance ζ between two non-contiguous disclination segments). Sources of non-collinearity - the cores of disclinations - extend only up to ξ (the restoring length for the matter field, forced to vanish at the center of the disclination core). The first physical examples of this geometrical vortex state or punctured continuum were type II superconductors - alloys (mostly) with useful properties - whose existence was postulated by Abrikosov in 1957.[37] In a representation of glass as an elastic continuum where every atom is a full sized tetrapod, the core of any vortex cannot exceed the size of the shortest ring, so that $\xi\approx a$, the interatomic spacing. Glasses are therefore as much in the extreme type II limit as they are continuous. Their vortices are the odd lines or 2π-disclinations, with a line tension energy given by the contributions to F (9) of the inner surfaces of the punctured space Σ.

equation). In the absence of superconductivity ($\rho=0$), the gauge field has two, massless (with infinite decay length), transverse modes, the photon with its two polarizations. In the absence of a gauge field, one would have one massless phase mode (spin wave in planar (XY) magnets which have also a complex function as order parameter), and, of course, no local (gauge) invariance. Minimal coupling combines these three massless modes in one massive vector boson (one single finite penetration depth for the magnetic induction in superconductors. This is Higgs's (-Kibble-Brout-Anderson...'s) mechanism.[38]

To obtain this result, simply find $\mathbf{C}$, a linear combination of gauge and matter fields which is gauge invariant, namely

$$\mathbf{C}(x) = \mathbf{A}(x) - (h/2\pi e^*) \text{ grad } \theta(\mathbf{x})$$
$$\mathbf{A}' = \mathbf{A} + \text{grad } \chi(\mathbf{x}) \tag{11}$$
$$\theta' = \theta + (2\pi e^*/h) \chi(\mathbf{x})$$
$$\mathbf{C}' = \mathbf{C} \ .$$

Then, the free energy reads

$$f(\mathbf{x}) = f_0 [\rho] + [(\text{curl } \mathbf{C})^2 + (\mathbf{C}/\lambda)^2]/(2\mu_0) \ , \tag{12}$$

where $1/\lambda$ is the mass of the mode $\mathbf{C}$ (a massive vector boson which has now three polarizations, and combines the original phase and photon modes); alternatively,

$$\lambda = \rho^{-1}(h/2\pi e^*)/\sqrt{(\mu_0\gamma)} \tag{13}$$

is the (now finite) penetration depth, the characteristic length of the gauge field. λ is, as anticipated, inversely proportional to the magnitude of the superconducting order parameter $\rho = \sqrt{n_s}$, with n_s, the superconducting density. This Higgs mechanism has been known in superconductivity since 1933 under the name of Meissner

(-Ochsenfeld) effect. The magnetic induction F = curl A = curl C is indeed gauge-invariant.*

The phase of the matter field appears as a gradient in the free energy and in the London-Josephson equation, curlcurl $C = - C/\lambda^2$. For a given superconducting configuration, this phase must change by an integral multiple of 2π when carried around a closed contour in space. Integration of the equation on a closed contour O inside the superconductor about a puncture or any normal region, leads to quantization of the fluxoid** in units of the flux quantum $\Phi_0 = h/e^* = 2 \times 10^{-15}$ Wb. Bloch[39] has shown that this quantization is a general property of a system whose free energy is gauge-invariant, with a homotopically non-trivial gauge group (here, $\pi_1(SO(2)) = Z \neq 1$). The free energy is then a periodic function of the external flux inducing the gauge transformation.

For example, consider a superconducting ring Σ. The free energy within Σ is gauge-invariant, the magnitude of the matter field is uniform within Σ, and the new configurations associated with multi-connectivity of base space Σ are degenerate, and characterized by quantized fluxoid. The degeneracy is only lifted when the electromagnetic free energy outside Σ (due to magnetic field trapped inside the ring's hole) is added. The multiple connectivity of S implies that there are several possible classical ground states, all degenerate because they are related to each other by large gauge transformations. This is to be contrasted with the unique ground

------

*   The coherence length ξ (length scale of matter field fluctuations) is easily found by expanding $V(\rho)$ about its minimum at ρ_0, and rescaling $\rho^* = \rho/\rho_0$ in (10) or (12). Then,

$$f_0[\rho] \cong (1/2)V''(\rho_0)\, \rho_0^2\, \{\xi^2\, (\text{grad } \rho^*)^2 + (\rho^*-1)^2\}\ ,$$

and $\xi = \sqrt{[\gamma/V''(\rho_0)]}$.

**   The fluxoid is $\Phi + (2\pi e^*/h) \int_O dx\ \lambda^2$ curlcurl A . Φ is the magnetic flux through O. Well away from the punctures, the second term (circulation of the supercurrent curlcurl A/μ_0) vanishes, and the flux is quantized in units of Φ_0

state in simply-connected space, which is one of the cornerstones of classical (crystalline) solid state physics.

Note that the phase θ is defined everywhere, except at the core of vortices and outside the superconductor, where the magnitude ρ of the order parameter vanishes. It does not contribute to the curvature (curl $\mathbf{C}$ = curl $\mathbf{A}$), except upon encircling a vortex, when it changes by 2π. θ is therefore uniquely defined everywhere, arcwise from a given origin within the superconductor, through the analogue in the continuum of the spanning tree of electrical network theory.[35,9] This continuum spanning tree is the space Σ, cut from every vortex puncturing it. The same geometrical construction is used to define the phase of the matter field in glasses. In the extreme type II limit, $\lambda >> \xi \approx a$, magnitude of ϕ and penetration depth λ (13) are constant within Σ (namely, within the superconductor punctured by the vortices).

Change of spanning tree is a large (discrete) gauge transformation: The phase is changed at a point $\mathbf{x}$, influencing its surrounding through the gradient, with suitable counteraction in $\mathbf{A}$ to keep $\mathbf{C}$ invariant. (The level at which one finds oneself in a castle depends on the path from the entrance gate selected on the floor map, or on the choice of spanning tree).

The three physical results obtained above in the case of superconductivity (a minimally coupled, SO(2) (abelian) gauge theory), namely: i) finite penetration depth, ii) existence of large gauge transformations giving rise to flux or fluxoid quantization, and iii) periodicity of the free energy as a function of the applied flux generating the gauge transformation, responsible, among others, for the Josephson effect[39], are in fact general consequences of minimally-coupled gauge theories. (Table I). To demonstrate this result in the general case of a non-commutative gauge group like SO(3) in glasses requires some care since the phase of the matter field is path-dependant even in a region of the material free of vortices or punctures. The following subsections are formal and can easily be omitted.

Table I.   Physical consequences of minimal coupling for gauge theories.

Gauge theory	Superconductivity (SO(2))	Glass (SO(3))
Higgs mechanism	Meissner effect (finite penetration depth)	finite energy $k \approx 0$ rotation modes
configurations classified by sectors	vortices (ground state in punctured geometry or elementary excitations)	odd lines or 2π-disclinations (ground state in punctured geometry, or elementary excitations)
large gauge transformations, free energy periodic in cranking flux	flux or vortex quantization, Josephson effect	2 configurations (2π, 4π) about an odd line, tunneling modes

---

3.2.1 The mathematical consequences of minimal coupling (Table I)

One proceeds, formally, as follows.[31] First, decompose the matter field in amplitude (its norm, $\rho = \sqrt{\mathrm{Tr}\,\phi^+\phi}$) and phase, with only the latter affected by gauge transformation:

$$\phi(\mathbf{x}) = [\rho(\mathbf{x})/\sqrt{d}]\,\Omega(\mathbf{x})\,, \qquad \Omega^+ = \Omega^{-1}\,, \qquad \rho^* = \rho, \qquad (14)$$

(Stueckelberg decomposition), with $d = \mathrm{Tr}\,\Omega^+\Omega$,

$$\Omega'(\mathbf{x}) = \mathbf{T}(\mathbf{x})\,\Omega(\mathbf{x})\,\mathbf{T}^{-1}(\mathbf{x})\,, \qquad \rho'(\mathbf{x}) = \rho(\mathbf{x}). \qquad (15)$$

In the extreme type II limit, the amplitude ρ is constant everywhere in Σ. It only vanishes inside the punctures. The potential term V in the free energy is then an additive constant without physical significance. On the other hand, the contribution of the integration of f($\mathbf{x}$) over the surfaces of the punctures is physical. It measures the tension energy of the frustration, odd or vortex lines.

The phase operator Ω can be defined everywhere: One proceeds arcwise from its value $\Omega_0 = \Omega(\mathbf{x}_0) = (\Omega_0^+)^{-1}$ at an arbitrary origin in space, without ever enclosing a puncture (vortex or odd line). Thus, on a tree spanning the punctured space Σ, Ω is given by

$$\Omega(\mathbf{x}) = \mathbf{W}(\mathbf{x})\, \Omega_0\, \mathbf{W}^{-1}(\mathbf{x}) \tag{16}$$

with the operator defined as an ordered sequence of operations (rotations in internal space) along a specific path from point $\mathbf{x}_0$ to point $\mathbf{x}$. Then,

$$\partial_i\Omega = [(\partial_i\mathbf{W})\mathbf{W}^{-1},\Omega] = \mathbf{W}\,[\mathbf{W}^{-1}(\partial_i\mathbf{W}),\Omega_0]\,\mathbf{W}^{-1} \tag{17}$$

and $\mathbf{X}_i = \mathbf{W}^{-1}(\partial_i\mathbf{W})$ is the phase shift along the spanning tree.[*] Like $\partial_i\theta$ in superconductors, its contribution to the curvature vanishes identically, $\partial_i\mathbf{X}_j - \partial_j\mathbf{X}_i + [\mathbf{X}_i,\mathbf{X}_j] = 0$, as long as no puncture is enclosed. A gauge transformation,

$$\Omega' = \mathbf{T}(\mathbf{x})\,\Omega\,\mathbf{T}^{-1}(\mathbf{x}) = (\mathbf{TW})\,\Omega_0\,(\mathbf{TW})^{-1}\,, \tag{18}$$

simply replaces $\mathbf{W}$ by $(\mathbf{TW})$, so that

$$\mathbf{X}_i' = (\mathbf{TW})\partial_i(\mathbf{TW})^{-1} = \mathbf{X}_i + \mathbf{W}^{-1}\mathbf{T}^{-1}(\partial_i\mathbf{T})\mathbf{W}\,. \tag{19}$$

The covariant derivative of $\mathbf{W}$ is given by

$$D_i\Omega = \partial_i\Omega + g\,[\mathbf{A}_i,\Omega] = \mathbf{W}\,[(\mathbf{X}_i + \mathbf{W}^{-1}\,g\mathbf{A}_i\,\mathbf{W}),\Omega_0]\,\mathbf{W}^{-1} \tag{20}$$

and this suggests introduction of a gauge-invariant linear combination of phase shift $\mathbf{X}_i$ and gauge field $\mathbf{A}_i$ (each of which are individually gauge- (and path-, or spanning tree-) dependant),

---

[*]   Measured in a rotating frame, through $\mathbf{W}$ and $\mathbf{W}^{-1}$ in (17).

$$C_i(x) = X_i(x) + W^{-1}(x)\, gA_i(x)\, W(x) \ , \qquad C_i^+ = -\, C_i \qquad (21)$$

$$C_i{}' = X_i{}' + W^{-1}T^{-1}\, gA_i{}'\, TW = C_i \ \ ,$$

by (20) and (5), and recalling that, by unitarity, $(\partial_i T)T^{-1} = -\,T(\partial_i T^{-1})$. As in eq.(11), the gauge transformation adds to $W^{-1}A_i W$ the same term it subtracts to X_i, and leaves C_i invariant.

The density of topological disorder is given by

$$\mathbb{F}_{ij} \equiv gW^{-1}F_{ij}W = \partial_i C_j - \partial_j C_i + [C_i, C_j] \ \ . \qquad (23)$$

It is gauge-invariant, and physically observable, as are all the terms in the free energy density,

$$f(x) = (1/2)\gamma\,\{(\partial_i\rho)^2 + (\rho^2/d)\mathrm{Tr}[C_i,\Omega_0][C^i,\Omega_0]^+\}$$

$$+\ (1/2\mu_0 g^2)\ \mathrm{Tr}\{\mathbb{F}_{ij}\mathbb{F}^{ij+}\} + V(\rho)\ \ , \qquad (24)$$

which has exactly the form of the superconducting free energy (12), its gauge invariance, and the same physical properties. (Set $\mu_0 = 1$ to relate (24) to (9)).

3.2.2 Application: Superconductivity

The general formalism can be applied to superconductivity, with a dictionary translating operations on a complex function into operations on an operator. We shall use Cayley-Klein operators (Pauli matrices) σ_α, with T and A_i always pointing in the same direction in internal space $\alpha = y$, say, perpendicular to the plane (xz) in which the gauge group (SO(2) in superconductors) acts. Then,

$$A_i \equiv -\, i\sigma_y\, A_i(x)/2\ \ , \qquad\qquad T \equiv \exp\{i\sigma_y\, g\chi(x)/2\} \qquad (25)$$

define the real functions $A_i(x)$ and $\chi(x)$.*

---

* One recovers the ubiquitous half-angles. This is because the phase angle (along the fiber) is half the orientation angle in quantum

The superconducting order parameter is constrained in the (xz) plane of internal space. Let us take $\Omega_0 = i\sigma_z$ along z,

$$\Omega = W \Omega_0 W^{-1} , \qquad\qquad W = \exp\{i\sigma_y \theta/2\} \qquad (26)$$

Note that Ω is antihermitian $\Omega^+ = -\Omega$, as well as unitary. Then,

$$D_i\Omega = -i\rho [\sin\theta \, \sigma_z + \cos\theta \, \sigma_x] [\partial_i\theta - g A_i] \qquad (27)$$

and

$$C_i = i\sigma_y [\partial_i\theta - g A_i]/2 \equiv -i\sigma_y g \, C_i(\mathbf{x})/2 \qquad (28)$$

$$\mathbf{F}_{ij} = (1/g) \, \mathbb{F}_{ij} = -i\sigma_y [\partial_i C_j - \partial_j C_i]/2 = -i\sigma_y (\partial_i A_j - \partial_j A_i)/2 \ ,$$

so that the free energy density (24) is exactly (12), with

$$g = 2\pi e^*/h \ , \qquad (29)$$

since $\mathrm{Tr}\{\mathbf{F}_{ij}\mathbf{F}^{ij+}\} = (\mathrm{curl}\ C)^2$, after summing over Latin indices and taking the trace over Greek indices. Gauge transformation affects $A_i(\mathbf{x})$, $\theta(\mathbf{x})$ and $C_i(\mathbf{x})$ as in (11), with $C_i'(\mathbf{x}) = C_i(\mathbf{x})$, gauge-invariant. Note that $\Omega' = i[\cos(\theta+g\chi)\sigma_z - \sin(\theta+g\chi)\sigma_y] = (\mathbf{TW})\Omega_0(\mathbf{TW})^{-1}$, illustrating the part played by $g\chi(\mathbf{x})$ as a phase shift.

3.3 Ground State Configurations, Tunneling Modes

The ground state configurations are are solutions $\{\Omega, A_i\}$ of the Euler-Lagrange equations minimizing the (gauge-invariant) free energy (9) or (24), or, equivalently, $\{W, A_i\}$ on a given spanning tree, since W parametrizes Ω (with $W(x_0) = 1$) as in eq.(16). C_i is a gauge-invariant, linear combination of the phase shift X_i and the gauge field A_i, neither of which are gauge-invariant. If one configuration can be found, all the others (degenerate with the first) will be obtained by large gauge transformations.

---

mechanics,[5] or in rotating an object with an internal phase, like the ribbon on the turntable of Fig.1.[8]

Let us now close a contour encircling a puncture, or reach the same point $\mathbf{x}$ through a path on the other side of the puncture. Ω must be rotated by a multiple of 2π, with 4π equivalent to the identity. We obtain therefore two possible "classical" configurations of Ω (path-ordered products of $\mathbf{W}(\mathbf{x})$), $|0>$ and $|2\pi>$, corresponding to rotations by 0 or 2π upon closing the contour. But since $\mathbf{C}_i$ is gauge-invariant, $\mathbf{W}$ also parametrizes the gauge field $\mathbf{A}_i$. Because there are only two distinct configurations of the matter field per puncture, up to continuous deformations, there are also two configurations of the gauge field, and therefore two ground state configurations $\{\Omega,\mathbf{A}_i\}$ per odd line, related by a large gauge transformation $\mathbf{G}$ (a rotation by 2π at $\mathbf{x}$). They have the same energy by gauge-invariance of the theory.

Neither "classical" configuration $|0>$ or $|2\pi>$ is gauge-invariant. Each is transformed into the other by a large gauge transformation $\mathbf{G}$, $\mathbf{G}|0> = |2\pi>$, $\mathbf{G}|2\pi> = |0>$. But the physical configurations must be gauge-invariant,

$$|\pm> = (1/\sqrt{2}) \ [\ |0> \pm \ |2\pi>]\ , \tag{30}$$

with one sign for the ground state and the other for the first excited state. Tunneling, however slow, must take place to restore gauge invariance. These quantum mechanical superpositions, formed by tunneling between the two classical potential wells or topological sectors of Fig.4, and split by $h/2\pi$ times the tunneling rate, have been observed in SQUIDS and in glasses: They are the celebrated tunneling modes, whose full identity has been revealed by a combination of four different types of experiments, specific heat, thermal conduction or sound propagation, saturation or (non-linear) ultrasonic attenuation and echoes.[47] See §4.1.

3.4 Gauge, Matter Fields and their Associated Lengths ($\propto$ Mass^{-1})

The formal manipulations above may appear tedious, but they illustrate the generality of minimal gauge coupling, and of its physical consequences. Note the existence of a purely geometrical curvature, quantized, concentrated in the punctures of Σ (the cores of defects, frustration or odd lines), additional to the (infinitesimal) contribution of the gauge field. There is a geometrical arbitrariness remaining in the choice of the spanning tree, and therefore in the definition of $\mathbf{C_i}$, through the rotation operator $\mathbf{W}$. Change of spanning tree produces a large gauge transformation, which is represented as a finite shift along the fiber of the bundle (Figs.7 and 2). The bundle is twisted around the punctures of Σ: it cannot be seen globally as the direct product of base space by fiber, although locally, it behaves as such. A large gauge transformation can also be made on a given spanning tree: One remains with the same gauge-invariant $\mathbf{C_i}$, and the same value of the free energy, but with a different $\mathbf{W'} = \mathbf{T(x)W}$ causing a finite shift in the fiber at point $\mathbf{x}$. Indeed, this transformation corresponds to a rotation of the matter field by 2π or 4π about a closed contour O, $P_s \exp \{\int_O d\mathbf{x}^i \, \mathbf{W'}^{-1}(\partial_i \mathbf{W'})\}$ (solution of (17), with P_s a path-ordering operator), and it leaves invariant the circulation of $\mathbf{C_i}$, $P_s \exp\{\int_O d\mathbf{x}^i \, \mathbf{C_i}\}$.

What is the physical content of the fields ϕ, $\mathbf{A}$, and $\mathbf{F}$? The identification of the matter field ϕ in the continuum was one of the important results of Toulouse[34] and of Dzyaloshinskii and Volovik[14] in spin glasses.* It cannot simply be a vector representing the spin or magnetization density, a natural choice in discrete magnets, since this would lead to line defects of the wrong kind (vortices ($\pi_1(SO(2))$ = Z) for planar (XY) spins, none ($\pi_1(S_2)$ = 1) for Heisenberg spins instead of frustration ($\pi_1(?)$ = Z_2)). It is the magnetization of a very small volume of the material (of the order of a coordination polyhedron of one atom and its magnetic neighbours, or of the diameter of a plaquette), namely a bunch of spins necessarily hirsute (shaggy) or non-collinear because of frustration, whose representative archetype is a rigid body, which gives rise to the correct line defect, $\pi_1(SO(3))$ = Z_2. In glasses, again, it is a shaggy bunch of tetrapods rather than a single tetrapod. (The fundamental

---

* What is a physical quantity in a continuum ? It must be an average over a "physically infinitesimal" volume element. (See ref.[4], §1, quoted in full in section 1.1).

group of the tetrahedral group T, $\pi_1(T)$ is non-abelian,[40] and that would have prevented its defect lines to cross each other in that imaginary material (even in its liquid phase), leading to topological obstruction at all the temperatures at which tetrapods retain their identity. This may be a desirable property for a glass, but awkward in the liquid phase of exactly the same material).

$\mathbb{F}$ measures the density of non-collinearity, that is the density of disorder. It is gauge-invariant, hence observable, in practice, through $\mathbf{W}$. But $\mathbf{W}$ is path-dependant, and only measurable on a spanning tree constructed on the punctured space Σ.

By rescaling the fields ϕ and $\mathbf{A}$ in (24), one obtains typical lengths and energy scales: In covalent glasses, where every atom supports a full-sized tetrapod, cores of odd lines have a radius smaller or equal to the interatomic spacing a. At that distance from the odd line, the matter field has reached its full amplitude, so $\rho=\rho_0$ everywhere in Σ. The coherence length ξ is the range of fluctuations in the magnitude of the matter field. It is also the core radius of odd lines. Thus,

$$\xi = a \, , \tag{31}$$

in the extreme type II limit. Let $\mathbf{C}_i{}^*=\lambda\mathbf{C}_i$, $\rho^*=\rho/\rho_0 \cong 1$. The free energy density (21) reads,

$$f(\mathbf{x}) = (1/2)\gamma\rho_0{}^2/\lambda^2 \; \mathrm{Tr}\{([\mathbf{C}_i{}^*,\Omega_0][\mathbf{C}^{i*},\Omega_0]/d) +$$

$$(1/\mu_0\gamma g^2\rho_0{}^2)(\partial_i\mathbf{C}_j{}^*-\partial_j\mathbf{C}_i{}^*+(1/\lambda)[\mathbf{C}_i{}^*,\mathbf{C}_j{}^*])^2\} + \mathrm{cst} \, . \tag{32}$$

Then, the length λ associated with fluctuations in the gauge field $\mathbf{C}^*$ is, clearly,

$$\lambda = 1/[(\sqrt{\mu_0\gamma})g\rho_0] \tag{33}$$

in agreement with the value (13), (29) for the penetration depth in superconductors. The only energy scale is,

40

$$\gamma \rho_0^2 / \lambda^2 \cong kT_0 \, ,$$

(34)

where T_0 is a typical glass temperature (at which the viscosity is infinite; see footnote this page). We will obtain another expression for this energy scale, in terms of elastic constants, in §6.5 (eq.6.)

A third, geometrical length ζ, measures the distance between disclinations or punctures in Σ. It is the average distance between one element of a disclination line to the nearest element found on a perpendicular plane (the nearest element which is not obviously following the first). If the two elements belong to the same loop, $\zeta=\eta$, the average diameter or girth of the disclination loop. If they belong to different loops, $\zeta=\delta$, the average distance between neighbouring loops. Three cases can be distinguished (the classification is that of polymer solutions), i) the dilute limit (alphabet soup): $\eta<<\delta$, ii) the melt (plate of spaghettis): $\delta<<\eta$, and iii) the semi-dilute case (spaghettis cooking in hot water) where $\delta\cong\eta\cong\zeta$ are indistinguishable on average, and one geometrical length suffices. Averaging over a distribution of loop sizes, maximum entropy arguments,[16,61,65] and inspection of hand-built[41] and iteratively decurved models,[18] all suggest that odd lines or 2π-disclinations in glasses are semi-dilute. In all cases, ζ is a natural cut-off length for the strain caused by one single disclination loop, so $\partial_i\mathbf{C^*} \cong \mathbf{C^*}/\zeta$. The non-linear term in $\mathbf{F}_{ij}$ (31) is negligible beside the gradient in the semi-dilute configuration $\lambda>\zeta$. It dominates the gradient in the dilute limit. ζ should therefore be compared to the extent λ of non-collinearity: If $\zeta>\lambda$, loops add as a superposition of individual sources of non-linear strain. If $\zeta<\lambda$, they act collectively, but the elastic strain is linear.*

---

*   Mobility of 2π-disclinations characterizes the transition between viscous liquid and frozen glass. In the glass, disclinations are frozen punctures, whose core is excluded from the space Σ in which matter is put. In the liquid, 2π-disclinations are mobile excitations with core within Σ. One can define a density of mobile disclinations $\rho_m=(a/\zeta)^2$ as an "order" parameter measuring the transition. ρ_m decreases with decreasing temperature in the liquid,

ζ is also the length over which the "rotation flux quantum" must move in tunneling modes (30). One obtains[61,58] a tunneling rate proportional to $\exp[-cst \, (\zeta/a) \, \ln \, (\zeta/a)]$, which is non-negligible because the rotation flux quantum is a short (size ζ), fat (range λ) object which has to move very little (over a distance $\zeta \ll \lambda$). The ln is indication of semi-dilution. By contrast, for a macroscopic superconducting ring, the tunneling rate goes as $\exp[-cst \times size \times distance]$ and is utterly negligible.

Minimally-coupled gauge theories are therefore highly automatic and geometrical. Once the gauge transformation (1) and the defects (sources in base space of non-triviality of the bundle) have been identified, one remains, in the extreme type II and semi-dilute regime, with two lengths ζ and λ (33), and one energy scale kT_0 (34).

Kerner[42] was the first to translate the ideas of defect[12,23], curvature[23] and gauge invariance[31] in glasses in the general fiber bundle formalism.

4. GAUGE INVARIANCE IN DISCRETE SPACE

4.1 Discrete Gauge Invariance in Spin Glasses

Gauge invariance in disordered condensed matter was first mentioned in a paper by Toulouse[10] dealing with spin glass on a lattice, described by the Edwards-Anderson Hamiltonian,

$$H = \Sigma_{(ij)} \, J_{ij} \, S_i \, S_j \tag{1}$$

---

to vanish at and below the Vogel-Fulcher temperature T_0 where the viscosity is infinite. At very high temperatures, ζ is small, and disclinations act collectively as sources of strain. As the temperature decreases, ζ increases, but disclinations keep acting collectively, until T^* ($>T_0$) (given by $\zeta \cong \lambda$) where they become individual sources of strain, and the viscosity crosses from Vogel-Fulcher behaviour (collective) over to Arrhenius (individual). T^* lies very close to T_0 for "fragile" liquids (e.g. o-terphenyl), further away for "strong" liquids (e.g. SiO_2) in the classification af Angell.[67].

42

where $S_i=\pm 1$ (Ising spins on lattice points i), and the coupling between nearest neighbour spins (ij) is $J_{ij}=\pm J$, according to some probability distribution given a priori. The {J} are said to be quenched. Hamiltonian (1), and thus the physics of the system, are invariant under the local transformation

$$S_i{}' = \tau_i \, S_i \, , \qquad J_{ij}{}' = \tau_i \, J_{ij} \, \tau_j \, , \tag{2}$$

parametrized by $\tau_i=\pm 1$ (Z_2 gauge transformation). This is a local transformation, involving variables which live on vertices (dynamical variables) and on edges (connections); it is therefore a gauge transformation. Transformation (2) is exact. It is also not very useful, since it involves two variables of physically different status: The spins S_i are dynamical variables, allowed to reach thermodynamic equilibrium within a canonical ensemble (Boltzmann distribution). By contrast, the couplings J_{ij} are fixed (quenched) when the sample under investigation was made. They are only random in the ensemble of different, but physically equivalent, realizations of similar spin glasses. It is the free energy of every particular spin glass realization which is averaged over the couplings. In short, half of the gauge transformation $J \to J'$ is not physically realizable on a given sample.

Invariants under transformation (2) include not only the Hamiltonian, but also the face (or plaquette) index $\Phi_f = \prod_{e \in f} J_e$ and therefore the geometrical frustration (or oddness in the language of §2.2), sig $\Phi = -1$. (It is easy to show that frustration, like oddness, forms close loops or lines terminating on the surface of the material by using the same demonstration as in §2.2). Frustration is a geometrical property, independent of the matter field S_i, so that geometry (and essential, topological disorder) is preserved under gauge transformation. Moreover, the partition function for a given sample, $Z[\{J\}]$, and thus its physics, is independent of all the details of the distribution of couplings {J} apart from those which are gauge invariant:[43]

$$Z[\{J'\}] = Z[\{J\}] \tag{3}$$

Thus, the tiling (warping, curvature, oddness,...) $\{\Phi\}$ characterizes completely the geometry of the system, and is gauge invariant. It is the base space of the fiber bundle (Fig.2). Frustration lines are the loci of non-triviality of the bundle, like the odd lines in random networks discussed in §2.2. As a corollary, the unfrustrated or Mattis model ($J_{ij}=J_i J_j$, sig $\Phi=1$), has the same statistical mechanics as a ferromagnet. Φ is also called Wilson's loop integral.

Transformation (2) cannot meaningfully be generalized to continuous (XY or Heisenberg) spins, because the couplings J_{ij} are essentially real numbers. By going from a lattice to the continuum, and from a microscopic (spins) to a semi-microscopic description of the matter field (hedgehog of spins on a plaquette: $\phi \in SO(3)$), a full exploitation of gauge invariance in disordered condensed matter (Yang-Mills theory) becomes possible (§3 and ref.[34,14]).

This simple example (2) of a spin glass on a lattice emphasizes the main characteristics of gauge invariance in disordered systems:

a) Gauge invariance is an exact symmetry of H. *

b) It preserves essential geometrical ingredients (frustration or odd lines).

c) It enables us to recover as much generative homogeneity as is compatible with (b), for example, by treating the couplings $\{J\}$ as

---

* It is amusing to see what happens to gauge invariance within the replica formalism.[44] For simplicity, consider the infinite range model for N spins,[45] with gaussian distribution of couplings. Then,

$$<\ln Z> \;\rightarrow\; <Z^n> = \Sigma_{\{S\}} \exp[\beta^2 J^2/(2N)\, \Sigma_{ij}\, \Sigma_{\alpha\beta}\, S_{i\alpha}\, S_{j\alpha}\, S_{i\beta}\, S_{j\beta}],$$

which is invariant under two types of transformations: Either a local spin flip $S_i{}^{\alpha'} = \tau_i\, S_i{}^{\alpha}$, in all replicas, or a global rotation of each replica independently, $S^{\alpha'} = R(\theta_\alpha)\, S^{\alpha}$. The Edwards-Anderson order parameter $q^{\alpha\beta} = <S_i{}^{\alpha}S_i{}^{\beta}>$ is not invariant under the latter transformation, which expresses the fact that the symmetry between replicas is broken ($<q_{\alpha\beta}>\neq 0$). The system is trapped in one of the many valleys (replicas) in configuration space, and hydrodynamic modes are associated with this broken symmetry.[46]

dynamical variables, and including the geometrical frustration (b) as constraints (source terms).*

The fiber bundle representation of an Ising spin glass has been given in Fig.2.

4.2 Classical Ground State Configurations in Random Networks

4.2.1 Potential valleys in configuration space

Glasses have anomalous physical properties - specific heat, phonon transport, saturability and echoes (coherence) - below 1K,[47] which can be modelled by tunneling modes between pairs of potential valleys in configuration space, represented schematically (but accurately as far as the energy scale and the physical properties are concerned) in Fig.4. Apart from the fact that one does not know precisely what tunnels, the presence of nearly degenerate (to within less than 10^{-4} eV) classical ground states (potential minima) in a system with no obvious symmetry to impose the degeneracy, is astonishing: Bulk condensed matter usually has one single ground state, and its potential energy, one single minimum in a many-dimensional configuration space. Excitations about this minimum are phonons, which can also be heard in glass. By contrast, condensed matter with punctures can accommodate several, nearly degenerate, low-energy configurations. For example, the energy of a superconducting ring trapping n magnetic flux quanta only depends on n through the contribution of the magnetic field energy outside the ring. The energy inside the material is identical for all n.

Observation of echo (analogous to spin echo, but generated by sound pulses) suggests that different tunneling modes - different pairs of potential valleys in Fig.4 - are uncoupled enough to preserve phase coherence between excitation and echo (~10μs at 20 mK).

---

*   From eq.(3), $Z = \Sigma_{\{J\}=\pm J} Z[\{J\}]$ δ(frustration). An integral representation of the delta function restores the full gauge invariance, at the price of a more complicated effective Hamiltonian.[43]

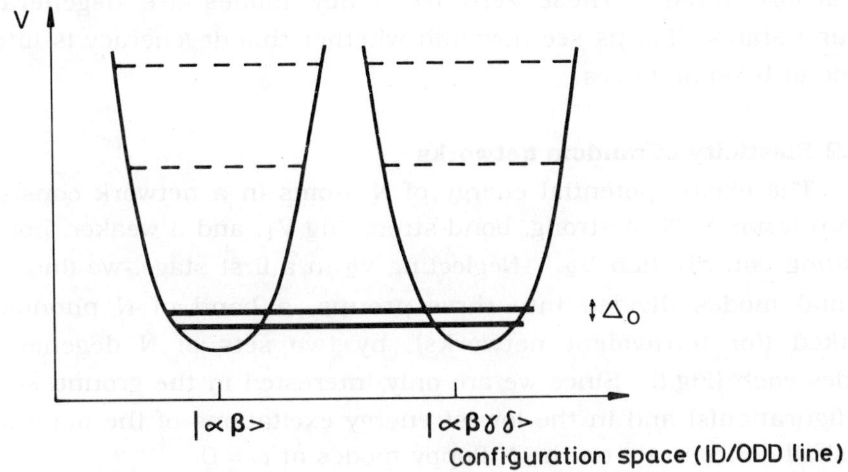

Fig.4.   Potential valleys in configuration space, giving rise to tunneling modes. The configuration space is a direct product of one-dimensional subspaces, each associated with one odd line. Only the topology of the valleys is necessary to obtain qualitatively ground state and elementary excitations in glass.[9]

The elastic energy spectrum of continuous random networks about odd lines is that of Fig.4[9] Even rings have standard spectrum, with one single potential minimum. By contrast, odd lines have (in a z=4 network) two alternative ground state configurations, degenerate in energy. The degeneracy is due to gauge invariance.

To demonstrate these assertions, consider a continuous random network. The elastic energy is carried by the bonds (as in spin glasses (1)). There are two types of potential energies, bond-stretching and bond-bending. Bond-stretching is very much the stronger of the two, but neglecting bond-bending altogether leaves the network underconstrained (wobbly), with as many zero frequency modes as there are atoms if the network is tetracoordinated. These zero frequency modes are degenerate ground states. Let us see how and whether this degeneracy is lifted by bond-bending forces.

4.2.2 Elasticity of random networks

The elastic potential energy of N atoms in a network consists of two terms:[48,9] A strong, bond-stretching V_1, and a weaker, bond-bending contribution V_2.* Neglecting V_2 in a first stage, we find the normal modes divided into three groups, a band of N phonons, flanked (for tetravalent networks), by two sets of N degenerate modes each (Fig.5) Since we are only interested in the ground state configuration(s) and in the lowest energy excitations of the material, we shall concentrate on the N floppy modes at $\omega = 0$.

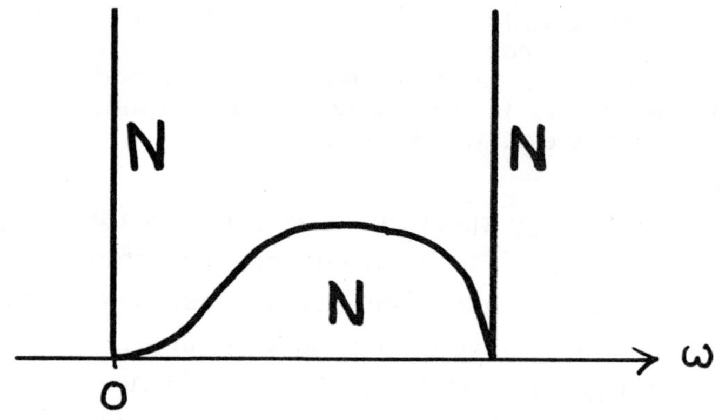

Fig.5. Schematic spectrum of the normal, bond stretching modes of a z = 4 covalent network representing a-Si.[37,9]

---

* V_1 can be written in terms of the relative displacement of two neighbouring atoms only. V_2 requires the direction towards a third atom as well. It is still a two-body potential, but the "bodies" are tetrapods.

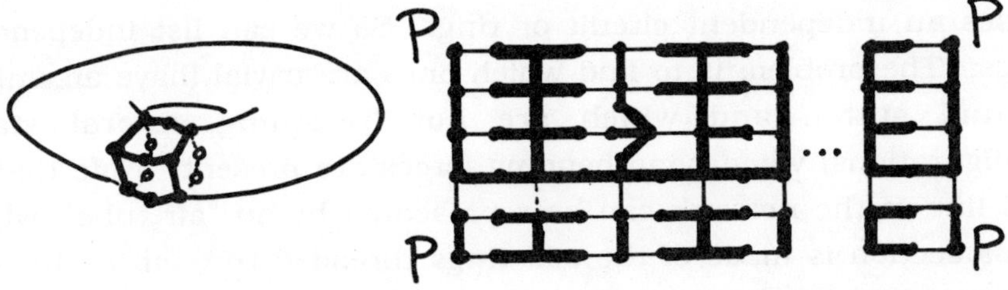

Fig.6.   The "air tube" subgraph of the random network surrounding an odd line. The subgraph can be laid down flat by making two cuts, one longitudinal, the other transverse (and imposing periodic boundary conditions). All the rings on the plane are even (they are on the surface of the tube), except those broken by the longitudinal cut, which are odd by definition of an odd line. (The rings broken by the transverse cut may also be odd, but would then be represented by another, thinner tube threading through this one).   On this subgraph, a spanning tree can be drawn (bold), and independent rings (closed by an edge not on the spanning tree) defined. All the independent rings are the same as on the plane (and even), except for the two (---- and ....) closing each cut. The independent ring ---- closing the longitudinal cut is odd, and is the representative of the odd line.

Accounting for these N floppy modes is the same as for the independent currents in an electrical network* (Kirchhoff):[35] They

---

*   Denote by $\mathbf{r}_i$ the displacement of atom i, by $\mathbf{r}_{i\alpha}$, that of its neighbour along direction α, and by $\mathbf{u}_{i\alpha}$ the direction of bond $i\alpha$ before displacement ($\alpha=1,\ldots,4$). Bond $i\alpha$ remains unstretched if $q_{i\alpha} = \mathbf{r}_i \cdot \mathbf{u}_{i\alpha} = \mathbf{r}_{i\alpha} \cdot \mathbf{u}_{i\alpha}$. So, an $\omega=0$ mode is characterized by an edge (bond) variable $q_{i\alpha}$, like a current in an electrical network.   Moreover, equilibrium requires $\sum_\alpha \mathbf{u}_{i\alpha}=0$, thus, $\sum_\alpha q_{i\alpha}=0$, Kirchhoff's current law.   There is also a ring closure relation analogous to Kirchhoff's voltage law, but it is unnecessary to count independent modes.[35,9] With $E=(z/2)N$ edges variables $q_{i\alpha}$, and N current law constraints, the number of $\omega=0$ modes is N for $z=4$. (Alternatively, there are $R_1=E-N+1$ independent circuits, where R_1 is the cyclomatic, or first Betti number, a topological invariant of the graph. The additional mode is the $\omega=0$ mode of the phonon band (rigid translation)).

are on edges not on a spanning tree of the network, each of which closes an independent circuit or ring. So we can list independent rings. The problem is to find which ones are trivial (have an unique ground state), and which are not (retaining several stable configurations) when bond-bending forces are present. Note that an odd line in the network can be represented by an "air tube", whose cross section is made of the odd rings threaded through by the odd line, and with even rings on its surface. (Fig.6).

Let us now include the bond-bending energy V_2, and see how the N+1 floppy modes tighten. Specifically, how many of these floppy modes survive as ground states? In order to answer this question, we must describe a configuration and measure its energy. By construction, floppy modes have only bond-bending energy. For a given bond, this energy is measured by comparing the orientations of the two tetrapods which it links, or, equivalently, through a congruent transformation of the tetrapod, from its orientation at i to that at iα. This is a generalization of the bond energy $S_iJ_{ij}S_j$ in spin glasses, which is given by comparing the directions of S_i and S_j, or by the congruent transformation (flip or identity) J imposes on the spin: Here, the congruent transformation is a rotatory-reflection, because bond α is common to the two tetrapods and imposes a mirror reflection. (The rotation part of the transformation is irrelevant to our argument).

The configuration of a n-sided ring is the product of n rotatory-reflections. In a given configuration, the tetrapod must be returned to its original orientation after being carried around the ring. This product of rotatory reflections is thus a covering transformation of the tetrapod, namely a permutation of its legs, or of the bond labels. If the ring is even, the permutation, a product of n reflections, is even. If the ring is odd, the permutation is odd. A ring is therefore a path in the discrete fiber bundle, returning to the same fiber (it is closed in base space), but not necessarily to the same point on the fiber. The permutation is the large gauge transformation moving along the fiber.

It is not the permutation itself which labels the configuration, but only its class. Suppose that we were to go around two different rings in succession, recording permutation R through one of them, permutation Q through the other. The total permutation is P = Q.R, or P' = R.Q, depending on the order of the circumnavigations, and, in general, P and P' are different. They are related by

$$P' = R.P.R^{-1} \tag{4}$$

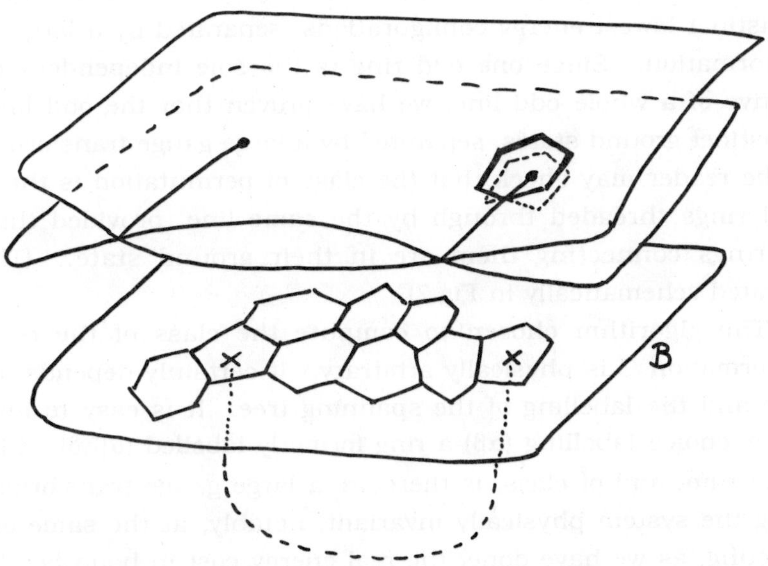

Fig.7.   The fiber bundle for the ground state of an elastic random network with z=4.   Here, the total space represents schematically the class of the covering transformation (permutation) of the tetrapod carried around a ring (in base space). The fiber consists of two points. ---- = odd line, x = odd ring, B = base space. Note the similarity with Ising spin glasses (Fig.2), and the difference: Fig.2 describes all spin configurations, ground state as well as excitations (the fiber represents the direction of the spin).

50

and belong to the same class of the permutation group. The physical configuration, independent of the order of circumnavigations made to measure it, is labelled by the set which includes P, P', etc., namely by the <u>class</u> of the permutation group to which they belong. The physical fiber bundle identifies in the fiber permutations belonging to the same class. (Fig.7).

The ground state of even rings clearly belongs to the identity class of the permutation group of degree $z=4$, S_4. It is non-degenerate, and even rings are dynamically trivial. The configurations of odd rings are labelled by the two, isomorphous classes of odd permutations of S_4 $((\alpha\beta)$ and $(\alpha\beta\gamma\delta)$, typically), so that odd rings have two distinct lowest-energy configurations, separated by a large gauge transformation. Since one odd ring is the sole independent representative of a whole odd line, we have proven that the odd line has two distinct ground states, separated by a large gauge transformation. But the reader may check that the class of permutation is the same for all rings threaded through by the same line, provided that the even rings connecting them are in their ground state. (This is illustrated schematically in Fig.7).

The algorithm chosen to compute the class of the covering transformation C is physically arbitrary. It certainly depends on the choice and the labelling of the spanning tree. It is easy to imagine another choice labelling $(\alpha\beta)$ a ring formerly labelled $(\alpha\beta\gamma\delta)$. Change of algorithm, and of class, is therefore a large gauge transformation, leaving the system physically invariant, notably, at the same energy (neglecting, as we have done, the real energy cost in bond-bending)*. Moreover, neither classical ground state configuration $(\alpha\beta)$ or $(\alpha\beta\gamma\delta)$ are gauge-invariant. The true, gauge-invariant ground state of an odd line is a tunneling, linear superposition of the classical configurations, $|+> = (1/\sqrt{2})[|\alpha\beta> + |\alpha\beta\gamma\delta>]$. There is also one low-lying excitation state, $|-> = (1/\sqrt{2})[|\alpha\beta> - |\alpha\beta\gamma\delta>]$. This fully confirms Fig.4, which is also the representation obtained in disordered elastic

---

* We have only investigated the geometrical consequences of a bond-bending energy, the reflection part of the rotatory-reflection connection.

continua (§3.3 and 6.5. See refs.[16,31]). It is also the picture of tunneling modes obtained by successive decurving operations.[18]

Note that the geometry of the base space (rings are either even or odd, odd rings form loops) does not match exactly the dynamics - geometry of the full bundle, or homotopy - (even rings have trivial ground states, whereas odd rings can be in either one of two states of twistedness or vorticity). Non-trivial geometry (frustration, odd line) is only the source of non-trivial dynamics, which also depends on the nature of the dynamical variables (tetrapods) through the connection, i.e. of the full bundle.

A final remark concerns a random network with z=3, which serves as a model for a-As. The odd permutations of S_3 belong to one class only: There can be no microscopic tunneling modes in a-As.

We have seen in this section (and in Figs.1.2 and 4.4), two examples of discrete fiber bundles which are non-trivial because of topological disorder (frustration or odd lines). Discreteness is a physical attribute of the material which they describe, and not only a useful mathematical artifice. Discrete fiber bundles (with discrete gauge group as fiber) have been introduced only recently in field theory,[49] but also as a guide in decurving the ideal state of glass, polytope {3,3,5}.[50]*

5.   DECURVING, PROJECTION AND DISORDER

5.1   Rolling, Pivoting and Projecting the Polytope

In this section,[17] we shall see how gauge invariance appears naturally within the programme of Kléman and Sadoc,[23] who suggested that an amorphous material could be represented as a crystal in some ideal, curved space, projected into our ordinary

---

*   Since {3,3,5} is a discrete scaffolding for the 3-sphere S^3, which is itself the total space of the (non-trivial: $S^3 \neq S^2 \times S^1$) Hopf bundle with gauge group S^1, one can illustrate concepts in the continuum in a small (120 points) discrete total space, and vice-versa. For example, the fibers are geodesics (circles) which wind around each other with winding number 1, indicating the non-triviality of the bundle.

Euclidean space. The crystal in curved space has generative homogeneity (it is generated by a discrete group of displacements). Its image, the glass in Euclidean space R3, has only overall, non-generative homogeneity, and disorder. By doing the projection explicitly, we shall see that the homogeneity of the glass and its disorder constitute, geometrically, a gauge symmetry. Specifically, we shall project a polytope (the framework of the ideal closed-packed metal) on its Euclidean tangent space. This guarantees that the local pattern of neighbouring frames is faithfully imprinted, albeit with a slight distortion. This is only a local procedure, and like the various maps of an atlas, the full glass structure is an assemblage of the all these local, slightly distorted, projections. The procedure guarantees that the strain is uniformly distributed throughout the glass, and that the local short-range order (the local frame in the polytope) is preserved by the projection.

5.2   Disorder as a Random Pivot

In order to understand how to carry out the projection of a scaffolding from curved space, consider the following problem: Suppose that we wish to imprint the surface of a soccer ball, complete with seams, on a plane of infinite extent in a consistent manner. A natural way to begin would be to roll the ball along one of its seams until a junction between seams is reached. At this point, an arbitrariness is introduced as one is confronted with a choice of directions in which to continue rolling. Moreover, we require closed loops (naturally pentagonal) of the soccer ball to be mapped into closed loops (naturally hexagonal) in the plane. A pivot is therefore needed to repeat the pattern consistently. In fact, at any point during the projection, a pivot can be added which is arbitrary, except for the requirement of satisfying compatibility conditions.

The seam corresponds to the local frame (tetrapod, short range order) in glass. It must be mapped into Euclidean space (the plane), which is tangent to the curved space. A local pivot (rotation of the frame, or of the tangent plane) is <u>necessary</u> to insure compatibility. Disorder implies that this pivot is arbitrary and random. This arbitrariness, in turn, is a gauge invariance.*

Projection also implies distortion, or strain, and the projection method is therefore related to classical elasticity theory (§6), in as much as both theories are mappings between spaces with frames of reference, subject to compatibility conditions. The tangent space is the natural state of the solid (defined in §6.1), with dislocations and disclinations, but stress-free. Classical elasticity theory uses a mapping between this natural state and the real, stressed configuration of the material, whereas the projection is its mapping from the ideal, perfect, but curved space. (See Fig.9).

5.3   Theory of Surfaces

The mathematical framework for the projection is provided by the classical theory of surfaces or differential geometry, which offers two alternative points of view.

a) The <u>intrinsic</u> theory views the surface from within, like a fly on a balloon, an inhabitant of Flatland, or, in one extra dimension, like ourselves within the universe. It is associated with the names of Riemann, Christoffel, Ricci, Levi-Civita, Einstein, etc.. The surface is described locally by a connection Γ, relating frames at different points on the surface, and its torsion (if any) and curvature are expressed in terms of Γ. This is also the method used in theory of continuum elasticity with defects (Kröner,[20] Bilby, Kondo, de Wit[19]), where torsion measures the density of dislocations, and curvature that of disclinations.

b) The <u>extrinsic</u> point of view is that of an observer detached from the surface, which is embedded in a space of at least one additional dimension (this may be the mathematical distinction between the theological notions of transcendence (b) and immanence (a)). It is associated with Gauss, Codazzi, Weingarten,

---

*   Projection from curved space implies that frame orientation in the tangent plane cannot be defined uniquely (remember the pentagons into hexagons). The physical properties of the material must be independent of such arbitrariness, and are invariant under local rotation of the tangent frame.

etc.. Plato's Myth of the Cave illustrates the relationship between the two points of view, with the ideal reality projected on the walls of a cave inside which us mortals are chained. Note that Plato already had the concepts of projection in a space of one less dimension, in which we are contingently stuck. After Plato, the only really novel idea of Kléman and Sadoc was that the ideal object was curved, but one cannot really expect Plato to have guessed non-Euclidean geometry, or reproach Kléman and Sadoc to have been born 25 centuries too late.

Consider a 2D surface from the extrinsic point of view, and embedded in Euclidean R3. The surface is described locally by the second fundamental quadratic form

$$b_{\alpha i}(\mathbf{r}) = - <\mathbf{e}_\alpha . \partial_i \mathbf{n}> \tag{1}$$

where $\mathbf{e}_\alpha$ lies in the tangent plane T at the point $\mathbf{r}$ and $\mathbf{n}$ its normal at $\mathbf{r}$, defined in the embedding space R3. $b_{\alpha i}(\mathbf{r})$ measures the deviation of the normal as one rolls along i

$$\partial_i \mathbf{n} = - b^\gamma_i \, \mathbf{e}_\gamma \, . \tag{2}$$

(See Fig.8). The surface is given objectively by its curvature at every point. The two points of view are linked by Gauss's equation (6), which states that the curvatures obtained in (a) or (b) are identical.

Figure 8 represents the surface S, embedded in R3. The surface is parametrized by intrinsic coordinates x_α, $\alpha = 1,2$, as $\mathbf{r} = \mathbf{r}(x_\alpha)$. This defines a natural tangent frame at $\mathbf{r}$,

$$\mathbf{e}_\alpha = \partial_\alpha \mathbf{r} \tag{3}$$

$(\partial_\alpha = \partial / \partial x^\alpha)$. Its variation in the direction i of R3 is given by the Gauss-Weingarten equation

$$\partial_i \mathbf{e}_\alpha = b_{\alpha i} \, \mathbf{n} + \Gamma^\gamma_{\alpha i} \, \mathbf{e}_\gamma \, , \tag{4}$$

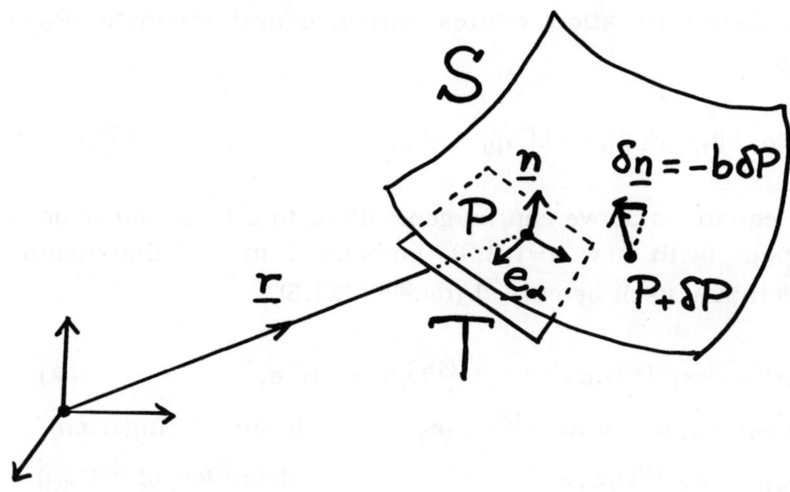

Fig.8.   Surface S, tangent plane T, local frame $\mathbf{e}_\alpha, \mathbf{n}$ and second fundamental quadratic form $b_{\alpha i}(\mathbf{r}) = -\langle \mathbf{e}_\alpha . \partial_i \mathbf{n} \rangle$.

where Γ is the connection. From the orthogonality condition $\langle \mathbf{n}.\mathbf{e}_\alpha \rangle = 0$, one obtains $\Gamma_{\gamma i \alpha} = \langle \mathbf{e}_\gamma . \partial_i \mathbf{e}_\alpha \rangle$, which is indeed an intrinsic quantity. We are distinguishing between Greek indices, labelling the frame $\{\mathbf{e}_\alpha\}$ in the tangent plane, and Latin indices labelling directions, because we shall later use the freedom to rotate or pivot the tangent plane, without changing the rolling direction. The pivot is the gauge transformation, and the Greek indices can refer to some internal parameter or isospin, which turns out here to have a geometrical interpretation. The main equations of surface theory [and their generalization] are classical: The Codazzi equation makes compatibility automatic,

$$\partial_j b_{\beta i} - \partial_i b_{\beta i} = \Gamma^\gamma_{\beta j} b_{\gamma i} - \Gamma^\gamma_{\beta i} b_{\gamma j} \ . \tag{5}$$

This point is extremely important because compatibility is the cornerstone of elasticity theory with defects,[19,20, see also §6.2)] and the homotopy theory of defects in crystals has been severely criticized for ignoring it.[51,52,13 (Mermin))] It has not escaped our attention here or in the other work on Z_2 defects in glasses.[53,12)]

The Gauss equation relates extrinsic and intrinsic ($R_{\alpha\beta ij}$) curvatures,

$$b_{\alpha i}\, b_{\beta j} - b_{\alpha j}\, b_{\beta i} = R_{\alpha\beta ij} \equiv (\partial_i \Gamma_{\alpha\beta j} - \partial_j \Gamma_{\alpha\beta i} + \Gamma_{\alpha\gamma i}\, \Gamma^\gamma{}_{\beta j} - \Gamma_{\alpha\gamma j}\, \Gamma^\gamma{}_{\beta i})\, . \qquad (6)$$

The equations above can be generalized to a three dimensional curved space (with now $\alpha=1,2,3$), embedded in a 4-dimensional space which may itself be curved (ref.[25], §21.5),

$$b_{\alpha i}(\mathbf{r}) = -\, \langle \mathbf{e}_\alpha . {}^{(4)}D_i \mathbf{n}\rangle \quad ; \qquad {}^{(4)}D_i\mathbf{n} = -\, b^\gamma{}_i \mathbf{e}_\gamma \qquad\qquad (7)$$

$${}^{(4)}D_i\mathbf{e}_\alpha = b_{\alpha i}\, \mathbf{n}\,/\langle \mathbf{n}.\mathbf{n}\rangle + {}^{(3)}\Gamma^\gamma{}_{\alpha i}\mathbf{e}_\gamma \qquad\qquad \text{(Gauss-Weingarten)}$$

$${}^{(3)}\Gamma_{\gamma\alpha i} = \langle \mathbf{e}_\gamma . {}^{(4)}D_i\mathbf{e}_\alpha\rangle \qquad\qquad \text{(definition of } {}^{(3)}\Gamma_{\gamma\alpha i})$$

$${}^{(4)}R_{\mathbf{n}\alpha ij} = (b_{\alpha i}|j - b_{\alpha j}|i)/\langle \mathbf{n}.\mathbf{n}\rangle$$

$${}^{(4)}R_{\alpha\beta ij} = {}^{(3)}R_{\alpha\beta ij} - (b_{\alpha i}\, b_{\beta j} - b_{\alpha j}\, b_{\beta i})/\langle \mathbf{n}.\mathbf{n}\rangle \qquad\qquad \text{(Gauss)}.$$

If $\mathbf{b} = 0$, both extrinsic and intrinsic spaces have the same curvature tensor, ${}^{(4)}\mathbf{R} = {}^{(3)}\mathbf{R}$, and the projection does not affect the curvature.

5.4 Random Pivot and Gauge Invariance

If the 4-dimensional space is Euclidean, the covariant derivatives ${}^{(4)}D_i$ (also denoted by ${}^{(4)}D_i b_{\beta j} = b_{\beta j}|i$) are replaced by ordinary ones ∂_i, and the curvature of the embedding space vanishes, ${}^{(4)}\mathbf{R} = 0$. Gauss's equation takes the simpler form,

$${}^{(3)}R_{\alpha\beta ij} = (b_{\alpha i}b_{\beta j} - b_{\alpha j}\, b_{\beta i})/\langle \mathbf{n}.\mathbf{n}\rangle \qquad\qquad (8)$$

similar to (6). Explicitly,

$$R_{\alpha\beta ij} = [\langle \mathbf{e}_\alpha .\partial_i\mathbf{n}\rangle\, \langle \mathbf{e}_\beta .\partial_j\mathbf{n}\rangle - \langle \mathbf{e}_\alpha .\partial_j\mathbf{n}\rangle\, \langle \mathbf{e}_\beta .\partial_i\mathbf{n}\rangle]/\langle \mathbf{n}.\mathbf{n}\rangle \qquad (9)$$

is valid in any dimension (compare with eq.6), is antisymmetric in $\alpha\beta$ and is covariant under gauge transformation (here, a pivot or rotation SO(D) of the frame $\{\mathbf{e}_\alpha\}$), like the Yang-Mills field intensity

(3.6) $\mathbf{F}_{ij}=\partial_i\mathbf{A}_j-\partial_j\mathbf{A}_i+g[\mathbf{A}_i,\mathbf{A}_j]$, $\mathbf{A}_i=A^\alpha{}_{\beta i}$, or the density of disclinations (6.10) to be introduced in the next section. Moreover, for the particular case with which we started, namely a 2D surface in R3, the antisymmetrical product can be rewritten by using vector products as

$$R_{\alpha\beta ij} = [(\mathbf{e}_\alpha \wedge \mathbf{e}_\beta) \cdot (\partial_i\mathbf{n}\wedge \partial_j\mathbf{n}>]/<\mathbf{n}.\mathbf{n}> , \qquad (10)$$

which is gauge-invariant under rotation (SO(2)) of the tangent plane, as is the magnetic induction $\mathbf{B}$ = curl $\mathbf{A}$ in electromagnetism, also a SO(2) gauge theory.

The theory has explicit gauge invariance, and the gauge transformation is a local (random) pivot. Its links with physically measurable quantities like strains are given by elasticity theory, through the connection and the curvature $\mathbf{R}$ itself, which acts as a source of strain.

Disorder, within the curved space framework of Kléman and Sadoc,[23] is a SO(3) gauge symmetry, which manifest itself as a random orientation of the local frames $\{\mathbf{e}_\alpha\}$. The sources of disorder are the 2π-disclination lines, which appear naturally in Euclidean space as a consequence of the triviality of the space group (absence of generative symmetry), but are also vestiges of the ideal curved space which had been introduced to express frustration. The 2π-disclinations are not active in decurving the ideal space,[18] and their density is proportional to the amount of non-collinearity in the local reference frames. Necessary arbitrariness constitutes the paradox of geometrical gauge invariance: Frames are necessary to determine the projection and to establish frustration, but their actual orientation is irrelevant physically, as is the variety of species and environment of the trees, necessary to establish the disorder of the forest.

Note that the extrinsic formalism was needed to exhibit gauge invariance. The intrinsic formulation is closer to classical continuum theory of elasticity with defects. We shall see in the next section that an extrinsic formalism, including the two states of the material and the mapping between them, will be needed to show the full invariance of elasticity theory.

6. THE DISORDERED ELASTIC CONTINUUM

6.1 Elasticity, Geometry and Defects

The prime suspect for gauge invariance is elasticity theory. It has, besides obvious and important defects (dislocations and disclinations), a motive (the equilibrium condition (45)), the environment (disorder) and the opportunity (geometry) to commit the deed.* We shall see that, in glass, the gauge group is much larger and has remarkable consequences, notably the presence of disclinations (rotation dislocations) in isotropic, three-dimensional solids.

The elastic continuum can be described by differential geometry (see [54,19,20] for excellent reviews), as a mapping between two states of the material, each characterized by its own set of coordinates:

1) The local, relaxed state - the tangent space of §5 - is characterized by a local frame $\mathbf{u}_\alpha$ (labelled by Greek index $\alpha=1,2,3$) and by non-holonomic (non integrable) Lagrange coordinates $\{dX^\alpha\}$. In units of the interatomic distance a, dX^α/a measures the number of steps along direction α of the frame, and serves as a ruler. The relaxed state is what Kröner[20] calls "natural (or intermediate) state". It is obtained from the actual, deformed configuration of the body when

---

* Unfortunately, the full gauge invariance of the theory has only been exploited recently. This is probably because full gauge invariance appears only in the presence of disclinations (cf. §6.3.2), and isolated disclinations do not occur in crystalline materials: They cost too much elastic energy. Friedel's is the only (ante 1965) classic book on dislocations in solids, where the possibility of disclinations is mentioned, to be dismissed for the reasons given above.[55] Liquid crystals, and colloidal crystals,[56] as well as glasses[16,23,12] have changed our preconceptions.

one allows the elastic strain to relax, or, alternatively, from the ideal state by plastic deformations.*

2) The other state of the material is the actual, stressed and deformed configuration of the body, described by frame $\mathbf{e}_i$ (labelled by Latin index i=1,2,3) and by Euler coordinates $\{y^i\}$, which are holonomic.

3) In addition, there is the ideal state of the crystal (the geometrical abstraction of a perfect single crystal), which can also be defined in glasses, but in curved space. The ideal state of amorphous metals and Frank-Kasper phases[59,60] is the polytope $\{3,3,5\}$.[23,24]

The state of the material defines the geometry of the space filled by its atoms. Both relaxed and actual spaces are macroscopically flat (Euclidean). Microscopically, the former concentrates curvature and torsion on the cores of line defects (dislocations and disclinations), while the latter has it spread out. The ideal space can be curved, as in glass.

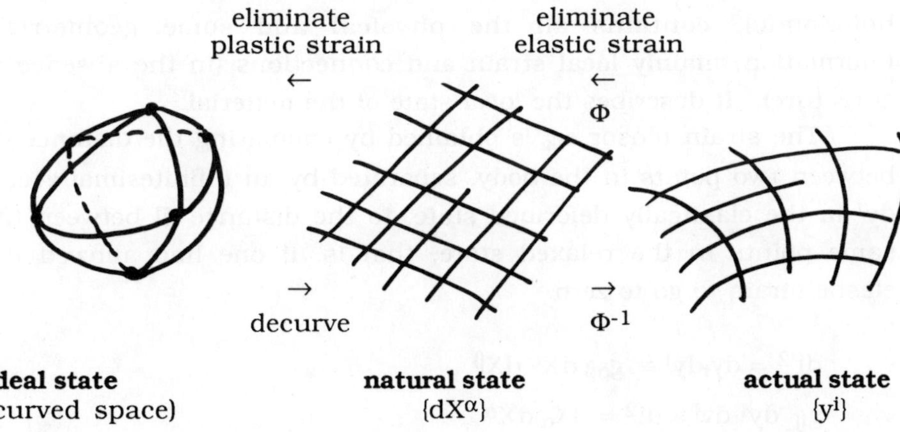

Fig.9.   Various reference states of an elastic material and mappings between them.

---

*   A nice example of natural state for close-packed metallic phases and glasses, is the corrugated space (a sort of three-dimensional egg carton), which is locally curved (portions of the ideal space, separated by negative disclinations), but globally flat.[68]

Glass can be regarded, on a semi-microscopic scale [when all relevant length scales (except dislocation core radius) are longer than the interatomic scale a] as an isotropic, elastic continuum with frozen-in elastic stresses. The stresses are due to entropy barrier preventing crystallization (to diamond or wurzite structure) when the covalent system is cooled from the melt, or to the fact that Euclidean space cannot be filled by packing configurations (regular tetrahedra) minimizing locally the energy. The presence of frozen-in stresses implies that the ideal state (on the left of Fig.9) of glass is only defined in curved space.[23] Curved space is a new feature in continuum elasticity theory, which has dealt traditionally with curved geometry in flat space.[19]

The mapping (Jacobian) matrix $\Phi^\alpha{}_i(\mathbf{y})$ from one state to the other,

$$dX^\alpha = \Phi^\alpha{}_i \, dy^i \tag{1}$$

(one cannot write $\Phi^\alpha{}_i = \partial X^\alpha/\partial y^i$ because the coordinates dX^α are not holonomic), contains all the physical and some geometrical information, mainly local strain and connections (in the absence of curvature). It describes the local state of the material.

The strain tensor e_{ik} is obtained by comparing the distance dl' between two points in the body, separated by an infinitesimal vector dy^i in the elastically deformed state, to the distance dl between the same points in the relaxed state, that is, if one had allowed the elastic strain to go to zero.

$$dl'^2 = dy_i dy^i \equiv g_{\alpha\beta} \, dX^\alpha \, dX^\beta$$

$$g_{ij} \, dy^i \, dy^j \equiv dl^2 = dX_\alpha \, dX^\alpha \tag{2}$$

The identity defines the metric g_{ij} of the relaxed state in the body (in the space described by the Euler coordinates $\{y^i\}$) and the metric

$g_{\alpha\beta}$ of the actual state in the space described by the Lagrange coordinate $\{dX^\alpha\}$.* Then,

$$dl'^2 - dl^2 \equiv 2\, e_{ij}\, dy^i\, dy^j = 2\, E_{\alpha\beta}\, dX^\alpha\, dX^\beta \tag{3}$$

defines the elastic strain tensor,

$$e_{ij} = (1/2)\,(\delta_{ij} - g_{ij}) = e_{(ij)} \tag{4}$$

Both e_{ij} and g_{ij} are symmetric in their indices.** They are also gauge invariant, whereas $g_{\alpha\beta}$ is covariant. Relation (4) is general and fundamental. One can also use (1) to obtain

$$g_{ij} = \Phi_{\alpha i}\, \Phi^\alpha{}_j \tag{5}$$

and relate the physical strain to the geometrical mapping (1).

The connection appears very simply in the construction of the Burgess vector b^α of a dislocation. Let C be a closed contour in the material $\{y^i\}$. Its image C' in $\{dX^\alpha\}$ is not closed, the missing amount being the Burgers vector

$$-\, b^\alpha = \int_{C'} dX^\alpha = \int_C \Phi^\alpha{}_i\, dy^i = -\iint [dy^i \wedge dy^j]\, 2\, T^\alpha{}_{ij} \tag{6}$$

where the dislocation density (a 2-form in $\{y^i\}$)

$$T^\alpha{}_{ij} = -\,(1/2)\,(\partial_i \Phi^\alpha{}_j - \partial_j \Phi^\alpha{}_i) = T^\alpha{}_{[ij]}\,, \tag{7}$$

---

*   By definition, the intrinsic metric of the relaxed state is $\delta_{\alpha\beta}$. The intrinsic metric of the actual state is δ_{ij} in linear elasticity (linearly in the strains), and also because we are in a continuum. The assumption of linearity in the strains[20] simplifies the algebra at no physical cost, and gives huge geometrical insight (see eqns.(20.2, 23, 30, 31 and 37)).

**   Notation: $s_{(ij)} \equiv (1/2)(s_{ij}+s_{ji})$, $a_{[ij]} \equiv (1/2)(a_{ij}-a_{ji})$.

is obtained from Stokes's formula. Transforming all indices in the material frame, one obtains the the torsion (or Cartan's) tensor

$$T^k{}_{[ij]} = (1/2) \, (\Gamma^k{}_{ij} - \Gamma^k{}_{ji}) \tag{8}$$

in terms of the connection (not a tensor)

$$\Gamma^k{}_{ij} = (\Phi^{-1}\partial_j\Phi)^k{}_i \, . \tag{9}$$

This connection is flat.[*] Here, the curvature (Riemann-Christoffel tensor, related to the density of disclinations)

$$R^a{}_{bij} = \partial_i\Gamma^a{}_{bj} - \partial_j\Gamma^a{}_{bi} + \Gamma^a{}_{ki} \, \Gamma^k{}_{bj} - \Gamma^a{}_{kj} \, \Gamma^k{}_{bi} \tag{10}$$

vanishes identically.

To introduce the curvature or frustration, and the disclinations required in glasses, one must start with the connection Γ (or the metric tensor g_{ij} (2)) rather than the mapping Φ (1), and define torsion and curvature tensors by eq.(8) and (10), respectively. The connection is constructed along a path in the body, by requiring that two overlapping neighbourhoods have local frames fitting together without any rotation (as one would in drawing the planisphere for an atlas). Parallel transport of any vector **v** between two points separated by **dy** along the path is given by the formal expression

$$\delta v^\alpha = - \, \Gamma^\alpha{}_{\beta i} \; v^\beta \; dy^i \tag{11}$$

which defines the connection. Connection here expresses the non-collinearity of nearby local frames.

One then uses the connection (11) to carry the vector **v** around a closed contour C in the body,

---

[*] It is called "pure gauge" by field theorists.

$$\Delta v^\alpha \;=\; -\int_C \Gamma^\alpha{}_{\beta i}\, v^\beta\, dy^i = -\iint [dy^i \wedge dy^j]\,\{\partial_i(\Gamma^\alpha{}_{\beta j}\, v^\beta) - (\partial_j(\Gamma^\alpha{}_{\beta i}\, v^\beta)\}$$

$$= -\iint [dy^i \wedge dy^j]\, R^\alpha{}_{\beta ij}\, v^\beta \tag{12}$$

(using Stokes's theorem and connection formula (11) again), where $R^\alpha{}_{\beta ij}$ is the curvature tensor (10) in the natural frame. It is gauge covariant.

Note that the local invariance appears differently in the torsion and in the curvature tensors. The first two indices of $\Gamma^\alpha{}_{\beta i}$ are internal and can be transformed independently of the third in the curvature tensor, whereas only the first index of $\Gamma^\alpha{}_{ij}$ is internal in the torsion tensor.

The complete problem of elasticity theory is to establish the relationship between stresses (σ) and strains ($\mathbf{e}$), in the presence of a given distribution of dislocations ($\mathbf{T}$) and disclinations ($\mathbf{R}$). The first task is to relate the physical strain to the geometrical connection, in the absence of equations (5) and (9). This was achieved by Kröner[20], with the introduction of an incompatibility tensor as source of strain. (Elastic strain $\mathbf{e}$ and stress σ are related by a constitutive equation, which is Hooke's law, in linear elasticity,

$$\sigma_{ij} = c_{ij}{}^{kl}\, e_{kl} \tag{13}$$

where $c_{ij}{}^{kl}$ are the elastic constants.)

6.2 Incompatibility

Continuum elasticity with defects is based on the concept of incompatibility. Compatibility (or integrability) sets single-valued, holonomic Lagrange coordinates for all points of the material; its failure defines topological defects (dislocations, disclinations) and is a source of elastic strain.

Compatibility is the necessary and sufficient condition for the existence of a single-valued, continuous vector field, the total displacement $\mathbf{u}^T$ of a point on the body,

$$\partial_{[m}\partial_{l]}\,u_k{}^T = 0 \; ; \qquad\qquad \partial_{[n}\partial_{m]}\partial_{[l}\,u_{k]}{}^T = 0 \qquad\qquad (14)$$

These two equations can be condensed into a single condition for the total strain $e^T{}_{ij} = \partial_{(i}\,u_{j)}{}^T \equiv (\partial_i\,u_j{}^T + \partial_j\,u_i{}^T)/2$,

$$0 = 4(\partial_n\partial_l e_{km}{}^T)_{[nm][lk]} \equiv \partial_n\partial_l e_{kn}{}^T - \partial_m\partial_l e_{kn}{}^T - \partial_n\partial_k e_{lm}{}^T + \partial_m\partial_k e_{lm}{}^T \quad (15)$$

(The identities define symmetry (ij) and antisymmetry [ij] in the indices i and j).

The total displacement and total strain are the sum of an elastic, physical component, and a plastic component, neither of which is single-valued on its own. The <u>incompatibility tensor</u> η, defined as the departure of the plastic strain from compatibility, is expressed in terms of the physical field, the elastic strain, as

$$\eta_{lknm} = -\,4\,(\partial_n\partial_l\,e_{km})_{[nm][lk]} \qquad\qquad (16)$$

In words, incompatibility is source of elastic strain.

Geometry and physics are linked through eq.(4) which relates the elastic strain **e** to the metric **g**. Geometry is carried along paths through the body by the connection Γ, and incompatibility will include a contribution of curvature **R**, torsion **T** (and also extra-matter in non-metric geometry, which will be defined presently and excluded immediately from further consideration). Connection and metric are related through covariant derivative of the latter,

$$D_m g_{lk} \equiv g_{lk;m} = \partial_m g_{lk} - \Gamma_{lkm} - \Gamma_{klm} = 0 \qquad\qquad (17)$$

which usually vanishes (Ricci lemma). It does so in the examples of §6.1, in glass, generally, in the absence of "extra-matter".*

---

* When the covariant derivative of the metric does not vanish, $D_m g_{lk} \equiv g_{lk;m} = -\,Q_{lkm}$, the right-hand side of (17) is the so-called Q-tensor. The corresponding defect is called extra-matter; it describes non-uniform thermal expansion or magnetostriction, vacancies, in short any additional source of strain, other than

and strain tensors in the body frame are manifestly invariant under gauge transformations, rotations of the local relaxed frame.

Geometrically, incompatibility is related to the connection. But, through the fundamental relation (4), it is only associated with the part of the connection containing the metric tensor explicitly, the Christoffel symbol Γ^0,

$$\Gamma^0_{ijk} = (1/2) (\partial_k g_{ij} + \partial_j g_{ik} - \partial_i g_{jk}) = \Gamma^0_{i(jk)} \tag{18}$$

Indeed, (16) reads

$$- \eta_{lknm} = \partial_n \Gamma^0_{lkm} - \partial_m \Gamma^0_{lkn} = R^0_{lknm} + O(\Gamma^0)^2 \tag{19}$$

This is the linear part (in the connection) of the Riemannian curvature tensor $\mathbf{R}^0$, defined by eq.(10) in terms of Γ^0 as the general curvature tensor $\mathbf{R}$ is to Γ. $\mathbf{R}^0$ has the following symmetries,

---

dislocation or disclination, which leaves the material unstressed. These do not occur in a homogeneous material like glass. We shall assume therefore eq.(17), and Ricci lemma to be valid from now on. Discussions of the role of extra matter in elasticity can be found in ref.[19,20(1958)], and, for its geometrical implications, in ref.[57].

Weyl's gauge ("calibration") theory,[2] whereby the length of a segment must preserved from one point to another, infinitesimally close to the first, is in fact an example of extra matter. Indeed, his equation (48) is our eq.(17), with $D_m g_{lk} = - \phi_m g_{lk}$, where ϕ_m plays the part of the gauge field for the infinitesimal transport of segments. Weyl calls *metric space*, a space equipped with this metric connection, in addition to the standard affine connection (which defines parallel transport (11), i.e. directional transport).

If ϕ_m is a gradient (integrable Weyl geometry), the calibration is uniquely defined everywhere and the effect of a non-vanishing $\mathbf{Q}$ can be absorbed (by a gauge transformation) into the Riemannian curvature $\mathbf{R}^0$ (19). Nevertheless, physical quantities (scalars, vectors,...) are scaled by a factor λ^n under change by $\lambda(\mathbf{x})$ of the standard of length. For a review of the cosmological consequences of local scale-, or calibration- invariance, in the case of integrable Weyl geometry, see Maeder.[69]

Magnetostriction (in an amorphous magnetic alloy) may provide a physical example of extra-matter, where $\mathbf{Q} \neq 0$ has a specific, additional contribution to the curvature (distance curvature, or "courbure segmentaire"[2]) and the calibration is path-dependent.

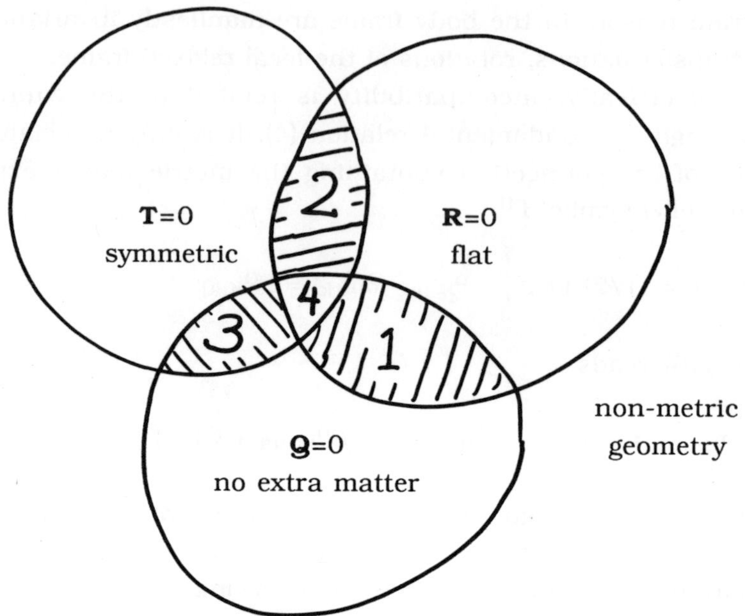

Fig.10.   Diagram showing that various geometries are defined by the non-vanishing fundamental tensors, **R** (curvature), **T** (torsion), and **Q** (associated with extra matter). Regions 1 (**R=0=Q**): realm of elementary dislocation theory. 3: Riemannian geometry. 4: Euclidean geometry. 2: an example of conformal geometry (cf. footnote preceding page).

$$R^O{}_{lknm} = - R^O{}_{lkmn} = - R^O{}_{klnm} = R^O{}_{nmlk} \qquad (20.1\text{-}3)$$

$$R^O{}_{lknm} + R^O{}_{lnmk} + R^O{}_{lmkn} = 0 \qquad (20.4)$$

and obeys Bianchi's identity

$$R^O{}_{lk[nm;p]} = 0 \ .^* \qquad (21)$$

Bianchi's identity and symmetry (20.1) are also obeyed by the full curvature **R** (obvious). Symmetry (20.2) is only obeyed by the linearized form of R (use representation (24) for Γ). Symmetry

---

*   Covariant derivative (;p) equals ordinary derivative (,p = ∂_p) in the linearized version of the theory.

(20.4) only holds for a symmetric connection like Γ^0, as does symmetry (20.3), which is a consequence of the other three.

The antisymmetry (20.1-2) in both couples of indices suggests, in 3D, a dual representation to lower the degree of the tensors, thus

$$\eta^{qp} = (1/4)\, \varepsilon^{qlk}\, \varepsilon^{pnm}\, \eta_{lknm} \; ; \qquad \eta_{lknm} = \varepsilon_{qlk}\, \varepsilon_{pnm}\, \eta^{qp} \qquad (22)$$

and the incompatibility tensor is (by (20.3)) a symmetric tensor of degree 2. Bianchi's identity becomes, in the linearized version,

$$\partial_p\, \eta^{qp} = 0 \qquad\qquad\qquad (23)$$

and incompatibility forms closed lines (cf. div $\mathbf{B} = 0$ in electromagnetism), carrying the index q. Finally, the last symmetry (20.4) of $\mathbf{R}^0$ reads $\eta^{[qp]} = 0$, and is redundant.[*]

The dual representation simplifies greatly the equations of the linearized version of elasticity theory[20] (linear in the connection Γ, large rotations Φ-1 are still permissible), which also become much more transparent (compare [20] and [57]).

Note that incompatibility, like the Christoffel Γ^0 (18), is firmly ensconced in the body frame, where the arbitrariness of the local relaxed frame (gauge invariance) cannot be exploited.

6.3 Differential Geometry and Defects in an Elastic Continuum

6.3.1 In the body frame

---

[*]   One can verify that if any two indices are identical, (20.4) reduces to one of the other three identities. It thus only brings new information if its 4 indices are different, which occurs in spaces of dimension no smaller than 4. The number of independent components of $\mathbf{R}^0$ equals $(1/2)\binom{D}{2}[\binom{D}{2} + 1] - \binom{D}{4} = (1/12)D^2(D^2-1)$. (The first term counts the number of independent components given the first three symmetries, and the second, the number of new relations brought about by (20.4)). This number is 6 for D=3, which is the number of components of the symmetric tensor η. It is 1 for D=2, where the incompatibility is a scalar.

The starting point is the definition of the connection from the metric (17).

There are two ways of expressing the connection. The simplest uses the symmetry of its first two indices,

$$\Gamma_{lkm} = (1/2)\ \partial_m g_{lk} + \Gamma_{[lk]m} \tag{24}$$

Both terms in the right hand side are basic elastic fields, the metric related to the elastic strain by (4), and

$$\Gamma_{[lk]m} \equiv -\kappa_{lkm} \tag{25}$$

where κ is called elastic bend-twist.

The second uses departure from Christoffel symmetric connection Γ^0,

$$\Gamma_{lkm} = \Gamma^0{}_{lkm} + (T_{lkm} + T_{kml} - T_{mlk}) \tag{26}$$

where Γ^0 has been defined in (18), and

$$T_{lkm} = \Gamma_{l[km]} \tag{8}$$

is the torsion or Cartan tensor. It measures the density of dislocations. The second term in (26), sometimes called contortion tensor, is antisymmetric in lk.

The curvature, or Riemann-Christoffel tensor **R**, defined in (10), measures the disclination density. This is all that is required to determine completely the physical problem, which consists of defect densities (**R** and **T**) sources of elastic fields (**e** and κ). One obtains (with some labour in the nonlinear case - rather unrewarding as the equations obtained are far from transparent - see [57]) with appropriate changes in notation) two classes of relations; a) continuity equations, identities between sources, which specify the topological nature of the defect, and b) field equations, relating the elastic fields to their sources.

In the linearized version of the theory, the continuity equations are

Bianchi's identity $\qquad R_{lk[nm,s]} = 0$ $\hfill$ (27)

Torsion identity $\qquad R_{l[knm]} + 2\, T_{l[kn,m]} = 0$ $\hfill$ (28)

(the latter replacing (20.4) for $\mathbf{R}^0$). The physical meaning of these equations is manifest in the dual representation. Notice first that the linearized $\mathbf{R}$ is antisymmetrical in both pairs of indices, since

$$R_{lknm} = \partial_n\, \Gamma_{lkm} - \partial_m\, \Gamma_{lkn} = \partial_n\, \Gamma_{[lk]m} - \partial_m\, \Gamma_{[lk]n} \qquad (29)$$

by using the first representation (24) of the connection. One introduces the dual $R^{qp} = (1/4)\, \varepsilon^{qlk}\, \varepsilon^{pnm}\, R_{lknm}$, and Bianchi's identity then read,

$$\partial_p\, R^{qp} = 0 . \qquad (30)$$

Like incompatibility, disclinations (labelled by index q) form closed loops. There is also a natural dual representation for the torsion, $T_{lkn} = (1/2)\, \varepsilon_{pkn}\, T_l{}^p$; $T_l{}^p = \varepsilon^{pkn}\, T_{lkn}$, and the torsion identity becomes

$$\varepsilon_{lqr}\, R^{qr} - \partial_r\, T_l{}^r = 0 \qquad (31)$$

so that dislocation lines (labelled by index l) may end on a disclination.*

---

*   Alternatively, a dislocation is a "current dipole" of 2 disclinations. Recall the geometrical identity of Poincaré, stating that the boundary ∂ of a boundary vanishes: $\partial.\partial \equiv 0$. It has a dual expression in terms of external derivative d: $d.d \equiv 0$. For example, div curl (vector) $\equiv 0$. (Here, duality is defined as in Stokes's theorem: The integral over a n+1-dimensional region of the external derivative of a n-form is equal to the integral of the n-form over the region's boundary). Differential geometry permits thus only three tiers of

The field equations are

$$R_{lknm} = -2\,\kappa_{lk[m,n]} \tag{32}$$

$$T_{lkm} = \Gamma_{l[km]} = -e_{l[k,m]} - \kappa_{l[km]} \tag{33}$$

or, in the dual representation, with $\kappa^q{}_m = (1/2)\,\varepsilon^{qlk}\,\kappa_{lkm}$,

$$R^{qp} = -\varepsilon^{pnm}\,\partial_n\,\kappa^q{}_m \tag{34}$$

$$T_l{}^p = \kappa^p{}_l - \varepsilon^{pkm}\,e_{lk,m} - \delta^p{}_l\,\kappa^q{}_q \tag{35}$$

Because strain only, and not bend-twist, enters the Hookian elastic free energy of a solid (unlike, for example, liquid crystals), the essential field equation relates strain to a source made of a combination of curvature and torsion, which is none other than the incompatibility. Using representation (26) for the connection, we obtain after some labor a relation between curvature $\mathbf{R}$, torsion $\mathbf{T}$ and Riemannian curvature $\mathbf{R}^0$ or incompatibility η, which reads in the dual representation

$$-\eta^{qp} = R^{qp} + \varepsilon^{pnm}\,\partial_n\,T_m{}^q - (1/2)\,\varepsilon^{qpn}\,\partial_n\,T_s{}^s \tag{36}$$

The last term is antisymmetric, whereas η is symmetric in its indices, so that

$$-\eta^{qp} = R^{(qp)} + (1/2)\,[\varepsilon^{pnm}\,\partial_n\,T_m{}^q + \varepsilon^{qnm}\,\partial_n\,T_m{}^p]$$

$$\equiv -\eta^{qp(R)} - \eta^{qp(T)} \; . \tag{37}$$

With the field equations in the right-hand side of (37), one recovers the fundamental field equation

$$\eta^{qp} = -\varepsilon^{qkl}\,\varepsilon^{pnm}\,\partial_n\,\partial_k\,e_{ml} \tag{38}$$

---

exterior derivatives. The density of dislocations $\mathbf{T} = d\phi$ is one level above than the density of disclinations $\mathbf{R} = d\Gamma$, which is the source of T as expressed in the torsion identity (31) or (43).

Incompatibility is source of elastic strain.

6.3.2 Gauge invariance ?

We have reviewed in this section the relationship between the linear theory of defects in elasticity, and differential geometry. All the equations have been written in the body frame. But most equations have an internal symmetry, exhibited if some of their indices are transposed to the natural frame with its local invariance, except those relying on symmetry (20.3) between the two indices in the dual representation, on the Christoffel symbols (18), or on the symmetry of the strain tensor, which forces all indices to be in the same frame.

For example curvature (disclination density) generates a complete gauge field theory, with $R_{\alpha\beta nm} \equiv \mathbf{R}_{nm}$ as source of the elastic field $\kappa_{\alpha\beta m} = - \Gamma_{[\alpha\beta]m} \equiv \kappa_m$.[*] The gauge field is the connection $\Gamma_{\alpha\beta m} \equiv \Gamma_m$. It has continuity equation (Bianchi identity)

$$\mathbf{R}_{[nm,s]} = 0 , \qquad \text{or} \qquad \partial_p R^{\gamma p} = 0 \qquad\qquad (39)$$

in the dual representation, and field equation

$$\mathbf{R}_{nm} = - 2 \kappa_{[m,n]} , \qquad \text{or} \qquad R^{\gamma p} = - \varepsilon^{pnm} \partial_n \kappa^{\gamma}_m . \qquad (40)$$

Gauge invariance is useful in this case, and will be exploited in the next section.

Torsion (dislocation density) generates its own gauge field theory, with torsion $T_{\alpha km} = \Gamma_{\alpha[km]} \equiv \mathbf{T}_{km}$ as source of elastic field $e_{\alpha k} \equiv \mathbf{e}_k$, defined by mapping half of the strain tensor in the natural frame. Its symmetry (and its fundamental relation (4) to the metric) has been lost through this schizophrenic manipulation. Here, the

---

[*]   Recall that the brackets () and [] surrounding indices represent complete symmetry () or antisymmetry [] of these indices, as defined in eq.(4) or (15).

72

gain of gauge invariance is not significant. Furthermore, the dislocation track is irrelevant when dislocations are not topologically stable defects, as in disordered materials where the space group is trivial.*

As expected, and unfortunately, there is no way to combine the two tracks into a single incompatibility-elastic strain relation in this mixed frame representation. The reason is that incompatibility is a Riemannian curvature which relies, like the Christoffels, on the symmetry between all its indices. In disordered solids, however, torsion (dislocation) is not a topologically stable defect, and serves only as a fluctuation to screen the 2π-disclinations, which are the only line defects surviving in the absence of any generative symmetry (trivial space group). The disclination track alone is relevant.

To exploit gauge invariance in elasticity, either disclinations, or dislocations should contribute to the incompatibility, but not both. This is the case in defect theory in 3D crystals, where disclinations are eliminated a priori, because their energy cost is prohibitively

---

*   For reference, though, here are the continuity (torsion identity) and field equations,

$$R_{\alpha[knm]} + 2\, T_{\alpha[kn,m]} = 0 \tag{41}$$
$$T_{\alpha km} = -\, e_{\alpha[k,m]} -\, \kappa_{\alpha[km]} \tag{42}$$

(strain and bend-twist have lost the symmetry and antisymmetry in their indices). A new dual representation must be introduced: $R^*_\alpha \equiv \varepsilon^{knm}\, R_{\alpha[knm]} = [\varepsilon^{knm}\, R_{\alpha knm}]/6$; $\kappa^*_\alpha{}^p = \varepsilon^{pkm}\, \kappa_{\alpha km}$, and the torsion identity reads,

$$(1/2)\ R^*_\alpha + \partial_p\, T_\alpha{}^p = 0 \tag{43}$$

while the field equation is

$$T_\alpha{}^p = -\, \varepsilon^{pkm}\, e_{\alpha k,m} -\, \kappa^*_\alpha{}^p \tag{44}$$

The torsion identity is an equation involving a three-form $R^*_{[knm]}$, (similar to Maxwell's equation div $\mathbf{D} = \rho$ in electromagnetism). An additional identity could be obtained by equating to zero its external derivative, a 4-form. But, in three dimensions, a 4-form vanishes identically, and the identity is empty.[58]

high (see eq.(50) below), through the ad hoc assumption of distant or absolute parallelism. This is also the case in glasses, where dislocations are not topologically stable defects in a material with space group, and can be treated simply as elastic fluctuations screening the disclinations. But then, η is neither symmetric, nor a tensor of degree 2 any longer (see §6.5).

The complete problem of linear elasticity, i.e. the relationship between stresses (σ) and strains (**e**) in the presence of line defects (dislocations, disclinations, incompatibility), is therefore summarized in the strain-incompatibility relation (38), the definition of incompatibility (37), and the constitutive relation between stress and strain, Hooke's law (13).* In addition, the equilibrium equation,

$$\partial_i \sigma^{ij} = 0 , \tag{45}$$

implies that stress forms uninterrupted lines, like disclinations (30, 39) and incompatibility (23). Analogy between eq.(45) and Maxwell's equation div **B** = 0 in electromagnetism motivated the introduction of a "vector potential" for stress, the stress function tensor, and a useful, but superficial use of its "gauge" invariance to solve the problem.[54,62] As we have suggested already, gauge invariance has a much broader range in elasticity, which we shall exploit fully in glass.

6.4 What is Different in Glass ?

Glass, like crystal, is an elastic continuum. The essential difference is that crystals have generative symmetry described by the space group, whereas such symmetry is completely broken in glasses. Glass is therefore the opposite of the Volterra continuum, an old favorite of dislocation theorists, which enjoys all rotation and translation symmetries.

---

*   The solution proceeds in two steps: i) Write down the elastic free energy as a functional of the strain field, by using the constitutive equation. ii) Integrate over all possible strain field configurations, subject to the restriction that incompatibility is source of strain (38).

Because glass contains frozen-in stresses and topological "defects", it is best described geometrically in its natural state (Fig.9), where the local reference frame is the configuration of a small neighborhood of **x**, cut out from the rest of the body, in which all elastic strains are relaxed. In order to compare the actual configurations of the glass at two different points **x** and **x'**, it is necessary to establish a connection, that is to define parallelism in the two configurations. The local frames are determined arcwise, by requiring that two overlapping neighborhood fit together without any rotation.* However, a finite circumnavigation does not necessarily restore the frame to its original orientation. It only does so in the absence of curvature, or of disclinations (Fig.3). In glasses, the source of curvature is concentrated in the core of isolated disclinations.

Let **v** be a vector undergoing parallel displacement. The connection Γ (11) is, in general, not symmetrical and warrant torsion (dislocations) as well as curvature (disclinations).

The orientation of the local reference frame is ambiguous in the presence of disclinations, whenever the connection is not flat (hence the need for compass and sidereal navigation). This arbitrariness is limited to some extent in elasticity by compatibility conditions: The frame specifies a configuration of the solid, and, like the tetrapods in network glass (§4), it must be returned to a physically equivalent orientation after circumnavigation. At any rate, the arbitrariness of the local reference frame implies that physics (strain tensor, free energy, etc.) and geometry (compatibility conditions) are invariant under any automorphism testing it. Here, automorphisms are local rotations of the frame, and continuum elasticity has SO(3) gauge invariance. The non-generative homogeneity of glass is tested not by translation, but by rotation of the local frames, and its symptom is their non-collinearity.

---

* The local frames provide a geodesic system of coordinates: In that frame, the (contravariant) components and length of an arbitrary vector at point P are unchanged when this vector undergoes an infinitesimal parallel transport. This requirement is expressed by eq.(11), together with (4) and (17).

Distant parallelism (equivalent to zero curvature or definition of a global reference orientation), a standard assumption in three-dimensional crystals, justified by the high energy cost of disclinations and by their conspicuous absence in any crystal (Frank-Kasper phases excepted[59,60,32]), destroys gauge invariance at the onset. It does not apply to glasses, whose local frame has arbitrary orientation, and where disclinations do not cost the prohibitive amount of energy that they have in crystals (see §6.5).

To identify a line defect like a dislocation or a disclination, one surrounds it with a Burgers circuit, which is the analogue of Gauss's or Ampère's contour used in electromagnetism to locate a charge or a current. The Burgers vector (6) (Nabarro vector (12) in the case of a disclination), which measures the mismatch of the local reference frame before and after circumnavigation, must belong to the symmetry group of the material (its space group), because both reference frames correspond to the same physical configuration, and cannot be physically distinguishable. In crystals (Fig.11), dislocations and disclinations may exist, are topologically stable, and are labelled by elements of the space group. In a Volterra continuum, whose space group contains all rotations and all translations, H=T(3)∧SO(3), arbitrary dislocations and disclinations are permitted.

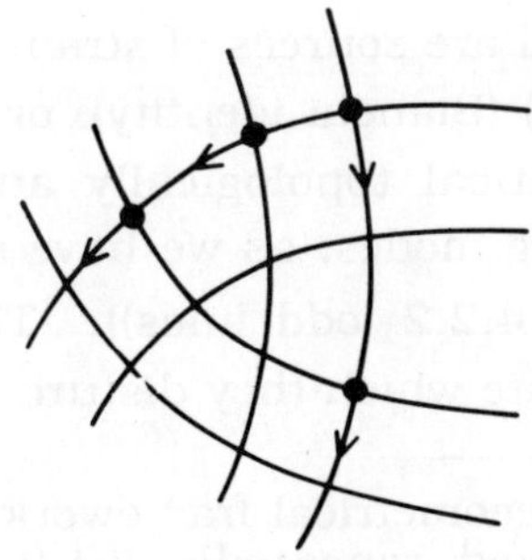

Fig.11.   In a crystal, rotation of the reference frame through a closed contour must be an element of the space group, which labels the disclination.

By contrast, glasses, whose generative symmetry is completely broken, have trivial space group. Nevertheless, one Z_2 line defect survives, because the rotation group SO(3) is not simply connected: a rotation by 2π, while it returns the object to its original orientation, entangles its connections with the rest of the material, and is therefore not continuously deformable to the identity. As the image of the Burgers contour by the mapping $\Phi(\mathbf{x})$, it cannot be continuously contracted to zero, even if the Burgers contour itself becomes shorter. Only a rotation by 4π restores object and connections: $\pi_1(\mathrm{SO}(3)) = Z_2$, as for the ribbon of Fig.1. All other operations associated with elasticity (translations, dilatations) are simply connected and topologically trivial (π_1 is called the first homotopy group).

Consequently, topologically stable 'defects' exist in amorphous materials with completely broken symmetry. They are disclination lines associated with rotation by 2π, oddness (in networks), or Z_2 algebra ($\{1,r\}$, $r^2=1$), and their topological stability is due to the multi-connectivity of the rotation group. Neither dislocations nor point defects ($\pi_2(\mathrm{SO}(3))=1$) are topologically stable in elastic continua. The same results hold for any SO(3) gauge theory of glasses, like the Yang-Mills theory written down a priori in §3.[31]

2π-disclinations have clearly the same algebra as the odd lines in random networks which we encountered in §2. Both are uninterrupted lines, and are sources of strain which satisfy a conservation law of the type (39) (Bianchi identity), or its discrete analogue. They are therefore identical topologically and in their physical effects (including tunneling modes, as we have seen by comparing §3.3 (2π-disclinations) and 4.2.2 (odd lines)). The only difference is in the geometrical substrate which they disturb, the base space.*

---

* The general, geometrical framework is that of a fiber bundle, as we have emphasized repeatedly (§1.2, 4.2. See also ref.[5,42]): Defects, incompatibilities or frustration are the loci in base space at which the bundle is non-trivial. The fibers are the orbits of the action of the gauge group on the field. Connections are defined as the bonds, links, or neighbourhood in base space, lifted on the bundle. Defects are sources of curvature in the bundle, whereby a closed path in base space around the defect is lifted in a path in the

6.5   Elastic Energy of Line Defects

Let us investigate now the physical consequences of line defects in glass. Specifically, we wish to know why disclinations are ruled out from crystalline solids because of their high energy cost, but can occur in disordered materials. To do so, we must express Hooke's law (13) in terms of a free energy functional of the strain. In linear elasticity, this is simply

$$F[e] = \int d\mathbf{x}\, \sigma^{ij}(\mathbf{x})\, e_{ij}(\mathbf{x}) \ . \tag{46}$$

The elastic "constants" c_{ijkl} are indeed constant throughout the elastic continuum. In an isotropic material, they are given in terms of the two Lamé constants, λ and μ, as

$$c_{ijkl} = \lambda\, \delta_{ij}\, \delta_{kl} + \mu\, (\delta_{ik}\, \delta_{jl} + \delta_{il}\, \delta_{jk}) \ . \tag{47}$$

One also introduces the Poisson ratio, $\nu = \lambda/[2(\lambda+\mu)]$.

Because stresses satisfy an equilibrium condition (45), and form lines, it is convenient to introduce a "vector potential for stress", and its analogue for strain, the strain function tensor χ.[20,62] There is a weak gauge invariance, associated with ambiguity on χ, which can be tracked down to the fact that stresses form lines. It is consistent with, but poorer than the full gauge invariance in the curvature $\mathbf{R}$ (40). By the fundamental equation (16) or (38), incompatibility, as source of strain, is also source for χ. In the "Coulomb" gauge, $\partial_i\chi^{ij} = \partial_j\chi^{ij} = 0$, the strain-incompatibility relation (38) becomes biharmonic,

$$\eta_{ij} = \partial^k\partial_k\partial^l\partial_l\, \chi_{ij} \equiv \nabla^4\chi_{ij} \ . \tag{48}$$

bundle which is no longer closed. The resulting shift along the fiber is an observable change in an internal parameter of the gauge field, which, among other effects, costs energy.

Both η and χ are symmetric tensors. The linear elastic free energy (46) can be written as a functional of χ,

$$F[\chi] = \int d\mathbf{x} \ \{2\mu[\chi_{ij} + (\nu/1-\nu) \ \delta_{ij} \ \chi^k{}_k] \ \nabla^4\chi^{ij}\} \tag{49}$$

Consider now two incompatibility lines L and L'. The Green's function of the biharmonic equation (48) is $|\mathbf{x}-\mathbf{x}'|$, so that, in an isotropic continuum, bare incompatibility lines interact through a potential increasing linearly with the distance,

$$F[\chi] = - (\mu/4\pi) \int d\mathbf{x} \int d\mathbf{x}' \ \{\eta_{ij}(\mathbf{x})+(\nu/1-\nu) \ \delta_{ij} \ \eta^k{}_k(\mathbf{x})\} \ |\mathbf{x}-\mathbf{x}'| \ \eta^{ij}(\mathbf{x}') \ , \tag{50}$$

which is catastrophically costly. Disclinations cannot occur therefore in isotropic 3D crystals or Volterra continuum because their strain energy is prohibitively large.[55]*

The situation is completely different in a disordered solid, where dislocations are not topologically stable, because the space group is trivial, as we have seen in §2, and again, in §6.4. They are not constrained to conserve a Burgers vector, and can be generated freely, as part of the heat bath, like strain fluctuations. In turn, free dislocations screen the elastic strain of disclinations, as they do in 2D systems, whether in the hexatic phase, in 2D cellular patterns (where dislocations are dipoles of disclinations[21]), in the skeleton of radiolarias[16], or, in 3D, at the interface (a thin amorphous layer) of mismatched films like silicon-on-sapphire.[63] **

The elastic energy can be obtained from the partition function Z, evaluated for a fixed distribution of the disclination part $\eta^{(R)}$ of the

---

* One remains with dislocation lines alone ($\mathbf{R} = 0$ in (37) and in the continuity equations (31) or (43)). and skip one differential level in the strain-incompatibility relation. Dislocation lines then interact through the Ampère potential $|\mathbf{x}-\mathbf{x}'|^{-1}$.

** The most efficient way to absorb the stress caused by structural mismatch is through an amorphous layer of roughly 7-8 interatomic distances in thickness. By contrast, mismatched crystalline interfaces induce extended defects (microtwins, stacking fault).[63,64,41] incompatibility, that is a given configuration of disclination loops, which are the only topologically stable "defects" in glass. The elastic configuration of the glass is given by the strain field $\chi(\mathbf{x})$ and the dislocation density $\mathbf{T}(\mathbf{x})$, and Z sums over all configurations, thus

$$Z = \int [D\chi] \, [D\mathbf{T}] \, \exp(-F/kT) \, P[\mathbf{T}] \, \delta(\eta^{(R)}+\eta^{(T)}-\nabla^4\chi) \, , \qquad (51)$$

with two weights, the Boltzmann distribution $\exp(-F/kT)$ (k is Boltzmann's constant, T is the temperature, F, the elastic free energy (50)) controlling strains, while fluctuations in incompatibility caused by the lack of topological stability of dislocations are governed, as a heat bath,* by a gaussian distribution,

$$P[T] \propto \exp \{ - (\gamma/kT) \int d\mathbf{x} \, T_{ij}(\mathbf{x}) \, T^{ij}(\mathbf{x})\} \, . \qquad (52)$$

$\gamma a \, (\approx kT_0)$, a typical elastic energy of the glass, is in fact its only energy scale (a result already obtained in §3). Calculation of Z is straightforward:** Beyond a moderate distance $(\sqrt{\gamma/\mu})$, the interaction potential is $|\mathbf{x}-\mathbf{x}'|^{-1}$ instead of the catastrophic $|\mathbf{x}-\mathbf{x}'|$ between bare

---

* The torsion identity (31) relates $\mathbf{T}$ to the antisymetrical part of $\mathbf{R}$. $\mathbf{T}$, and $\eta^{(T)}$ are therefore independent of the disclination part of the incompatibility, $\eta^{(R)}$, which measures the symmetrical part of $\mathbf{R}$. See (37).

** Use auxiliary fields to express the delta function as another functional integral. This reduces Z to a multiple gaussian integral. Then, the elastic energy is given by,

$$E \equiv - [\partial/\partial(1/kT)] \ln Z \qquad (53)$$

$$= (\gamma/4\pi) \int d\mathbf{x} \int d\mathbf{x}' \, [R^{ij}(\mathbf{x})v(|\mathbf{x}-\mathbf{x}'|)R_{ij}(\mathbf{x}') + R^i{}_i(\mathbf{x})w(|\mathbf{x}-\mathbf{x}'|)R^j{}_j(\mathbf{x}')] \, ,$$

with $v(r)=2[1-\exp(-r/\delta)]/r$, $w=[\exp(-r/\delta)-\exp(-r/\delta')]/r$, $\delta=\sqrt{(\gamma/\mu)}$, $\delta'=\sqrt{\{(1-v)/(1+v)\}}\delta$. At short distances $(r<\delta)$, the potentials v and w are both proportional to r, and one recovers the unscreened interaction (50) between disclinations. But $v(r)\approx 2/r$ and $w(r)\approx 0$ if $r>\delta$, and their interaction potential is reduced to the Ampère form $|\mathbf{x}-\mathbf{x}'|^{-1}$.

disclinations. Thus, disclinations can and do exist in glass and other disordered materials with broken translational symmetry, because dislocations are then free to screen their self- and mutual interactions.[16)]

The same result is obtained from the model free energy (3.8) obtained a priori for a minimally-coupled, SO(3), gauge theory. The only necessary assumptions are, i) extreme type II, and ii) semi-dilute configuration of 2π-disclinations. The non-linear terms in the gauge field in $\mathbf{F}_{ij}(\mathbf{x})$ (3.6) are negligible compared to the gradients, in the semi-dilute limit $\zeta<\lambda$ (the "penetration depth" (3.33), not the Lamé constant), and one recovers a free energy similar to that of vortex loops in extreme type II superconductors, or to the induction of current loops in electromagnetism, namely one ruled by an Ampère potential $|\mathbf{x}-\mathbf{x}'|^{-1}$.

The entropy of a collection of loops can also be calculated, and the thermodynamics of a viscous fluid above the glass transition can be obtained by that method. This result will not be discussed here, since it is only remotely connected to gauge geometry, and has been described elsewhere.[16,31,61)]

7.   CONCLUSIONS

This chapter has presented several variations on the same theme, that disordered condensed matter can be represented by a non-trivial fiber bundle, which is the mathematical support of a gauge theory.

The base space is the structural scaffolding, a random network or a disordered continuum, on which are put the physical quantities; the "real" space. It contains the minimal, simplest conceivable line "defects", the odd lines (frustration lines or 2π-disclinations), which are the geometric locus of non-triviality of the bundle. Disorder manifests itself in the base space, at two levels: The distribution of odd lines is random, and, it is argued, semi-dilute. Moreover, any generative symmetry has disappeared. The space group is trivial, and even the involution associated with the bonds (connection, J_{ij} or reflection in §4) never adds up to an uniform action. It is indeed the action of this involution which differentiates odd and even rings, frustrated and flat plaquettes.

It is in the fiber that the physical characteristics of the material are to be found, and one would have expected solid state physicists to take charge, and flood the subject with words, acronyms and particular cases. Not at all. Large gauge transformation (topological curvature), the discrete shift along the fiber which is the physical manifestation of non-triviality of the bundle, is, like its source in base space, universal. It is a Z_2 twist which is its own antitwist. It has spectacular physical consequences in the low temperature behaviour of glass.[47]

Even the connection, which requires identification of the physical (matter) field, turns out to be automatic and general in all our examples of disorderd condensed matter. This is to be contrasted with the case of the swimming paramecium, which can only move by selecting an efficient connection. The connection may be "natural" for the animal, but not for the mathematicians describing it[7] and for their readers.

One has a feeling of balance and simplicity obtained not through development from axioms, but by reducing, through essential experiments, a collection of particular phenomena to its bare essentials.[47,10] Gauge theory has helped us, as Mark Kac has put it in the postscript of his autobiography,[66] "to steer a straight course between the dead ends of extreme pragmatism and empty and useless abstraction". Kac then illustrates this balance and its inherent conflicts by quoting from Plutarch: "Eudoxus and Archytas had been the first originators of this far-famed and highly prized art of mechanics, which they employed as an elegant illustration of geometrical truths and as a means of sustaining experimentally, to the satisfaction of the senses, conclusions too intricate for proof by word and diagrams... But what with Plato's indignation at it,... to ask help (not to be obtained without base supervisions and deprivation) from matter; so it was that mechanics came to be separated from geometry, and, repudiated and neglected by philosophers, took its place as a military art."

The simplest physical model of a bundle is the Ising spin glass (Fig.2 and §4.1): A harmless-looking square lattice for base space, a fiber consisting of two points (spin up and down) and an obvious connection (the exchange interaction J_{ij}). A circuit about a frustrated plaquette ($\Pi J_{ij}=-1$) flips the spin on the fiber. Unfortunately, the model is too simple because it does not distinguish between an ordinary excitation (spin flip) and a large gauge transformation between topological sectors (tunneling mode). By contrast, the topological sectors (classical configurations) are evident in network glasses (Figs.4,7 and §4.2), and distinct from ordinary elementary excitations.

In continuous models, gauge transformation produces a connection automatically, by means of the covariant derivative. Compatibility, and the concept of a single-valued physical configuration, suffice to define the fiber bundle entirely. Compatibility forces one to distinguish between models with generative symmetry (non-trivial space group), and those (like glass) which have none. In this automatism, one may forget the essential part played by the local reference frame. This is why the projection method of §5, with its necessary and physically irrelevant pivot of the local frame, offers the most direct tutorial model for gauge invariance, and its geometrical representation of disorder.

The reader may feel that there is a certain amount of wishful thinking in the argumentation of §5 that, since there is necessity for a pivot, and because this pivot is manifestly not systematic or ordered, it must be completely random, restricted only by compatibility relations. It is not at all evident that this "all or nothing" treatment of generative symmetry and space group describes all the physical cases, and that interesting intermediate situations, poor in, but not entirely devoid of generative symmetry, could not occur. This is certainly what has happened in homotopy theory of defects.[51] This is why a classification of the algebraic modes of producing disorder, as commutations of successive decurving operations on an initial polytope in curved space[18,28], should be especially rewarding, but such a study remains almost entirely to be done.

We concede therefore that the link between gauge invariance and the notion of disorder has not been rigorously established. But what is disorder? It is only defined in negative terms (eg. trivial space group, avoidance of the niceties of adjustment[22]), apart from the existence and maximum entropy configuration of odd lines in base space. Disorder is not necessary to ensure non-triviality of the bundle or even some degree of gauge invariance (witness fully-frustrated models like the triangular antiferromagnet, which can be treated as in §4.1), but the local arbitrariness it imposes (to avoid it, one must carefully respect "the niceties of adjustment" everywhere) is an additional, gauge symmetry.

We remain with a single picture of remarkable simplicity: Disordered condensed matter is represented by a non-trivial fiber bundle, with a base space which reproduces the structure of the glass (as seen, for example, by neutron diffraction) and contains elementary (odd) lines as geometrical locus of non-triviality. Each line creates two topological sectors on the fiber, linked by a large gauge transformation, and between which the system tunnels. Gauge invariance, and the existence of large gauge transformation, impose the remarkable degeneracy of the two topological sectors (observed in the low-temperature specific heat of glass) without invoking a non-existent generative symmetry. Finally, the theory is linear in the strains.

Acknowledgements

I would like to thank Gérard Toulouse for several influential seminars, Jean-François Sadoc, André Katz, Alain Comtet, Dorothy Duffy, Anthony Lawrence, Natanael Rohr da Silva, Helena Gilchrist, Daniel Bovet, André Koch, Alf. Shapere and John Madore, for directing me towards many useful references, and for helpful discussions, Miguel Jorand, for technical assistance, and my co-authors for all of the above, besides their important contributions. The hospitality of the Institut de Physique Expérimentale, Université de Lausanne, where this chapter was completed, has been appreciated. Last, but not least, I am most grateful to the Editor, and to the other authors of this volume, for awaiting patiently the completion of this chapter, and for their encouragement.

REFERENCES

1)   *Glass and W.E.S. Turner*, E.J. Gooding and E. Meigh, eds., Soc.. of Glass Technology, 1951.

2)   Weyl, H., *Temps, Espace, Matière*, Blanchard, Paris 1922, §16. [*Space, Time, Matter*, Dover, NY 1952, §16].

3)   Yang, C.N. and Mills, R.L., Phys. Rev. $\underline{96}$, 191 (1954).

4)   Landau, L and Lifshitz, E., *Electrodynamique des Milieux Continus*, Mir, Moscow 1969, §1.

5)   Bernstein, H.J. and Phillips, A.V., "Fiber Bundles and Quantum Theory", Scient.Amer. $\underline{245}$, 94 (7/1981).

6)   Jaynes, E.T., "The Well-Posed Problem", Found. of Phys. $\underline{3}$, 477 (1973).

7)   Shapere, A. and Wilczek, F., Am.J.Phys. $\underline{57}$, 514 (1989);
-     -   J.Fluid Mech. $\underline{198}$, 557 (1989);
Wilczek, F., Phys.World $\underline{2}$, 36 (1989).

8)   Stong, C.L., Scient.Amer. $\underline{233}$, 120 (12/1975). The antitwister mechanism described there is due to D.A. Adams.

9)   Rivier, N., Adv.Phys. $\underline{36}$, 95 (1987).

10) Toulouse, G., Comm.Phys. $\underline{2}$, 115 (1977), where the label "frustration" is attributed to P.W. Anderson.
-     , in *Modern Trends in the Theory of Condensed Matter*, A. Pekalski and J. Przystawa, eds., Springer 1980, p.195.

11) Rivier, D., Rev.Théol.Phil. $\underline{122}$, 15 (1990).

12) Rivier, N., Phil.Mag.A $\underline{40}$, 859 (1979).

13) Toulouse, G. and Kléman, M., J.Physique Lett. $\underline{37}$, 149 (1976);
Volovik, G.E. and Mineev, V.P., Zh.E.T.F.Pis'ma $\underline{23}$, 647 (1976);

Rogula, D., "Large Deformation of Crystals, Homotopy and Defects", in *Trends in Application of Pure Mathematics to Mechanics*, G. Fichera, ed., Pitman, NY 1976.

Reviews by Mermin, D., Rev.Mod.Phys. 51, 591 (1979), and by Michel, L., *ibid.* 52, 617 (1980).

Digest by Toulouse, G., in *Modern Trends in the Theory of Condensed Matter*, A. Pekalski and J. Przystawa, eds., Springer 1980, p.189.

14) Dzyaloshinskii, I.E. and Volovik, G.E., J.Physique 39, 693 (1978).

15) Venkataraman, G. and Sahoo, D., Contemp.Phys. 27, 3 (1986).

16) Duffy, D.M. and Rivier, N., J.Physique Coll. 43, C9-475 (1982);

Duffy, D.M., Thesis, University of London (1981);

Rivier, N., in *Amorphous Materials: Modelling of Structure and Properties*, V. Vitek, ed., AIME, New York 1983, p.81.

17) Rivier, N. and Lawrence, A., J.Physique Coll. 46, C8-409 (1985).

18) Sadoc, J.F. and Rivier, N., Phil.Mag.B 55, 537 (1987).

19) De Wit, R., "A View of the Relation between the Continuum Theory of Lattice Defects and Non-Euclidean Geometry in the Linear Approximation", Int.J.Eng.Sci. 19, 1475 (1981).

20) Kröner, E., Erg.Angew.Math. 5, 1 (1958);

- , "Plastizität und Versetzungen", in *Mechanik der Deformierbaren Medien*, by A. Sommerfeld, 5th.ed., Akad. Verlagsg., Leipzig 1964, ch.9.

- "Continuum Theory of Defects", in *Physics of Defects*, R. Balian, M. Kléman and J.P. Poirier, eds., North Holland, Amsterdam 1981, p. 215.

21) Weaire, D. and Rivier, N., Contemp.Phys. 25, 59 (1984).

22) Lewis, F.T., Am.J.Botany 30, 74 (1943).

23) Kléman, M. and Sadoc, J.F., J.Physique Lett. 40, 569 (1979).

24) Sadoc, JF., J.Non-Cryst.Solids 44, 1 (1981);

Sadoc, J.F. and Mosseri, R., J.Physique 46, 1809 (1985).

25) Misner, C.W., Thorne, K.S. and Wheeler, J.A., *Gravitation*, Freeman, San Francisco 1973.

26) Christ, N.H., Friedberg, R. and Lee, T.D., Nucl.Phys.B 222, 89 (1982).

27) Wooten, F. and Weaire, D., Sol.St.Phys. $\underline{40}$, 1 (1987).

28) Rivier, N. and Sadoc, J.F., J.Non-Cryst.Solids $\underline{106}$, 282 (1988).

29) Rivier, N., Weaire, D. and de Römer, R., J.Non-Cryst.Solids, $\underline{105}$, 287 (1988).

30) Rivier, N., Phil.Mag.B $\underline{52}$, 795 (1985).

31) Rivier, N.and Duffy, D.M., J.Physique $\underline{43}$, 293 (1982).

32) Shoemaker, D.P. and Shoemaker, C.B, Acta Cryst.B $\underline{42}$, 3 (1986).

33) Joannopoulos, J.D. and Cohen, M.L., Sol.St.Phys. $\underline{31}$, 71 (1976).

34) Toulouse, G., Phys.Rep. 49, 267 (1979).

35) Kirchhoff, G., Pogg.Ann.Physik, $\underline{72}$, 32 (1847); Biggs, N., *Algebraic Graph Theory*, Cambridge Un.Press, 1974.

36) Tinkham, M., *Introduction to Superconductivity*, McGraw-Hill, NY 1975.

37) Abrikosov, A.A., Zh.E.T.Fiz. $\underline{32}$, 1442 (1957) [JETP 5, 1174 (1957)].

38) Coleman, S., "Secret Symmetry: An Introduction to Spontaneous Symmetry Breakdown and Gauge Fields", in *Laws of Hadronic Matter*, A. Zichichi, ed., Acad.Press 1975, p.138.

39) Bloch, F., Phys.Rev. B2, 109 (1970).

40) Nelson, D.R. and Widom, M., Nucl.Phys.B $\underline{240}$, 113 (1984).

41) Major, A., Taylor, N., Lawrence, A., Rivier, N. and Sadoc, J.F., Phil.Mag.B $\underline{55}$, 507 (1987).

42) Kerner, R., Phil.Mag.B $\underline{47}$, 151 (1983).

43) Fradkin, E., Huberman, B.A. and Shenker, S.H., Phys.Rev.B $\underline{18}$, 4789 (1978).

44) Edwards, S.F. and Anderson, P.W., J.Phys.F $\underline{7}$, 965 (1975); Kac, M., Ark.Det.Fys.Sem.Trondheim $\underline{11}$, 1 (1968).

45) Sherrington, D. and Kirkpatrick, S., Phys.Rev.Lett. $\underline{35}$, 1792 (1975).

46) Fert, A. and Hippert, F., Phys.Rev.Lett. $\underline{49}$, 1508 (1982); Gullikson, E.M., Fredkin, D.R. and Schultz, S., Phys.Rev.Lett. $\underline{50}$, 537 (1983).

47) Hunklinger, S. and Raychaudhari, A.K., Progr.Low T.Phys. IX,
     D.F. Brewer, ed., North Holland, Amsterdam 1986, p.265;
     Phillips, W.A., ed., *Amorphous Solids. Low-Temperature
     Properties*, Springer, Berlin 1981.

48) Keating, P.N., Phys.Rev. 145, 637 (1966);
     Alben, R., Weaire, D., Smith, J.E.Jr. and Brodsky, M.H.,
     Phys.Rev.B 11, 2271 (1975).

49) Manton, N.S., Comm.Math.Phys. 113, 341 (1987).

50) Nicolis, S., Mosseri, R., and Sadoc, J.F., Europhys.Lett. 1, 571
     (1986).

51) Trebin, H.R., Adv.Phys. 31, 195 (1982).

52) Ovidko, I.A. and Romanov, A.E., Phys.Stat.Sol.(a) 104, 13 (1987).

53) Rivier, N., Phil.Mag.A 45, 1081 (1982).

54) Lardner, R.W., *Mathematical Theory of Dislocation and
     Fracture*, Univ. Toronto Press 1974.

55) Friedel, J., *Dislocations*, Pergamon, Oxford 1964, §2.1.6.

56) Rothen, F. and Pieranski, P, La Recherche 176, 312 (1986).

57) Schouten, J.A., *Ricci-Calculus*, Springer, Berlin, 1954.

58) Rivier, N., in *Topological Disorder in Condensed Matter*,
     F. Yonezawa and T. Ninomiya, eds., Springer, Berlin 1983,
     p.14.

59) Rivier, N. and Sadoc, J.F., Europhys.Lett. 7, 523 (1988).

60) Sadoc, J.F., J.Physique Lett. 44, 707 (1983);
     Nelson, D.R., Phys.Rev.Lett. 50, 983 (1983).

61) Rivier, N., Rev.Bras.Fis. 15, 311 (1985).

62) Nabarro, F.R.N., *Theory of Crystal Dislocations*, Clarendon,
     Oxford 1967.

63) Smith, D.J., Freeman, L.A., McMahon, R.A., Ahmed, H., Pitt, M.G.
     and Peters, T.B., Inst.Phys.Conf.Ser. 67, 83 (1983).

64) *Physics of Granular Media*, D. Bideau and J.A. Dodds, eds., Nova
     Science, NY 1990. (§§ by J. Desrues, J.M. Georges and
     conclusion).

65) Rivier, N., in *Structure of Non-Crystalline Materials 1982*, P.H.
     Gaskell, E.A. Davis and J.M. Parker, eds., Taylor and
     Francis, London 1983, p.517.

66)   Kac, M., *Enigmas of Chance*, Univ. of Calif. Press, Berkeley 1987.

67)   Angell, C.A., J.Non-Cryst.Solids $\underline{102}$, 205 (1988).

68)   Gaspard, J.P., Mosseri, R. and Sadoc, J.F., Phil.Mag.B $\underline{50}$, 557 (1984).

69)   Maeder, A., "La cosmologie de Dirac et l'invariance d'échelle dans la théorie de la gravitation", in *Théories Cosmologiques*, AVCP (course XX), EPFL Lausanne 1978, p.257.

Chapter II
MORPHOLOGY OF STRATIFIED FLUIDS

Jean CHARVOLIN and Jean François SADOC

MORPHOLOGY OF STRATIFIED FLUIDS

J. Charvolin and J.-F. Sadoc,
Laboratoire de Physique des Solides,
associé au CNRS (LA 02),
bat. 510, Université Paris-Sud,
91405 Orsay, France.

ABSTRACT

In the classical view of a fluid the degrees of freedom of the molecules, positions, orientations and velocities, are randomly distributed. There exist however situations in which they are periodically modulated. For example, the positions can be modulated by the existence of interfaces in the lamellar and smectic phases of lyotropic and thermotropic liquid crystals, the orientations can be modulated by a constant chiral twist in the layered organization of cholesteric phases and the application of an external field can induce the building of a coherent velocity field in periodic convective rolls. We call these organizations stratified fluids, we shall discuss here the two first cases only, as they are those corresponding to equilibrium structures. Their organization in flat layers periodically stacked along a one-dimensional lattice, where the constant distance imposed by the interactions between interfaces or the pitch of the chiral twist is kept everywhere, is the most commonly observed and, indeed, the easiest to understand. However, when some thermodynamical parameter is varied and a symmetry broken, the systems can eventually build more complex organizations along two or three-dimensional lattices, hexagonal and cubic phases or "blue" phases, where the above constraint is not uniformly respected. We propose here to consider all these organizations in systems of very different natures on the same basis. In our view they would result from a conflict, or frustration, arising between the requirement of constant distance and the appearance of a new symmetry in the stratifications introduced by a lateral area difference between different layers of the stratification or a double twist. We show that the frustration appears because of the flatness of the Euclidean space in which the systems are embedded, and would be relaxed if they were embedded in adequatly chosen curved spaces. This remark is the fundamental point of the search for possible geometrical configurations optimizing the frustrations in the Euclidean space. These can be obtained through a mapping of the curved space, containing the relaxed structure, onto the Euclidean space, by introducing defects of rotation, or disclinations, respecting the symmetry of the relaxed structure in the curved space. The resulting configurations have topologies and symmetries which agree well with those of the observed structures. This implies that the energy terms implicitly present in this purely geometrical approach, those associated with the compressibility of the periodic layering and the curvature elasticities of the fluid layers around some spontaneous curvature, are the dominant ones. Such an approach may be seen as the basis for establishing the crystallography of periodic systems of fluid films. Together with other chapters of this book devoted to solid materials, it illustrates the unifying power of the concept of defect in condensed matter physics by extending it to the understanding of mesoscopic structures formed by fluid materials.

I-EXAMPLES OF STRATIFIED FLUIDS

A fluid is classically considered as an assembly of molecules whose degrees of freedom, positions, orientations and velocities, are randomly distributed. It is however possible, by a suitable choice of the molecules in the two first cases or the application of an external field in the third case, to create situations in which the corresponding degrees of freedom are periodically modulated, i.e. to generate stratifications in fluids. The two first cases are well illustrated by the so-called mesoscopic structures of liquid crystalline materials, lyotropic as well as thermotropic, and the third one by the convective instabilities induced by applying a thermal gradient, an electric or magnetic field perpendicularly to thin slabs of isotropic and nematic liquids.

I-1.Stratification of Position in Lamellar and Smectic Systems

The best example is provided by the lamellar structures formed by amphiphilic molecules such as soaps, detergents, lipids and copolymers. These molecules are built in two parts having different affinities for different solvents, as shown in fig.1a for dodecyltrimethylammonium chloride (DTACl). In this simple case the paraffinic chain of the molecule has a good affinity for organic solvents, such as oil, but not for polar solvents, such as water, it is the opposite for the ionic group which is polar. When such molecules are put in presence of water their polar heads build interfaces which protect the chains from the contact with water. The shape and organisation of these interfaces depend upon temperature and concentration. Phase diagram [1] and structural [2] studies of various systems have shown that the lamellar structure shown in fig. 1b, where symmetric layers limited by flat interfaces at constant distances are periodically organized on a 1-D lattice with a characteristic parameter of a few tenths of Å, is of very general occurence.

Other structures, with curved interfaces, can be observed for other values of the parameters. The particular case of dodecyltrimethylammonium chloride provides a good illustration of the evolution of the structures[3]. Starting from the lamellar phase shown in fig. 1 and increasing the water content, one meets successively a "bicontinuous" cubic structure made of two interwoven labyrinths of amphiphiles separated by a film of water and organized on a 3-D lattice with space group Ia3d (it may be Pn3m or Im3m in other systems [4]) , a hexagonal structure made of infinite cylinders of amphiphiles separated by a connected film of water and organized on a 2-D lattice, a "micellar" cubic structure possibly made of small isometric and anisometric micelles organized on a 3-D lattice with space group Pm3n (it may be Fd3m in other systems [4]) and, finally, a disordered

[1] P. Ekwall, Adv. Liq. Cryst. $\underline{1}$,1 (1975), edited by G.H. Brown, Academic Press.

[2] V. Luzzati, Biological Membranes $\underline{1}$, 71 (1968), edited by D. Chapman, Academic Press.

[3] R.R. Balmbra, J.S. Clunie, and J.F. Goodman, Nature $\underline{222}$, 1159 (1969).

[4] V. Luzzati, P. Mariani and T. Gulik-Krzywicki, in "Physics of Amphiphilic Layers", Springer Proceedings in Physics $\underline{21}$, 131 (1987), edited by D. Langevin and J. Meunier and

P. Mariani, V. Luzzati and Delacroix H., J. Mol. Biology, $\underline{204}$, 165 (1988)

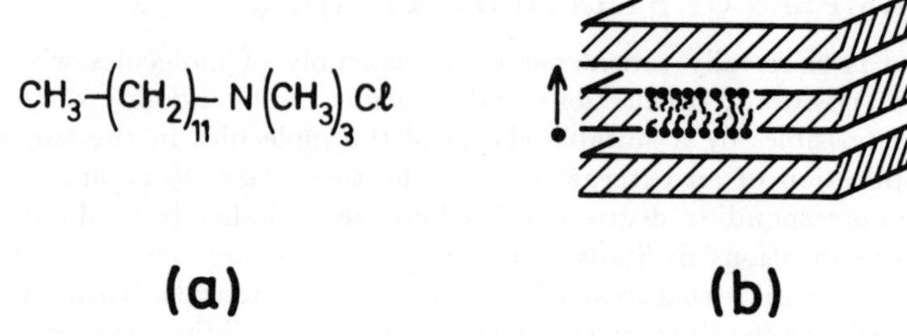

(a) **(b)**

fig. 1
An amphiphilic molecule DTACl (a), the lamellar structure with flat
interfaces formed by such a molecule in presence of a few per cent in weight
of water (b).

solution of micelles. Drawings of these structures are presented in fig. 2. Such a
sequence of structures is typical of the so-called polymorphism of lyotropic liquid
crystals and micellar phases . It is important not to forget that, although these
structures are strictly speaking crystalline, because of their long range order, the
local organization of the molecules within them is not crystalline, X ray scattering
[2] and nuclear magnetic resonance studies [5] have demonstrated their liquid-like
behaviour. Indeed, the long range order of these structures is not the result of
the propagation of a short range order of their molecules, as in ordinary molecular
crystals. They can be depicted, in a somewhat provocative manner, as ordered
entanglements of disordered liquids separated by interfaces. This polymorphism is
not limited to molecules having a medium molecular weight, it was recently shown
that amphiphilic copolymers, with much higher weight, exhibit similar sequences
of structures [6].

One may expect that a similar situation is provided by the thermotropic
polymorphism of liquid crystals formed by the so-called mesogenic molecules,
although their case is less well documented at the moment. These molecules
are also made in two parts, aromatic and paraffinic, as shown by the example
in fig. 3a, but they are not usually considered as amphiphilic molecules as the
difference of affinities between these two parts concerns organic solvents only and
is not much pronounced. However they also tend to build structures where the
two parts segregate on either sides of an interface. The most common of these

[5] J. Charvolin and A. Tardieu, Solid State Physics suppl. $\underline{14}$, 209 (1978), edited by L. Liebert,
F. Seitz and D. Turnbull, Academic Press.

[6] E.L. Thomas, D.B. Alward, D.J. Kunning, D.C. Martin, D.J. Handlin and L.J. Fetters,
Macromolecules $\underline{19}$, 2197 (1986).and

H. Hasegawa, H. Tanaka, K. Yamasaki and T. Hashimoto, Macromolecules $\underline{20}$, 1651 (1987).

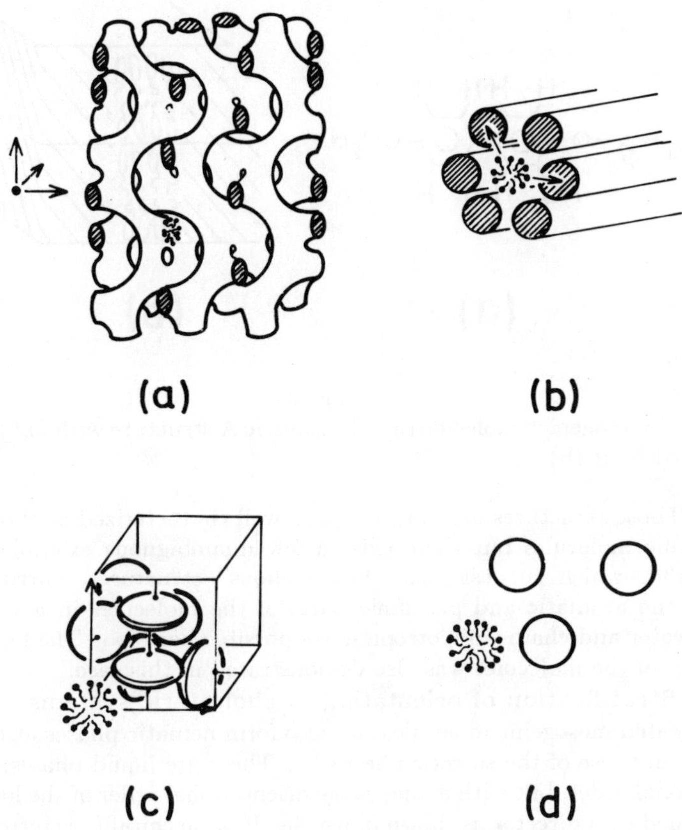

fig. 2

Structures with curved interfaces observed in DTACl, beyond the lamellar one, for increasing water content: cubic with space group Ia3D (a), hexagonal (b), a proposition for the cubic with space group Pm3n (c), disordered micellar (d). Constant interfacial distances can not be respected everywhere in such structures.

structures is the smectic one shown in fig. 3b [7], with flat interfaces periodically organized along a 1-D lattice with a parameter of the order of the molecular length. It is quite analogous to the lamellar structure formed by amphiphilic molecules.

In certain cases, when the temperature is varied or when particular mixtures of different mesogenic molecules are prepared, 3-D cubic structures can be observed

---

[7]   P.-G. de Gennes, The Physics of Liquid Crystals, Oxford University Press (1975).

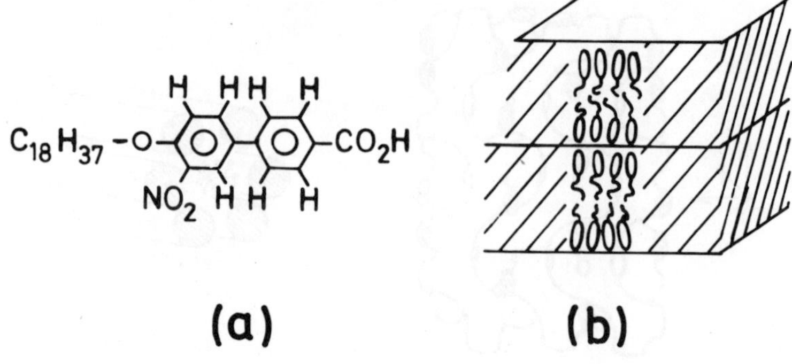

fig. 3

A "mesogenic" molecule (a), the smectic A structure with flat interfaces formed by it (b).

too [8] . Those structures are not always as well characterized as those formed by amphiphilic molecules but there exist a few unambiguous examples with space group Ia3d, which suggest that "bicontinuous" structures, partitioning space between the aromatic and paraffinic parts of the molecules in a way similar to that of water and chains in lyotropics, are possible here too. The liquid-like local behaviour of the molecules was also demonstrated in this case.

I-2. Stratification of orientation in cholesteric systems

Elongated mesogenic molecules can also form nematic phases at temperatures higher than those of the smectic phases [7] . These are liquid phases, without any translational order, but with a long range orientational order of the long molecular axes defined by a director as shown in fig. 4a. If some chirality is introduced in the nematic liquid, either on the molecule itself or on a solute, its director undergoes a helicoidal distortion of the type of that shown in fig. 4b. This is a cholesteric state, in which the half pitch of the helix defines a periodic layering of the system on a 1-D lattice with a characteristic parameter of a few thousands of Å, while the local molecular behaviour remains similar to that of the nematic [7] . This periodic change of orientation in an oriented liquid is a second example of stratification.

However, this is not the only type of organization possible for a cholesteric material. Recent studies have shown the existence of 3-D cubic organizations for

[8]   A. Tardieu and J. Billard, J. de Physique 37,C3-79 (1976),

J. Billard, C. R. Acad. Sc. Paris 305, serie II, 843 (1987),

G. Etherington, A.J. Leadbetter, X.J. Wang, G.W. Gray and T. Tajbakhsh, Liq. Crystals 1, 209 (1986),

N.H. Tinh, C. Destrade, A.M. Levelut and J. Malthète, J. de Physique 47, 543 (1986),

Y. Fang, Thesis, Orsay (1988).

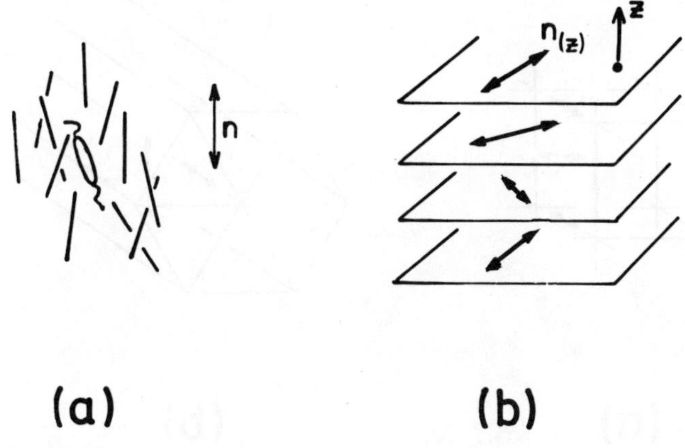

fig. 4

Nematic (a) and cholesteric (b) structures, the planes drawn in (b) are virtual.

some compounds, at temperatures higher than that of the 1-D cholesteric structure [9]. These are the so-called "blue" phases, with space groups $I4_132$ and $P4_232$ closely related to those of the "bicontinuous" cubic phases formed by amphiphilic and mesogenic molecules, but with larger parameters. Fig. 5a suggests the way the orientation changes in the case of the $P4_232$ structure. It has also been shown that the application of an external electric field on a blue phase can induce structural transformations above a certain threshold value, particularly one into a hexagonal structure organized on a 2-D lattice [10]. Fig. 5b suggests the orientation change in this structure. In the case of these structures it is clear that constant distances can not be kept between loci of parallel orientations.

I-3. Stratification of Velocities in Convective Rolls

The simplest example is that of a thin slab of an ordinary liquid, contained between two plates and heated from below. The lower layers of liquid, which are expanded and have a lower density, want to move up while the upper ones, which have a higher density, want to move down. This is an unstable situation and, above a threshold value of the temperature gradient, convective rolls settle, appearing as parallel lines at distance of a several microns fixed by the thickness of the slab, the liquid moving up along one line, down along the next, and again as

---

[9] H. Stegemeyer, Th. Blùmel, K. Hiltrop, H. Onusseit and F . Porsch, Liq. Crystals 1, 3 (1986).

[10] R.M. Hornreich, M. Kluger and S. Strikman, Phys. Rev. Lett. 54, 2099 (1985)

P.E. Cladis, T. Garel and P. Pieranski, Phys. Rev. Lett. 57, 2841 (1986)

G. Hepke, B. Jérome, H.S. Kitzerow and P. Pieranski, Proceeding of the $12^{t}h$. International Liquid Crystal Conference, Freiburg 1988, Liq. Cryst. to be published.

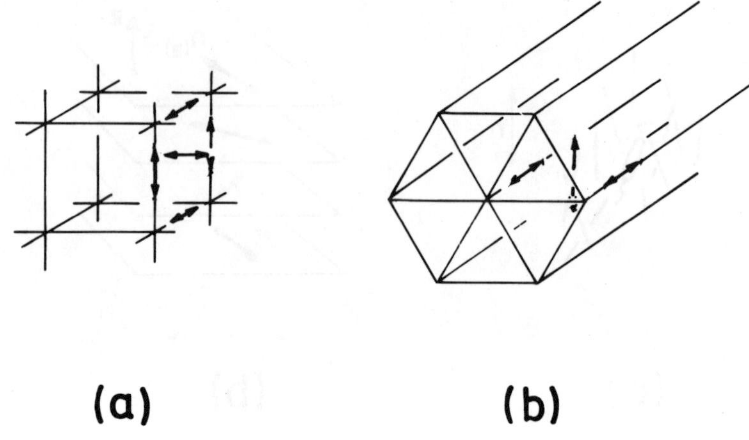

(a)   (b)

fig. 5

Schematic representations of a cubic blue phase (a), a hexagonal one (b), with part of the helicoidal twist.

shown in fig. 6. This is a system with a stratification of the velocity distribution which, because of the finite size of the slab of liquid, can be seen as a 2-D example of stratification. Under certain conditions the rolls may eventually form square or hexagonal patterns. Similar instabilities may be observed when magnetic or electric fields are applied along the normals of thin slabs of nematic liquids.

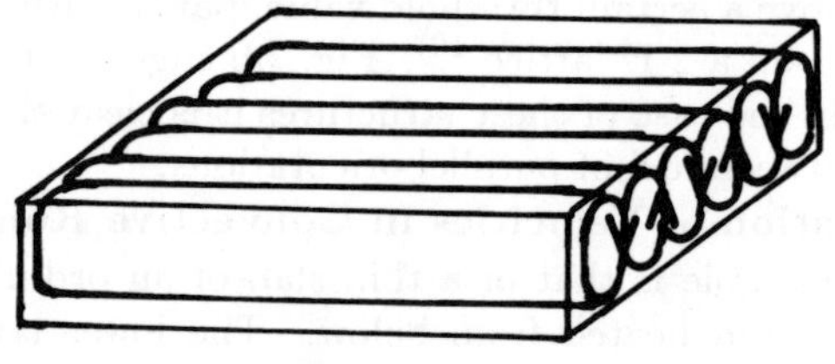

fig. 6

Parallel convective rolls in a thin slab of liquid.

II- PRESENTATION OF THE CHAPTER

II-1. Direction Followed

The presentation of the structural descriptions given above can be concluded in the following way. We have described periodic modes of organizations, on characteristic distances varying from a few tens to a few thousands of $\mathring{A}$ according to the nature of the system, found in liquids of molecules having very different chemical structures and sizes, put in various situations bearing no relation between them. Nevertheless, in spite of all these differences, the symmetries of these organizations are very close, lamellar for 1-D organizations, most often hexagonal p6m for 2-D organizations and cubic Ia3d or Pn3m for 3-D ones. Obviously, the fundamental reason for this can not be usefully traced back to the different chemical or environmental details. We propose here to look for it in the common existence of a stratification in films, as defined in figure 7. On this simple basis we shall try to show that all the structures formed can be understood within a common conceptual frame.

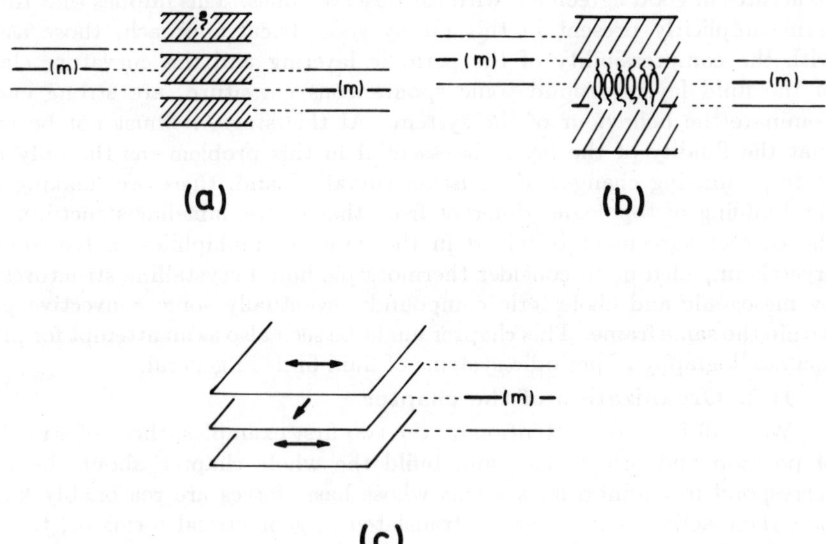

fig. 7
Definitions of the films and their middle surfaces (m), two definition are possible in the cases of stratifications by interfaces (left and right schema) ,in a lamellar structure (a), in a smectic structure (b), in a cholesteric structure (c).

The general idea of what will be developed in the chapter is the following. A perfect periodic stratification in symmetric layers or films is only possible in the Euclidean space if the planes of stratification are flat. If an internal constraints appears, which deforms the planes, the perfect periodic stacking at constant distance is no longer possible. There is a conflict, or frustration, between the two constraints, deformation and constant distance, and the frustrated system is forced to change its structure in order to find an acceptable compromise in the Euclidean space. It can only be found by introducing defects in the stratification. This is a geometrical approach to a morphological problem, without the explicit presence of energy terms, and this might be seen as a very serious limitation. However, if we refer to classical solid state physics, such a geometrical, or crystallographical, approach has indeed proved to be very efficient in providing the necessary basis for understanding the organizations of atoms or molecules. One might think that this success in solid systems holds to the strength of the interactions and that in, the fluid systems considered here, softer potentials allow the systems to escape the rigor of geometry. Nevertheless, we recently tried to analyze the well documented case of the structures of interfaces formed by amphiphilic molecules along these lines. We found the topologies [11] , symmetries [12] [13] and sequence [14] of structures in good agreement with the observed ones. This implies that the energy terms implicitly present in this purely geometrical approach, those associated with the compressibility of the periodic layering and the curvature elasticities of the fluid layers around some spontaneous curvature, are strong enough to dominate the behaviour of the system. At this stage, it must not be forgotten that the fluidity of the layers is essential in this problem, as the only physical state permitting changes of Gaussian curvature and, therefore, making possible the building of topologies different from that of the lamellar structure. Finally, the correct agreement obtained in the case of amphiphiles in lyotropic liquid crystals impelled us to consider thermotropic liquid crystalline structures formed by mesogenic and cholesteric compounds, eventually some convective patterns, within the same frame. This chapter might be seen also as an attempt for proposing a crystallography of periodic systems of fluid films in general.

II-2. Organization of the chapter

We shall focus our attention on the two first examples, those of stratifications of position and orientation, and build the whole chapter about them. They correspond to equilibrium systems whose basic forces are reasonably known, so that their actions can be easily translated in geometrical terms or, the relation between the forces and the organization in space can be easily perceived. This is not the situation for the third case, that of the stratification of the velocity field. It corresponds to dynamical systems which are out of equilibrium, where several phenomena, not always well-known or whose actions are not always well discriminated because of strong non linear couplings, intervene in the building of the patterns. In most situations, their formulations in geometrical terms are but a phenomenological description at the moment. For this reason we shall present this last case in a rather conjectural manner, in a brief paragraph at the end of the chapter.

---

[11] J.F. Sadoc and J. Charvolin, J. de Physique $\underline{47}$,683 (1986).

[12] J. Charvolin and J.F. Sadoc, J. de Physique $\underline{48}$, 1559 (1987).

[13] J. Charvolin and J.F. Sadoc, J. de Physique $\underline{49}$, 521 (1988).

[14] J. Charvolin and J.F. Sadoc, J. Phys. Chem. $\underline{92}$, 5787 (1988).

III- STRATIFICATIONS OF POSITIONS AND ORIENTATIONS IN LIQUID CRYSTALS

III-1. Stratified Fluids in Flat Space and Frustrations

In our approach of these stratified fluids the structural elements are no longer the molecules but the films defined by the unit of translation along directions normal to the layers. Examples of films are shown in fig. 7. The organizations of such films are governed by several forces whose actions are not always compatible in an Euclidean space, leading to situations of frustration as discussed below.

III-1.1. Frustration in lamellar and smectic systems

In these cases the stratifications concern the positions of the molecules. We present the problem in the case of systems of amphiphiles and lamellar phases only, similar considerations hold for mesogenic molecules and smectic phases. The film is made of two identical interfaces and several forces fix the inter- and intra-layers distances : Van der Walls forces, electrostatic interactions between polar groups at the interfaces, forces created by water polarization and charge distributions in aqueous layers, hydrophobic interactions which prevent the presence of water within the layers and fluctuation-induced forces [15]. They are not totally known at the moment, as well as their interplay . However it is reasonable to think that the role of the components normal to the interfaces is to maintain constant distances between them, if the interfaces are homogeneous, and that the role of the components parallel to them is to determine the interfacial curvatures, as they do not vary necessarily in concordance at different levels within the layer of the film. Owing to the symmetry of the film with respect to its middle surface it is clear that the fact that two facing interfaces may have symmetric curvatures is not compatible with constant distances between the interfaces, if a lamellar-type stacking is kept, as shown in fig. 8. The situation with flat interfaces is represented in fig. 8a and one situation of conflict with curved interfaces in fig. 8b. This is a typical case of frustration which has no solution in the 3-D Euclidean space R_3, the system must find other structures in which the forces can build acceptable compromises.

III-1.2. Frustration in cholesteric systems

In this case the stratification concerns the orientation of a director, which is submitted to a twist so that a period appears corresponding to half the pitch of the helicoidal reorientation. The layered situation of the classical cholesteric phases, described in the first paragraph and recalled in fig. 9a, corresponds to the case where the twist propagates along one direction only. But this is not the most general situation, and one may envisage an isotropic propagation of the twist along directions perpendicular to the axis of the molecule. In such a case, if two molecules are displaced from the origin O to point M following the two different paths indicated on fig. 9b, they have different orientations in M , and it is not

---

[15]   J.N. Israelachvili, "Intermolecular and surface forces", Academic Press (1985).

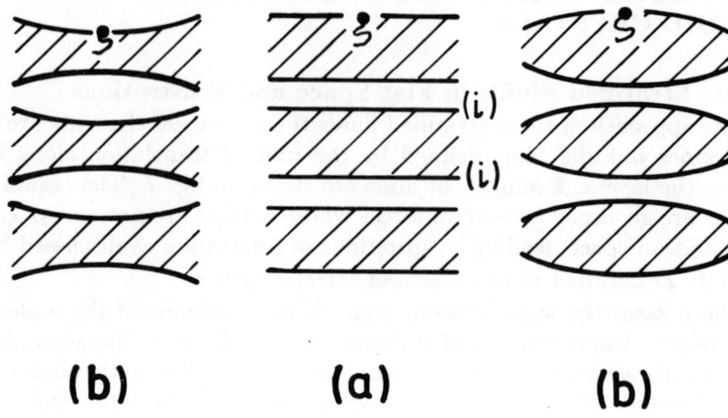

fig. 8

Schematic representation of a periodic system of films with flat interfaces (a), constant interfacial distances and zero curvature are compatible in R_3, the same with symmetrically curved interfaces (b), constant interfacial distances and non zero curvatures are no longer compatible in R_3 and the system becomes frustrated.

possible to find a coherent configuration of orientations over the whole space [16]. This is another example of geometrical frustration in the 3-D Euclidean space R_3.

III-2. Stratified Fluids in Curved Spaces

A frustration is directly related to the geometry of the space containing the system under study. As shown in fig. 8 and 9, where systems embedded in the flat 3-D Euclidean space R_3 are represented, the situations with flat films are the only situations compatible with a strict respect of the constraint of constant distance in this space. However, one might think that, if the frustrated systems were embedded in adequately curved spaces, a compatibility might be restored in a manner formally similar to the problems of bi-dimensional tilings with regular polygons and tri-dimensional packings of regular polyhedra [17]. We show in this paragraph that ideal stratified structures without frustration, i.e. conciliating the two antagonistic constraints, can indeed be built in curved spaces. Such ideal structures have of course no reality but we shall show in the next paragraph that they are needed, as starting points, to generate the possible configurations

---

[16] S. Meiboom, M. Sammon and W.F. Brinkman, Phys. Rev. A <u>27</u>, 438 (1983),

E. Dubois-Violette and B. Pansu, this book;

E. Dubois-Violette and B. Pansu, Mol. Cryst. Liq. Cryst. <u>165</u>,(1988).

[17] J.F. Sadoc and R. Mosseri, Phil. Mag. B <u>45</u>, 457 (1982),

J.F. Sadoc and R. Mosseri, Pour la Science <u>87</u>, 10 (1985).

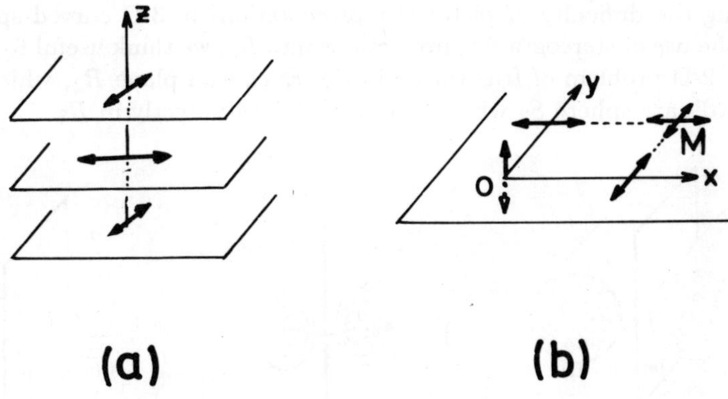

fig. 9
Schematic representation of a cholesteric system with a twist along one
direction only (a), the simple twist and the layering are compatible in R_3,
the same with the twist propagating isotropically (b), the double twist and
the layering are not compatible in R_3 and the system becomes frustrated.

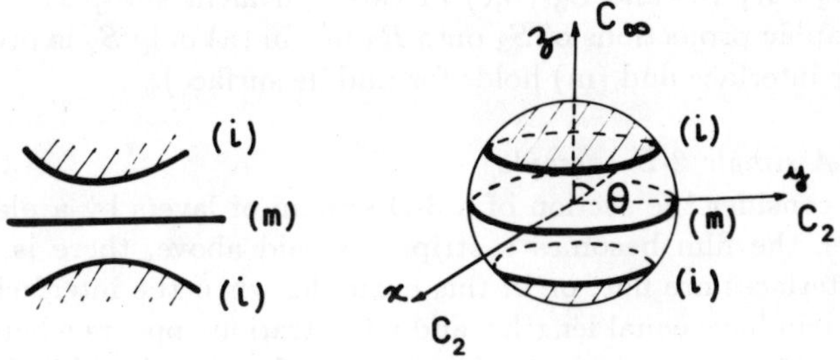

fig. 10
The 2-D periodic system of parallel strips, suppression of the
frustration in R_2 by transfering the strip onto S_2.

optimizing frustrations in the Euclidean space R_3.

We limit this first approach to curved spaces with homogeneous positive
curvatures. We want homogeneous curvatures, because we look in this first
approach for configurations which relax the frustrations equally everywhere, and
positive, because such spaces are finite so that the condition of periodicity can
be subtituted by cyclic conditions. These spaces are the hypercylinder $S_2 \star R_1$
and the hypersphere S_3. These spaces and some of their properties needed for the
following are described in the appendix.

104

Facing the difficulty of pictorial representations in 3-D curved spaces, which require the use of stereographic projections into R_3, we think useful to discuss first a simple 2-D problem of frustration in the Euclidean plane R_2, which is relaxed on the ordinary sphere S_2 which can be visualized directly in R_3.

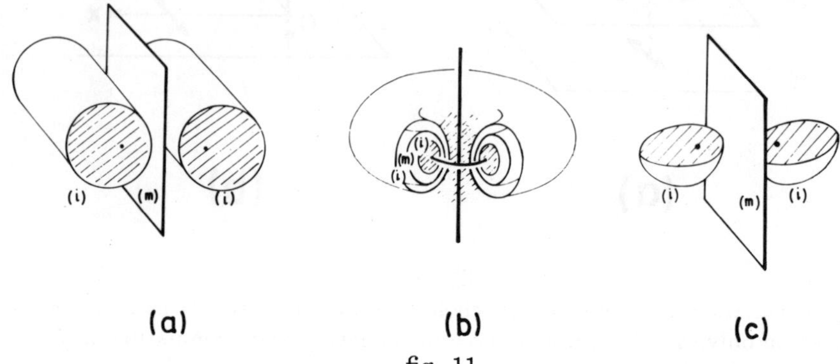

fig. 11

The 3-D periodic system of parallel films, suppression of the frustration in R_3 by transfering the film onto the surface separating the curved spaces $S_2 \star R_1$ (a) and S_3 (b,c) in two equivalent subspaces. ((b,c) are stereographic projections of S_3 onto R_3 but in (a) only S_2 is projected, (i) holds for interface and (m) holds for middle surface).

III-2.1. A simple 2-D example

We just consider the section of a 3-D system of layers by a plane R_2 normal to the layers, the film becomes a strip. As said above, there is no frustration when the interfaces are flat, or in this example, when the interfacial and middle lines of the strip have equal lengths, and a frustration appears when the interfaces become symmetrically curved, or when the interfacial and middle lengths become different. This situation can be homogeneously relaxed when the flat space R_2 is curved into the spherical space S_2 of radius R if, as shown in fig. 10, the middle line is placed on the equator and the shorter interfacial lines are placed on parallels at equal distances from the equator. The frustration is obviously suppressed, as the interfacial lengths $2\pi R \sin\theta$ are smaller than the middle length $2\pi R$, while the basic symmetry of the system is preserved, as the equator is a stationary line separating the sphere in two identical hemispheres. Moreover, when moving along meridians, which are great circles and therefore geodesics of S_2, the lines are periodically crossed, as when moving along the normals to the lines of the infinite system in R_2. This cyclic behaviour in the curved finite space S_2 makes the finite system of one strip in this space, equivalent to the infinite periodic system of an infinite number of strips without frustration in the infinite flat space R_2. Finally the symmetry axes of the relaxed structure in S_2, which we shall need in the next paragraph, are one C∞ axis normal to the equatorial plane and a family of C2 axes in this plane.

III-2.2 Lamellar and smectic systems in $S_2 \star R_1$ and S_3

This is the 3-D case where the frustration arises from area differences between the middle surface and the interfaces of the films. It requires the use of 3-D curved spaces. The way we proceed is quite similar to that followed in the preceding 2-D example. The situation can be homogeneously relaxed when the flat space R_3 is curved into the cylindrical space $S_2 \star R_1$ or the spherical space S_3. The middle surface of the film is placed on a particular surface of the curved space which divides it in two identical subspaces and is a stationary surface. Such a surface is minimal in the curved space because of the condition that this space is divided in two identical parts. In the case of $S_2 \star R_1$ such a surface is the usual cylinder $S_1 \star R_1$. In the case of S_3 there are two possible surfaces: a great sphere S_2 of S_3 and the so called spherical torus T_2. These surfaces are shown on the fig. 11 using a stereographic projection of the curved spaces on R_3; they are also described with more details in the appendix. The two interfaces are placed on equidistant, parallel cylinders, tori or spheres [11] [14]. In these situations the symmetry of the film is preserved, as the surfaces supporting its middle surface are stationary surfaces separating the spaces in two equal subspaces, the frustration is relaxed, as the interfacial area are smaller than the middle area, the distances between the interfaces are constant, as the surfaces of a family are parallel surfaces, and the periodicity of the stacking appears if we consider a continuous cyclic displacement of a local point of observation along geodesics normal to the surfaces. Finally, in view of their use in the next paragraph, the symmetry axes of relaxed structures are to be examined. They are C2 and C∞ axes. In $S_2 \star R_1$,the C2 axis are great circles normal to $S_1 \star R_1$ and generators of the cylinder; the C∞ axis are the lines parallel to to R_1 through the poles of S_2. In S_3 the C2 axis are geodesic great circles normal to T_2 or lying in the surfaces of T_2 and S_2; the C∞ axis are the torus axes.

III-2.3. Cholesteric systems in $S_2 \star R_1$ and S_3

In this 3-D case the frustration arises from the "double-twist" imposed to the orientation of the molecules by their chirality. This is a problem of director field which requires the use of 3-D curved spaces and their fibrations. The use of the fibration of S_3 was recently suggested [18] , we propose here to consider also the helicoidal fibration of $S_2 \star R_1$ defined in the appendix. The situation can be homogeneously relaxed when the flat space R_3 is curved into the cylindrical space $S_2 \star R_1$ or the spherical space S_3 if the frustrated director field is fitted

---

[18]   J.P. Sethna, D.C. Wright and N.D. Mermin, Phys. Rev. Lett. __51__, 467 (1983),

J.P. Sethna, Phys. Rev. Lett. __51__, 2198 (1983),

J.P. Sethna, Phys. Rev. B. __31__, 6278 (1985).

J.F. Sadoc, in "Physics of disordered materials", edited by D. Adler, H. Fritzsche and S.R. Ovshinsky, Plenum Press (1985).

M. Kleman, J. Phys Lett. __46__, L-723 (1985).

E. Dubois-Violette and B. Pansu, in this book.

with the fibration lines of the two spaces, as shown in fig.12. Here, as in the preceding case, displacements of an observation point along the geodesic normal to the surfaces supporting the fibrations are periodic, and the systems in curved spaces $S_2 \star R_1$ or S_3, although finite along certain or all dimensions, are equivalent to the infinite periodic system without frustration in the infinite flat space R_3. The symmetry axes of the relaxed structures are C2 and C∞ axes. The first are , in $S_2 \star R_1$, great circles normal to the family of parallel cylinders supporting the helicoidal fibration, in S_3 geodesic great circles normal to the family of parallel tori supporting the fibration. The second are the polar lines of the spaces.

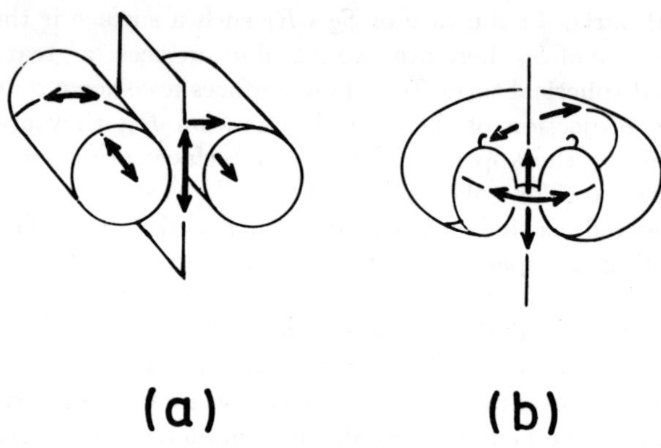

fig. 12

The cholesteric system with double twist, suppression of the frustration in R_3 by transfering it onto the helicoidal fibrations of $S_2 \star R_1$ (a) and S_3 (b). (Projections as defined in fig.11).

III-2.4. Remark

There is an important difference between the two physical problems of cholesteric and lamellar structures in the sense that it is always possible to find a homogeneous curved space to relax the frustration for the first case, while it is not for the other. Cholesteric systems are described by their pitch, which is related to the curvature radius of the required space by the very simple relation $p = \pi R$, so that it is always possible to find an adequate space whatever is the pitch. Lamellar and smectic systems, are described by two parameters, the area difference within the layers, or the interfacial curvature, and the distances between the interfaces. The curvature radius R is therefore defined by two relations and, from the elimination of R between them, a new relation appears between the two parameters of the systems. This imposes that it is possible to find spaces to relax the frustration in lamellar and smectic systems , only for particular values of the parameters. This is a geometrical constraint which is introduced by the use of spaces with homogeneous curvatures. Its role was examined in details in the case of lamellar systems of amphiphilic molecules and it was shown to be the major cause for the existence of a sequence of structures common to all phase diagrams [14]. We do not develop this aspect here.

III-3. Structures of disclinations in flat space

In the preceding paragraph we have shown that ideal stratified structures without frustration, i.e. conciliating the two antagonistic constraints, can indeed be built in curved spaces. Such ideal structures have of course no reality but are helpful to generate the geometrical configurations possible in the Euclidean space. For this it is necessary to map the curved space onto the flat space. A classical example of such a process with 2-D spaces is that of the mapping of a sphere onto a plane. A sphere can not be deformed into a plane without being torn, and if matter is added during the process. If a structure is present on the sphere the introduction of matter must respect the symmetry of this structure. This is generally known in condensed matter physics as a Volterra process and corresponds to the creation of discontinuities, or defects, in the structure which can be defects of translation (dislocations) and rotation (disclinations) [19] . As the direct displacements on a sphere are rotations the defects needed to map it onto a flat space are disclinations only. (Another argument would be to say that the sphere suffers from an angular defect intergrated on its surface of 4π relative to the plane, which can but be cured by a total amount of defects of rotation of -4π). In a similar way the mapping of the 3-D curved space onto the 3-D flat space is obtained by introducing a network of disclination lines. When there is no structure in the curved space disclinations of infinitesimal angle are introduced around each point of the space, when a structure is present disclinations of finite angles, respecting the symmetries of the structure without frustration in the curved space, must be introduced. In the case of films the disclination axes may be normal to the films or contained in them; their effects are shown in fig. 13. The density of the disclination network is determined by the curvature of the curved space [20]. Configurations obtained that way can be called structures of disclinations.

III-3.1. The 2-D system of strips

We come back to the case involving 2-D spaces only for a simple presentation of the process. The frustration in R_2 is relaxed by putting the strip on the sphere S_2, as shown in fig. 10. The mapping of S_2 onto R_2 is obtained introducing the adequate disclinations around the $C\infty$ and C2 axes of the relaxed structure. Disclinations around the $C\infty$ axis can be of any angle, their effect is to increase the area of the sphere and decrease its curvature locally without changing the topology of the strip, as shown in fig. 14. The result in R_2 is an infinite flat strip of constant thickness. This can not be got without unreasonable distorsions of the initial strip, involving stretching energy. Disclinations around C2 axes are

---

[19] J. Friedel, Proceedings of the Sixth General Conference of the European Physical Society, page 25, Prague (1984).

W.F. Harris, Sc. Am. **237**, 130 (1977).

[20] J.F. Sadoc and R. Mosseri, J. de Physique **45**, 1025 (1984).

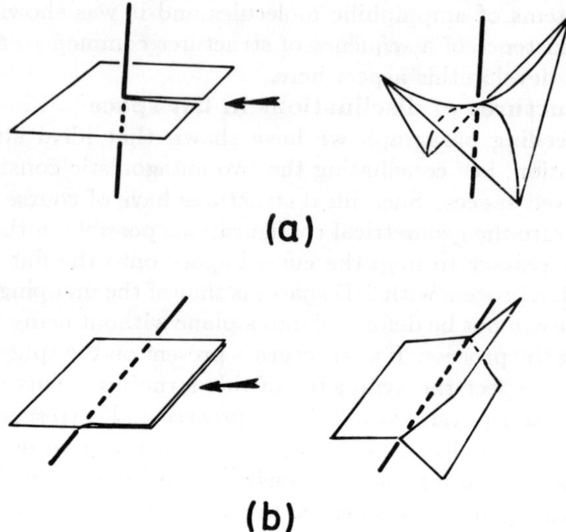

(a)

(b)

fig. 13

Disclinations normal (a) and parallel (b) to a film.

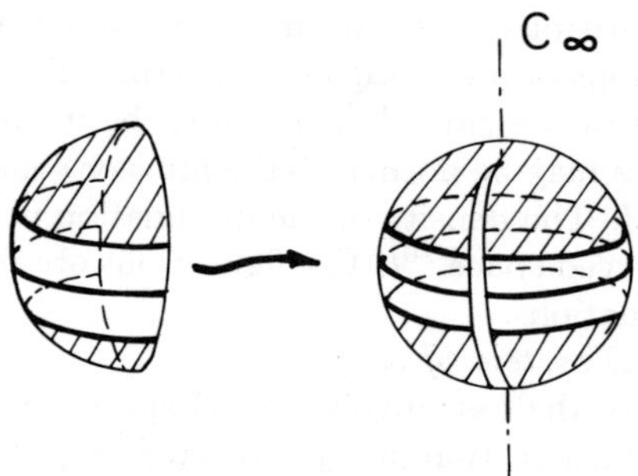

fig. 14

The 2-D periodic system of parallel strips, back to R_2 by introducing
disclinations around the C∞ axis, following a Volterra process.

$-\pi$ disclinations because of the C2 symmetry [19], they decrease the curvature
of the sphere locally and change the topology of the structure, as shown in fig.
15. The result in R_2 is a structure of closed cells delimited by a connected strip
with curved interfaces, this process involves curvature energy essentially. The two
processes are geometrically equivalent but, if one consider the energies associated

with the distorsions, they are different, the second is most likely preferable to the first because curvature energies are known to be smaller than stretching ones in liquid crystals. After having considered the topologies of the solutions, the last question about them concerns their possible organizations, or the way the disclinations might be ordered. In the new structure the organization of the strips separating the closed cells is connected and the topological structure of its 2-D flat pattern should obey the classical laws of topological stability telling that each cell must have in average 6 edges meeting 3 by 3 at each vertex [21]. That means, using Schlafli's notation [22] , that any organization must be a $\{6,3\}$ tiling of the plane, i.e. a hexagonal lattice or any distorsion of it, as represented on the right of fig. 15, if ordered.

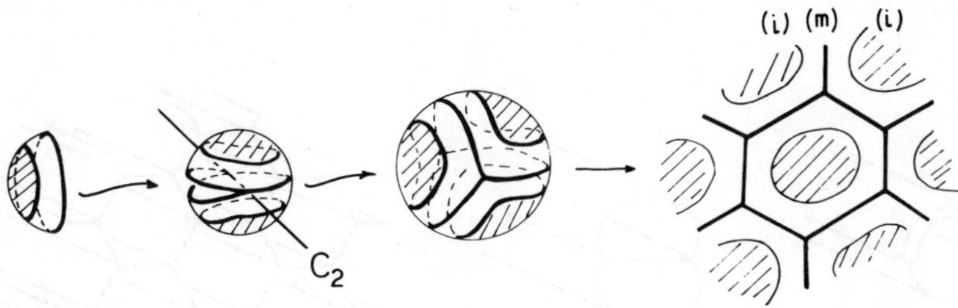

fig. 15

-The 2-D periodic system of parallel strips, back to R_2 by introducing $-\pi$ disclinations around C2 axes, following a Volterra process.

III-3.2. Non-lamellar and non-smectic structures

We now deal with a periodic stacking of interfacial films. The frustrations in R_3 are relaxed by transfering the stacking into 3-D spaces with homogeneous positive curvatures. Two such spaces have been used, the hypercylinder $S_2 \star R_1$ and the hypersphere S_3. The second has an isotropic curvature while the first has not it has no curvature along direction R_1 and all the curvature is concentrated on the sphere S_2 orthogonal to this direction .

In the three following cases a film has its mid-surface on a stationary surface of the curved space as described in III-2.2. Then the whole curved space is decurved using a disclination procedure.

III-3.2.1. Disclinations in $S_2 \star R_1$

The film is on a usual cylinder defined by the space product of a great circle of S_2 and R_1. All the curvature of the space results from that of S_2 (see the

---

[21] D. Weaire and N. Rivier, Contemp. Phys. **25**, 59 (1984).

[22] The Schlafli's notation $\{p,q\}$ for a tiling of polygons means that the polygons have p edges that meet q by q at vertices, the notation $\{p,q,r\}$ for a packing of polyhedra means that the polyhedra have faces with p edges, each polyhedron has q faces around one vertex and there are r polyhedra around one edge.

110

appendix) so the disclination process in $S_2 \star R_1$ [11] is formally equivalent to the one used just above in the case of the system of strips. The result of disclinations around the C∞ axis in R_2 is a film with flat interfaces but with important distorsions of the distances within the strip and, as already said, the process involves stretching energies. The result of $-\pi$ disclinations around C2 axes in R_2 is a structure of connected flat films delimiting infinite straight cells with curved interfaces, a process which involves curvatures energies essentially. Once again, the second process is chosen preferably to the first because curvatures energies are classically smaller than stretching ones in liquid crystals and, from then on, we shall not consider disclinations around C∞ axes any more. Finally, the structure optimizing the frustration corresponds to that of the hexagonal phase of lyotropic liquid crystals as shown in fig. 16.

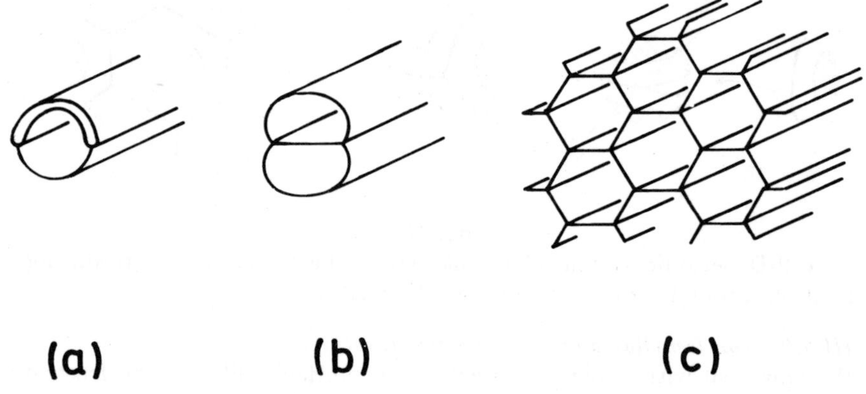

(a)   **(b)**   **(c)**

fig. 16

Introduction of two $-\pi$ disclinations around C2 axes of a great cylinder $S_1 \star R_1$ in $S_2 \star R_1$ following a Volterra process. (for topplogical reasons defects must be introduced by pairs). The cylinder is partly split in two sheets limited by the C2 axes (a), a space is introduced in between the sheets (b), the honeycomb structure obtained after the introduction of the necessary disclinations (c) (see also fig. 23)

III-3.2.2. Disclinations in S_3, the spherical torus

The film on the spherical torus is represented in fig. 11. We consider disclinations around C2 axes normal to the surface only, as it can be shown that disclinations around C2 axes on the surface must be accompanied by C∞ disclinations around a polar axis so that the final result correspond to the situation described just above [11]. A schematic representation of the introduction of a $-\pi$ disclination around a C2 axis normal to the surface, valid only as far as topology is concerned [12], is given in fig. 17. The torus of genus 1 is cut in half, the lips are separated and half a torus is inserted. The result is a torus of genus 2 which, like the original torus, separates the space in which it is embedded in two identical subspaces. When the set of disclinations needed to map S_3 onto R_3 is introduced, a complex surface of high genus is generated which keeps the important property of separating R_3 in two identical subspaces. The topology of such a configuration is obviously "bicontinuous" , as that of the cubic phases found in the immediate vicinity of lamellar and smectic phases.

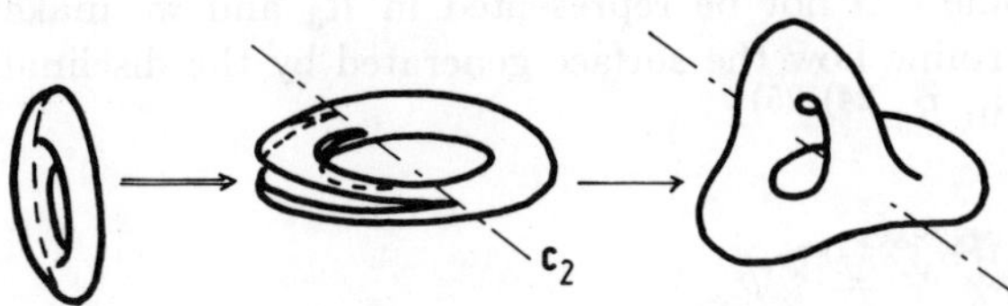

fig. 17

Introduction of a $-\pi$ disclination around a C2 axis of a torus, following a Volterra process. The validity of this figure in the T_2 case concerns the topology of the process only.

To find the possible ordered organizations of the disclinations, we make use of an important property of the spherical torus which is that it can be built by identification of the opposite sides of a square sheet and that its only regular possible periodic tiling is a $\{4,4\}$ tiling of squares [12] .

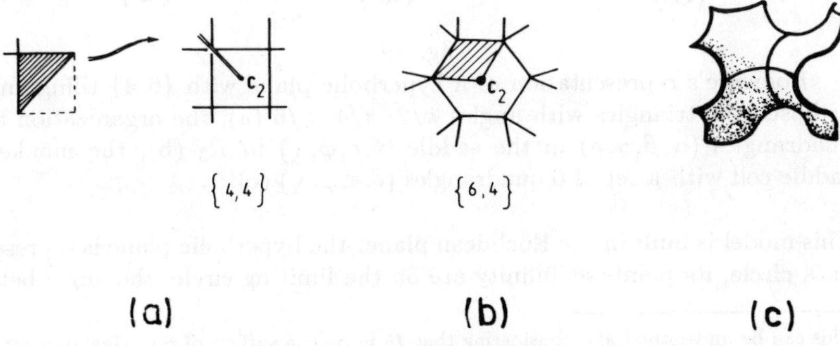

fig. 18

A $-\pi$ disclination in a square transforms it into a hexagon (a), the 4 connectivity is preserved at every vertex (b), if the original squares are in a plane with zero Gaussian curvature the plane is transformed into a hyperbolic surface with negative Gaussian curvature (c).

The homogeneous introduction of the disclination network must respect this tiling and its symmetry, so that the elementary aspect of the disclination process is the disclination of a square of the $\{4,4\}$ tiling. This is represented in fig. 18. The $-\pi$ disclination transforms the square into a hexagon without changing the coordinence of the vertices. The effect of the introduction of the total set of disclinations is therefore to transform the $\{4,4\}$ tiling into a $\{6,4\}$ tiling, which is obviously not Euclidean, as hexagons can only be organized on $\{6,3\}$ in R_2. The surface generated this way has the same local properties as a hyperbolic surface of constant negative Gaussian curvature, or hyperbolic plane [23]. The hyperbolic plane can not be represented in R_3 and we make use of Poincaré's model to determine how the surface generated by the disclination procedure can be embedded in R_3 [24] [25].

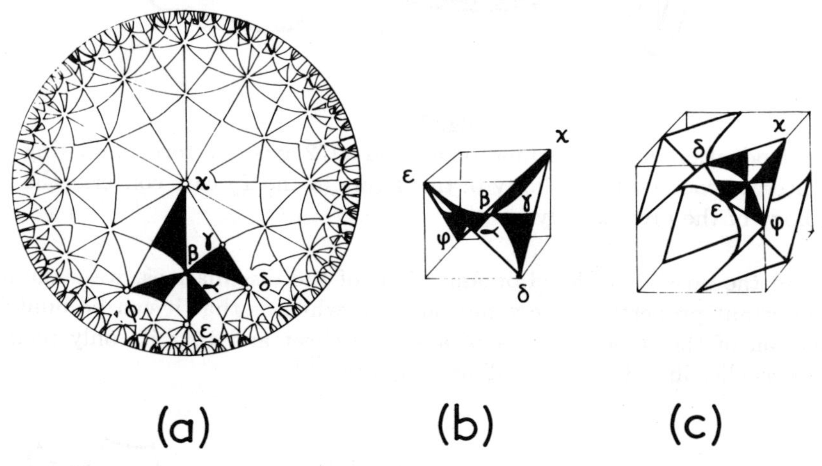

(a)   (b)   (c)

fig. 19
Poincaré's representation of a hyperbolic plane with $\{6,4\}$ tiling and orthoscheme triangles with angles $\pi/2$, $\pi/4$, $\pi/6$ (a), the organization of quadrangles $(\alpha, \beta, \gamma, \delta)$ in the saddle $(\delta, \varepsilon, \varphi, \chi)$ in R_3 (b), the monkey saddle cell with a set of 6 quadrangles $(\delta, \varepsilon, \varphi, \chi)$ (c).

This model is built in the Euclidean plane: the hyperbolic plane is represented within a circle, its points at infinity are on the limiting circle, the angle between

---

[23] This can be understood also considering that T_2 in S_3 is a surface of zero Gaussian curvature in a space of positive Gaussian curvature so that, when the curvature of the space is decreased to zero , that of the surface becomes negative.

[24] D. Hilbert and S. Cohn-Vossen, "Geometry and the Imagination", Chelsea publishing Company, New-York (1983).

[25] H.S.M. Coxeter, "Introduction to Geometry", John Wiley, New-York (1961).

lines are preserved, the metrics is chosen in order to represent that of the surface, i.e. the distance between two points representing a constant segment on the surface decreases when the two points approach the limiting circle [25]. The $\{6,4\}$ tiling of the hyperbolic plane we need is represented in fig.19a, together with its orthoscheme triangles which are the smallest cells, or assymmetric units, from which the whole surface can be constructed by reflections on its sides. It is clear on this representation that the hexagons and the triangles are not Euclidean. The hexagons have angles of $\pi/2$, which are angles of the squares before the introduction of the disclinations, so that the orthoscheme triangles have angles of $\pi/2$, $\pi/4$, $\pi/6$. An examination of the possible structures of these triangles, embedded in R_3, readily shows that the possible surfaces in R_3 fall into three categories. The surfaces for which the two sides of the right angle of the orthoscheme triangle are straight lines, those for which the hypotenuse only is a straight line and those for which all sides are curved. We just treat here of the first case, details concerning the other cases can be found in [12]. If the two sides of the right angle of the orthoscheme triangle are straight they build a network of intersecting straight lines whose element is a non-planar quadrangle $(\alpha, \beta, \gamma, \delta)$ with three angles of $\pi/2$ and one angle of $\pi/3$. Such elements can be assembled in a more symmetrical quadrangle $(\delta, \varepsilon, \phi, \kappa)$ with four angles of $\pi/3$ and four equal sides having the saddle shape shown in fig. 19b. These quadrangles are regularly organized in the hyperbolic plane and it was shown by Schoenflies [26] and Schwarz [27] that, when embedded in R_3, six of them build the network having the shape of a monkey saddle shown in fig. 19c, which is the translation cell of a cubic lattice. Schwarz also showed that this periodic network of straight lines is the support of an infinite periodic minimal surface, or IPMS, separating R_3 in two identical subspaces. The translation cell of the surface, together with the two congruent labyrinths separated by it, are represented in fig. 20. Notice that the translation cell of the surface shown in fig. 20 is not exactly that of the network represented in fig. 19c as, on the latter, the normal to the surface would rotate by π, and not 2π, when moving by a translation vector. This organization has the Pn3m symmetry of one of the "bicontinuous" cubic phase [4]. The two other cases of orthoscheme triangles, with straight hypotenuse or all sides curved, lead to organizations with Im3m and Ia3d symmetries which have been observed experimentally too [4].

This rather intuitive approach was recently supported by the systematic investigation of translation sub-groups in the $\{6,4\}$ hyperbolic plane [28]. For this we considered the intrinsic geometry of this particular hyperbolic plane and analyzed its symmetry groups, focussing our interest onto translation sub-groups and fundamental regions. The smallest fundamental region for generating surfaces of genus 3 is the dodecagonal cell containing 96 orthoscheme triangles shown in

[26]   A. Schoenflies, Compte Rendus 112, 478 (1882).

[27]   H.A. Schwarz, "Gesammelte Mathematische Abdhandlungen" band1, Springer Verlag, Berlin (1890).

[28]   J.F. Sadoc and J. Charvolin, Acta Cryst. A. 45, 10 (1989)

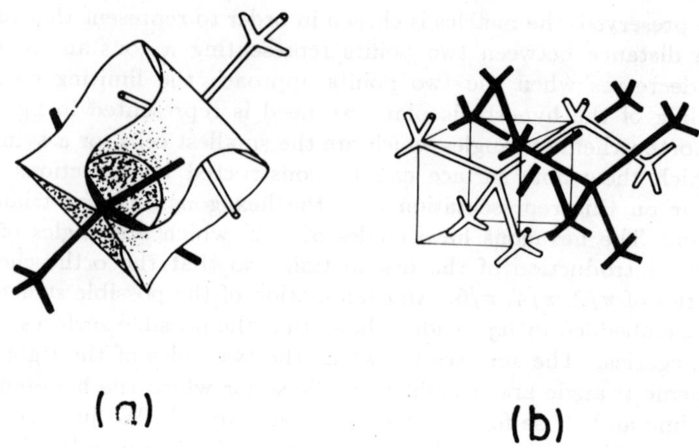

fig. 20
The translation cell for the F surface of Schwarz (a), and the labyrinths
of the Pn3m cubic structure separated by this surface (b).

fig. 21. Its periodic reproduction by a translation sub-group covers the whole
hyperbolic plane without overlap. Another aspect of the periodic behaviour,
usally known as the Born-Von Karmann conditions in classical crystallography,
is to identify the sides of an isolated fundamental region two by two, so that all its
vertices are gathered in one point, in order to built a toroidal surface with several
holes. In this case all the elements of the translation sub-group are change into
the indentity operation.

The periodical hyperbolic surface and the toroidal surface are embedded in non
trivial spaces. It is also possible to have an intermediate situation between these
two extreme aspects of the periodicity by restricting the number of translations
which are changed into the identity operation and we show that it is possible to
obtain periodical surfaces embedded in R_3, the IPMS. Schematic representations
of the identifications needed to build the cells of the 3 surfaces are given in fig.
22. This of course imposes some "surgery" in the hyperbolic plane, as some parts
of it have to be cut out in order to permit the identifications, and implies also
that the IPMS can not keep a constant Gaussian curvature, it is only its averaged
value which corresponds to that of the initial hyperbolic plane. These distortions
are to be related to the fact that the whole hyperbolic plane can not be embedded
in R_3 and the following simple argument can be used to illustrate this point. The
number of points reproduced by elements of the symmetry group is infinite in all
cases but we can look at the way it increases within a disk of increasing radius.
In the hyperbolic plane this variation is exponential but follows a power law in
R_3, thus the hyperbolic plane is too "large" to be put in R_3 without the cutting
out of some parts of its surface. More generally, this investigation also permits

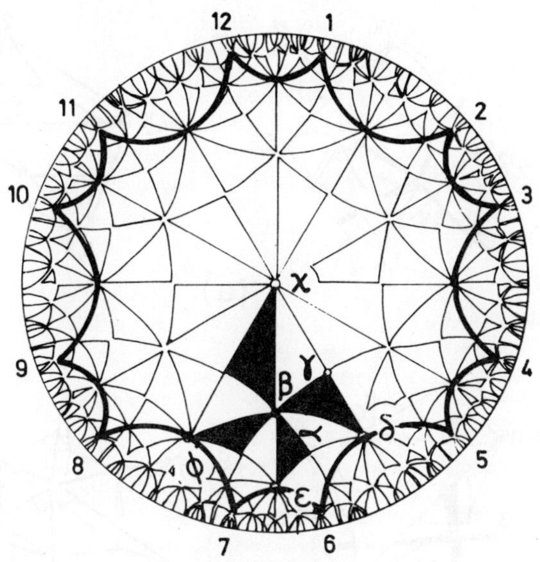

fig. 21
The dodecagonal translation cell in the 6,4 hyperbolic plane.

to study the cristallography of these complex surfaces from an intrinsic point of view, using operations of groups of symmetry defined by displacements on their surface. It transforms the 3-D problem of infinite periodic minimal surfaces into a 2-D problem and, although the latter is to be treated in a non-Euclidean space, provides a relatively simple formalism for the investigation of infinite periodic surfaces in general and the study of the geometrical transformations relating them.

III-3.2.3. Disclinations in S_3, the great sphere

The film on the great sphere is represented in fig. 11. The C2 axes are great circles of S_2 and the introduction of a disclination along one of these axes brings in a third finite subspace, as shown in fig. 23. The introduction of a set of disclinations just multiplies the number of identical finite subspaces. The topology created this way is therefore that of an infinite number of finite cells separated by a self-intersecting film [13]. This is the topology of the micellar phases and, most likely, that of the cubic phases found in the immediate vicinity of micellar phases. It is important to notice that the topological stability of the cells implies that the walls limiting them meet three by three along common edges which meet four by four at each vertex [21], [29].

Because of the identity of the cells the search for ordered organizations is that of eventual space-filling assemblies of regular polyhedra obeying this law. Such

[29] N. Rivier, Phil. Mag. B <u>52</u>, 795 (1985).

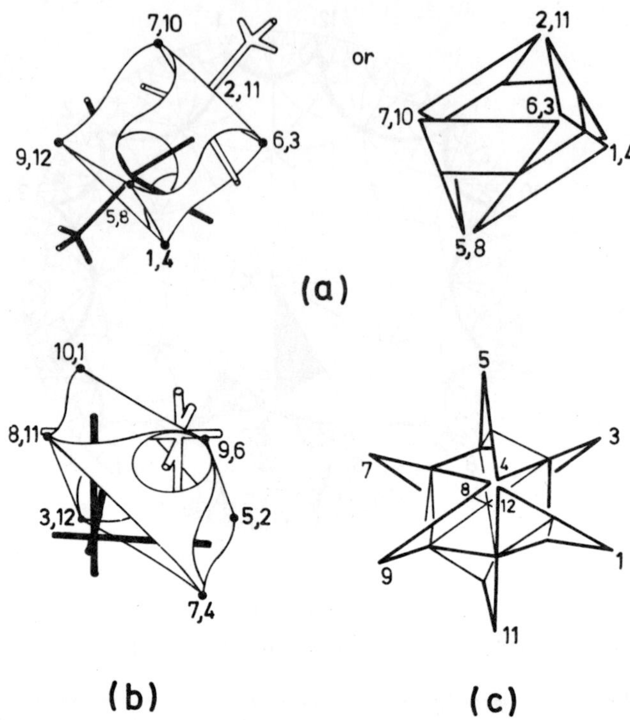

fig. 22

Identifications of vertices of the dodecagonal cell building the cells
with genus 3 of the 3 basic IPMS, P and D surfaces of Schwarz (a,b),
a topological analog of the G surface of Schoen (c). The number are those
of the vertices of the dodecagonal cell represented in fig.21.

assemblies are called polytopes. We must therefore search for polytopes having
three faces per edge and four edges per vertex or, having three faces belonging
to the same polyhedron around one vertex and three polyhedra around one edge.
This means, using Schlafli's $\{p,q,r\}$ notation [22], the polytopes of the $\{p,3,3\}$
family. They are four which, exist in curved spaces only[30] :

-in spherical spaces with constant positive Gaussian curvatures there are the
three $\{3,3,3\}$, $\{4,3,3\}$ and $\{5,3,3\}$ polytopes

-in a hyperbolic space with constant negative curvature $\{6,3,3\}$.

As there is no $\{p,3,3\}$ polytope in a flat Euclidean space our problem admits

---

[30] H. S. M. Coxeter, "Regular Complex Polytope", Cambridge University Press (1974).

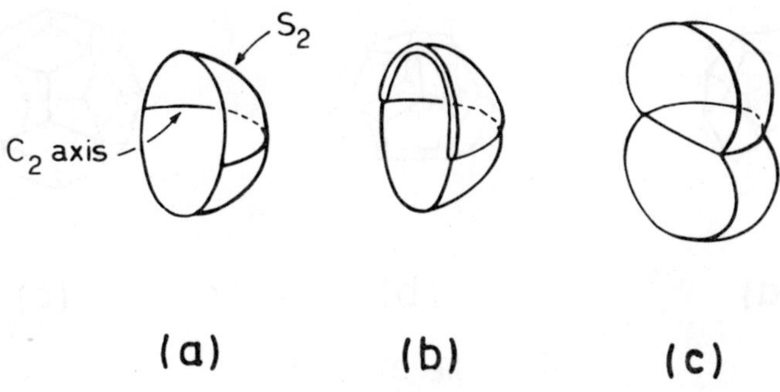

fig. 23

Introduction of a $-\pi$ disclination around a C2 axis of a great sphere S_2 in S_3, following a Volterra process. This is a stereographic projection of S_3 onto R_3, for the sake of clarity, only half a sphere S_2 is shown and the film supported by it is not represented. S_2 separates S_3 in two identical sub-spaces (a), S_2 is partly split in two sheets limited by the C2 axis (b), and a third sub-space, identical to the two first is introduced between the sheets.

no solution with identical regular polyhedral cells [31], indeed the only regular tesselation of R_3 is the polytope of cubes $\{4,3,4\}$. We are therefore driven to search for eventual non-regular but, nevertheless, periodic solutions. For this it is useful to consider the fact that the regular $\{p,3,3\}$ polytopes are found in spaces of decreasing curvatures and are therefore met one after the other during the progressive introduction of disclinations needed to map S_3 onto R_3. They are met at well defined steps of the process, when the curvature of the space is such that it can be tiled by a $\{p,3,3\}$ polytope. In between two steps, when the curvature of the space is not compatible with a regular polytope, the filling of the space can not be regular. This is the situation we shall be confronted with as our Euclidean space R_3 is not compatible with either $\{5,3,3\}$, which exists in the spherical space of lowest positive curvature for a $\{p,3,3\}$ polytope, or the following polytope $\{6,3,3\}$, which exists in a hyperbolic space of negative curvature. Indeed it can be shown, writting the curvature as a function of p and making it equal to zero, that, in R_3, p = 5.1 and is not an integer [32] . If non regular solutions exist in R_3 they should be assemblies of polyhedral cells with an average number of edges per face close to 5.1.

[31] Soap bubble froths are macroscopic examples of the impossibility to fill R_3 with $\{p,3,3\}$, the solution is random in this case see

W. D' Arcy Thompson, "On growth and form", abridged edition by J.T. Bonner page 88, Cambridge University Press (1975), and [29].

[32] H.S.M. Coxeter, Illinois J. Math. 2, 746 (1958).

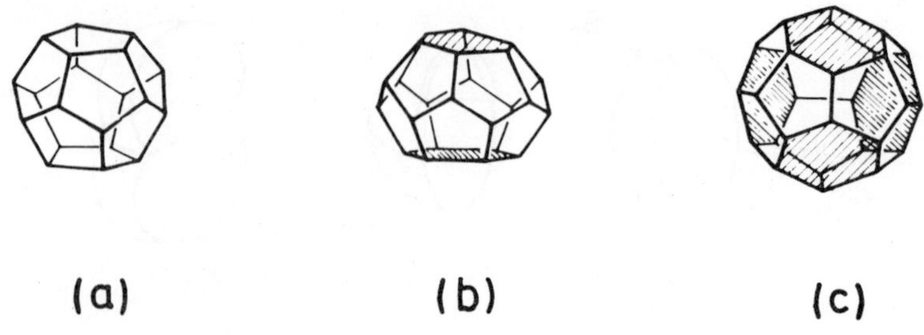

(a)   (b)   (c)

fig. 24

Transformation of a dodecahedron (a) into a tetrakaidecahedron (b) by one $2\pi/5$ disclination around an axis normal to two pentagonal faces, and into a hexakaidecahedron by four half disclinations (c). The hexagons are hatched.

The polytope of interest for us is obviously $\{5,3,3\}$ which exists in the space of lowest positive curvature. We must start from it and find the disclination process realizing the final mapping of S_3 onto R_3 which leads to ordered cellular structures. This question was previously adressed in the case of clathrate structures of water and silicon-sodium alloys and, in a related manner, in the case of the dual polytope $\{3,3,5\}$ to analyze the Frank and Kasper's structures of Laves phases and A15 alloys [33]. It was shown that disclinations normal to pentagonal faces of a dodecahedron transform it into the tetrakaidecahedron and hexakaidecahedron shown in fig. 24. These non-regular polyhedra are particularly interesting here as it is known that, when slightly distorted, they can be packed so that they build periodic space-filling assemblies of cells [34]. These assemblies are indeed the organizations of cages trapping host molecules in clathrates of water molecules, the vertices being occupied by oxygen atoms and the edges by hydrogen bonds. Coming back to our problem, these organizations should be those of the film and micelles permitted by the properties of our Euclidean space, the middle surface film being supported by the faces of the polyhedra and the interfaces being those of the micelle located in each polyhedron. Several structures can be built along the above principle and, among them, we can distinguish two large families according to the fact that their dihedral and edge angles stay close to 120°and 109°28' or show departures from these values. If we limit ourselves to the first family, whose angles are the closest to those of fluid films balancing their tensions, we are left with two relatively simple structures. One, type I structure, has space group Pm3n and its crystallographic unit cell contains 2 dodecahedra and 6 tetrakaidecahedra, the local arrangement of its polyhedra being shown in fig. 25. The second, type II structure, has space group Fd3m and its crystallographic unit cell contains

[33] J. F. Sadoc in reference [18].

[34] R. Williams, "The geometrical foundation of natural structures", Dover (1979).

16 dodecahedra and 8 hexakaidecahedra. If larger departures from the classical angles of films are accepted, other structures, involving also pentakaidecahedra, are possible [34]. Cubic structures having these symmetries have been observed in between hexagonal and micellar phases [4]. The topology of the structural models proposed therein to analyze these data is not of the "micellar" type discussed here. However NMR data support a topology of this type [35].

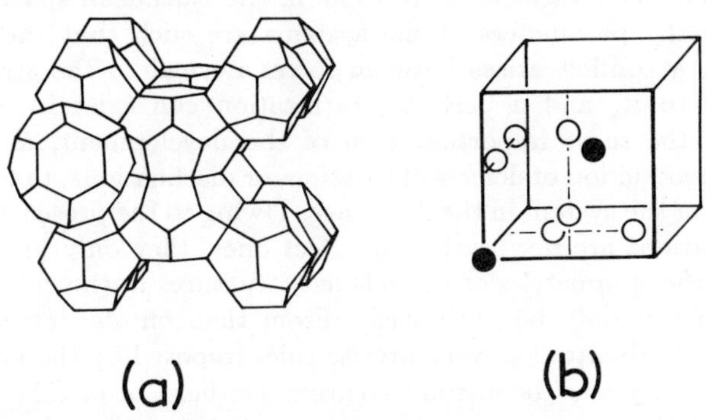

(a)   **(b)**

fig. 25

Periodic aggregation of slightly distorted 12-hedra and 14-hedra, on the right the position of the 12-hedra (•) and 14-hedra (o) in the cell of the Pn3m lattice, from reference [34].

III-3.3. "Blue" phases

The frustration due to the double twist was relaxed on the fibrations of the hypercylinder $S_2 \star R_1$ and the hypersphere S_3. Those fibrations are supported by the families of cylinders or tori, which were used in the cases of systems of interfaces, and it is easy to see that the C2 symmetry axes of interest in the two problems are the same. In the case of the fibration in the hypercylinder $S_2 \star R_1$ they are the generators of the cylinder $S_1 \star R_1$ with maximal area separating the hypercylinder in two identical subspaces, in the case of the fibration in the hypersphere S_3 they are the great circles normal to the tori. The introduction of the disclinations is therefore similar to the procedure used in the case of systems of interfaces and the structures built by the disclinations must have related symmetries. They are hexagonal p6m in the case of the mapping of $S_2 \star R_1$ onto R_3, and cubic Ia3d, Pn3m or Im3m in the case of the mapping of S_3 in which, as the chirality of the director field must be taken in account, mirror operations are to be suppressed; the symmetries are indeed those of the observed in "blue" phases. The point of view adopted here is also that which is used in another chapter by E. Dubois-Violette and B. Pansu to study the director field around a disclination line.

III-4. Summary and Comments

[35]   T. Bull and B. Lindman, Mol. Cryst. Liq. Cryst. <u>28</u>, 155 (1974).

Fluid systems of very different natures, with interfaces in lamellar and smectic liquid crystalline materials or periodically twisted in cholesteric materials, are able to build structures exhibiting similar symmetries, on very different scales, with very different molecular interactions. This indicates that the occurence of these structures is not to be looked for in the detailed nature of each system but, rather, in their common ability to build stratified layers.

The only perfect stratification possible in the Euclidean space is that of flat layers. When the parameters of the systems are such that the layers can no longer be flat, a conflict arises between forces normal to the stratification and forces parallel to it, and a perfect stratification can exist in a curved space only. This is the most important step of the development, as it is it which imposes the introduction of defects of rotation, or disclinations, to find the possible configurations of the system in the flat space. Owing to the presence of the defects, these configurations are necessarily imperfect ones, they only preserve locally in the flat space the symmetries of the relaxed structures in the curved spaces, and the frustration can only be optimized. From then on the introduction of the disclinations are submitted to very precise rules imposed by the symmetry of the problem. First, they must be introduced around either $C\infty$ or $C2$ axes. In the case of the $C\infty$ axes, the system is forced back to the flat situation, this is a very brutal solution. In the case of disclinations around $C2$ axes, the solutions proposed to the systems are more subtle as they imply topological changes. The first process may be valid for weak degrees of frustration, while the second may be preferred for stronger ones, as shown by the observations. Second, the modes of organizations of these disclinations, if any, are also imposed by the symmetry of the problem. Their systematic investigation provides symmetries for the structures of disclinations which match well with those of the most commonly observed structures.

The agreement obtained with the broad lines of the description is rather satisfying and suggests that the basis of this approach, the conflict between two types of forces, might be correct to understand these complex mesoscopic or macroscopic structures. One direction now would be its extension to the case of less frequent structures, quadratic, tetragonal and rectangular for example, which have been recently shown to exist in the systems discussed here [10] [36]. This certainly goes through the giving up of the hypothesis of isotropy of the film, implicitly contained in our approach, and a precise examination of the density of disclinations and of their modes of introduction. Another obvious extension is of course the consideration of energy terms to evaluate the relative stabilities of the structures. Particular terms appear at each step of the geometrical development. For instance, cubic and hexagonal structures, which have been shown to be equivalent solutions on the purely geometrical aspect of space filling [14] , should be different energetically. They differ at the first step, that of the embedding in a curved space, as the cubic structure is obtained from the isotropic space S_3

---

[36] S. Alperine, Y. Hendrikx and J. Charvolin, J. de Physique Lett. 46, L-27 (1985).
P. Kekicheff and B. Cabane, J. de Physique 48, 1571 (1987)

and the hexagonal one from the anisotropic space $S_2 \star R_1$. There is a cost of energy for moving from the first to the second as it is necessary to introduce a disclination around a $C\infty$ axis, transforming for instance the spherical torus into a cylinder. Then they differ at the second step, when disclinations around C2 axis are introduced as they are normal to the film for the cubic structure and contained in it for the hexagonal one. Normal disclinations change the Gaussian curvature, not parallel ones, and this implies different energy contributions [37]. These different terms are being evaluated in this precise case and it appears that the cubic structure is preferred for weak degree of frustration while the hexagonal structure ifs preferred for stronger ones [38]. This result agrees well with the order of appearance of the two phases in phase diagrams of amphiphilic molecules [1] and might help understanding, the cubic structure of "blue" phases must be submitted to a electric field for transforming into a hexagonal one [10] .

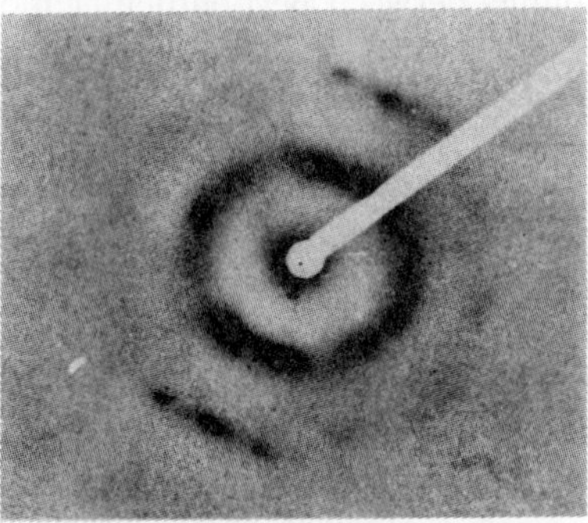

fig. 26
One particular plane of the reciprocal space of a model cubic structure Ia3d (a), observation of the same plane in a real structure (b).

Finally, our last comment is to insist again on the very nature of the considerations developed in this chapter. To introduce it in a novel manner we present two particular sections of the reciprocal space of a cubic structure Ia3d in fig. 26. One is that expected if the structure was that obtained following our approach, the other is that which was recently observed with a monocrystal of

[37] W. Helfrich, in "Physics of Defects", Les Houches summer school 1980, edited by R. Balian, North Holland (1981).
[38] J. Charvolin and J.-F. Sadoc, in preparation.

the cubic phase formed by an amphiphile/water system [39]. The two patterns are different, Bragg spots only are visible on the first, Bragg spots and rather intense diffuse scatterings are visible on the second. The latter show that the real structure is the seat of very intense structural fluctuations. The structures obtained in our approach are zero temperature structures, this approach is but a crystallographic investigation of the possible configurations of fluid films taking spatial constraints into account only.

IV- STRATIFICATION OF VELOCITY IN CONVECTIVE ROLLS

In the case of classical electrical instabilities of thin slabs of nematic liquids submitted to a perpendicular electric field, transitions from the 1-D lattice of parallel convective rolls to 2-D square lattices appear above critical values of the field. These transitions can be phenomenologically described in terms of defects [40] [19]. However, their analytical modelisation, which implies strong non linear couplings between perpendicular modes, does not call for any phenomenon of the type of those described in the two other cases of stratifications. This case of convective rolls does not seem to be relevant of the type of approach developed in this chapter. However we have been tempted to speculate on its eventual interest in one example of thermal instability.

In classical thermal instabilities adjacent rolls have opposite senses of rotation and determine upward and downward flows which have equal velocities. In this case the horizontal velocities in the regions in between those flows are parallel, there is no divergence in the field and this is perfectly compatible with a pattern of straight parallel rolls. This is indeed because the system is considered to have vertical symmetry. If the latter is broken, for instance by a non linearity in the expansion coefficient of the liquid, the upward and downward velocities are no longer equal and the flow conservation imposes a divergence of the horizontal velocities which is no longer compatible with a periodical stacking of straight parallel rolls in the plane of the slab. This might be treated as a case of frustration in a 2-D Euclidean space, quite equivalent to that discussed for the 2-D system of parallel strips. Its solutions might be looked for in the same way, with the formal difference that the divergence of the meridians from the poles of S_2 should be used rather than the difference of lengths between the parallels and the equator. The result is also a 2-D $\{6,3\}$ hexagonal pattern.

V- CONCLUSION

The observation of a unique morphological scheme, in a wide variety of stratified fluids of very different chemical natures, suggested that the origin of this scheme is not to be looked for in the particular details of the molecular

[39] Y. Rançon and J. Charvolin, J. de Physique 48, 1067 (1987).

[40] R. Ribotta and A. Joets, in " Cellular Structures in Instabilities", edited by J.E. Weisfred and S. Zaleski, Lecture Notes in Physics 210, Springer Verlag (1984).
A. Joets and R. Ribotta, idem.

structures. We proposed it to be in the existence of the stratification, whose modes of organization are imposed by the properties of the space in which the system is embedded. This is a purely geometrical formulation of the problem, which is rather unusual in the fields of physical chemistry and physics of fluids from where most of the systems studied here are issued. As a matter of fact this problem, particularly in the case of liquid crystalline structures, is mainly envisaged from the thermodynamical point of view at the moment. We think that the two points of view, geometrical and thermodynamical, are indeed complementary, the first providing the frame within which the second must be developed. Finally, the notions of frustration and defects intervening in our arguments were mostly developed for analyzing solid state physics problems related to departures from crystallographic order and also large cell crystals. Their success in giving account of the broad lines of a morphological problem in liquid state physics contributes to the demonstration of their unifying power in condensed matter physics. They may help to open new ways in the difficult study of the disordered states of these systems which, actually, suffers from some conceptual weakness.

APPENDIX: CURVED SPACES AND THEIR PROPERTIES

The development presented in this article requires the knowledge of some properties of curved spaces with constant intrinsic curvature.

I- S_2 or the 2-sphere

This is the classical sphere (its surface), which is a 2-D space, with constant, homogeneous and positive curvature. The curvature is defined by the intrinsic curvature $\kappa = 1/R^2$ which can be obtained by direct measurements on the surface itself: for instance the area of a geodesic triangle is $S = (2\pi - A - B - C)/\kappa$ where A,B and C are the angles of the triangle. A sphere is currently defined in Euclidean coordinates as $x_1^2 + x_2^2 + x_3^2 = R^2$ but displacements on the surface are defined by two parameters θ and ϕ and it is better to use spherical coordinates rather than Euclidean ones. They are related by

$$x_1 = R \sin\theta \cos\phi, \, x_2 = R \sin\theta \sin\phi, \, x_3 = R \cos\theta$$

This is an intrinsic representation of S_2.

Geodesic lines are great circles, any other circle is smaller. Two geodesics have two common points diametrically opposite on the sphere.

II- S_3 or the 3-sphere.

This is the hypersphere, it is a 3-D space of constant, homogenous and positive curvature. This space is usually defined by four coordinates in an Euclidean 4-D space as $x_1^2 + x_2^2 + x_3^2 + x_4^2 = R^2$.

However, important aspects of this space used in this study are better seen from an intrinsic point of view, i.e. described in the 3-D space with three parameters. For this reason we must use two other systems of coordinates, the toroidal and spherical systems, in which the coordinates of one point are expressed as functions of the radius of curvature R and of three adequately defined angles θ, ϕ and ω.

124

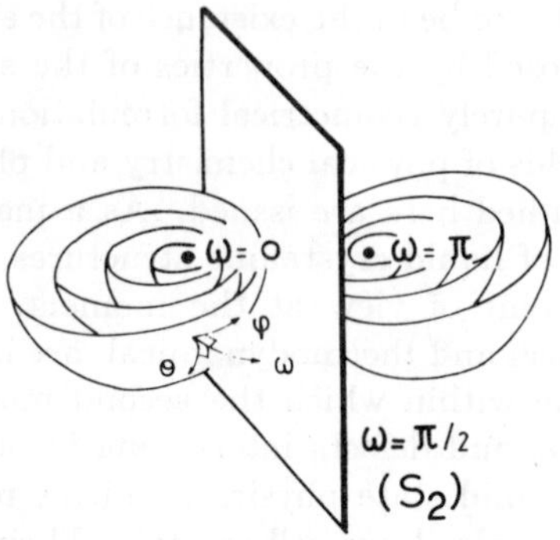

fig. 27

Stereographic projection in R_3 of a family of spheres with $\omega = $ constant in S_3, the great sphere S_2 corresponds to $\omega = \pi/2$, θ and ϕ are variables on the surfaces of the spheres.

This is indeed quite comparable to moving from Cartesian coordinates to spherical ones in the case of S_2 presented above.

It is possible to define spherical coordinates with two singularities at two poles. These coordinates are related to the 4-D Euclidean coordinates by:

$$x_1 = R \sin\theta \cos\phi \sin\omega, \ x_2 = R \sin\theta \sin\phi \sin\omega, \ x_3 = R\cos\theta \sin\omega, \ x_4 = R\cos\omega$$

In this system surfaces at constant ω are spheres S_2 with radii $R\sin\omega$ centered on the two points $x_4^2 = R^2$ obtained for $\omega = 0$ and π as shown in fig. 27. Among them the great sphere S_2, obtained for $\omega = \pi/2$, is at equal distances from the two points and separates S_3 in two equivalent sub-spaces. So, embedded in S_3, we can define great 2-spheres, dividing the space in two half hyperspheres, exactly like an equator divide the usual 2-sphere into two north and south hemispheres.

Any point on a 2-sphere is therefore determined by the variables θ and ϕ and a constant ω. Varying the two first by $d\theta$ and $d\phi$ leads to the orthogonal displacements $R\sin\omega d\theta$ and $R\sin\theta \sin\omega d\phi$ on the sphere and changing the third by $d\omega$ leads to a displacement $Rd\omega$ normal to the sphere. The element of area on the sphere ω is therefore $ds = R^2 \sin\theta \sin^2\omega d\theta d\phi$ and the element of volume in S_3 is $dv = R^3 \sin\theta \sin^2\omega d\theta d\phi d\omega$ so that the area of a small sphere in S_3 is $S = 4\pi R^2 \sin^2\omega$ and the volume enclosed $V = 2\pi R^3(\omega - (\sin 2\omega)/2)$.

More particularly, the great sphere S_2 with $\omega = \pi/2$ has an area $4\pi R^2$ and a volume $\pi^2 R^3$, which is exactly half that of the volume of the hypersurface S_3.

On a great sphere we can draw a great circle: great circles are geodesic lines of the 3-D spherical space.

It is also possible to define toroïdal coordinates which are related to Cartesian

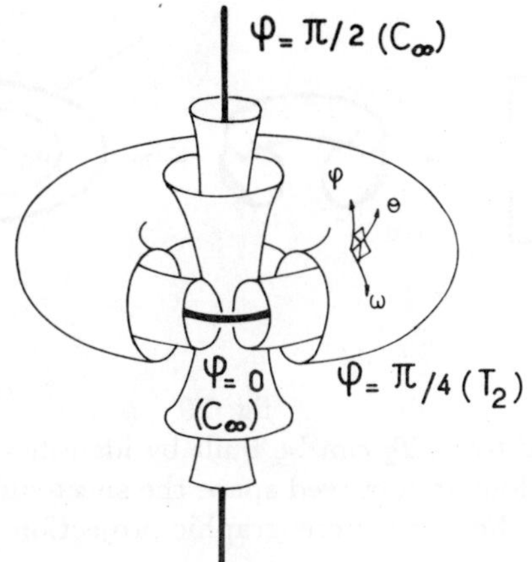

fig. 28

Stereographic projection in R_3 of a family of tori with $\phi = $ constant in S_3, the spherical torus T_2 corresponds to $\phi = \pi/2$, θ and ω are variables on the surfaces of the tori.

coordinates by:

$$x_1 = R\cos\theta\sin\phi,\, x_2 = R\sin\theta\sin\phi,\, x_3 = R\cos\omega\cos\phi,\, x_4 = R\sin\omega\cos\phi$$

In this system the surfaces at constant ϕ are tori organized around their $C\infty$ symmetry axes which are the two interlaced great circles, $x_1^2 + x_2^2 = R^2$ and $x_3^2 + x_4^2 = R^2$, obtained for $\phi = 0$ and $\pi/2$ as shown in fig. 28. Among them the spherical torus T_2, obtained for $\phi = \pi/4$, is at equal distances from the two axes and separates S_3 in two equivalent sub-spaces. Any point on a torus is therefore determined by the variables θ and ω and a constant ϕ . Varying the two first by $d\theta$ and $d\omega$ leads to the orthogonal displacements $R\sin\phi d\theta$ and $R\cos\phi d\omega$ on the torus and changing the third by $d\phi$ leads to a displacement $Rd\phi$ normal to the torus. The infinitesimal length dl on the spherical torus is given by

$$dl^2 = R^2(d\theta^2 + d\omega^2)/2$$

this is the metrics of an Euclidean plane and, therefore, the spherical torus has zero Gaussian curvature. Indeed it can be built by identification of the opposite sides of a square sheet of sides $\sqrt{2}\pi R$ as shown in fig.29 [30)] [11)]. Then the element of area on any torus ϕ (which is also a surface without Gaussian curvature) is $ds = R^2\sin\phi\cos\phi d\theta d\omega$ and the element of volume in S_3 is $dv = R^3\sin\phi\cos\phi d\theta d\omega d\phi$ so that the area of the torus is $S = 4\pi^2 R^2\sin\phi\cos\phi$ and its volume $V = 2\pi^2 R^3\sin^2\phi$

More particularly, the spherical torus T_2 with $\phi = \pi/4$ has an area $2\pi^2 R^2$ and an enclosed volume $\pi^2 R^3$, which is exactly half that of S_3.

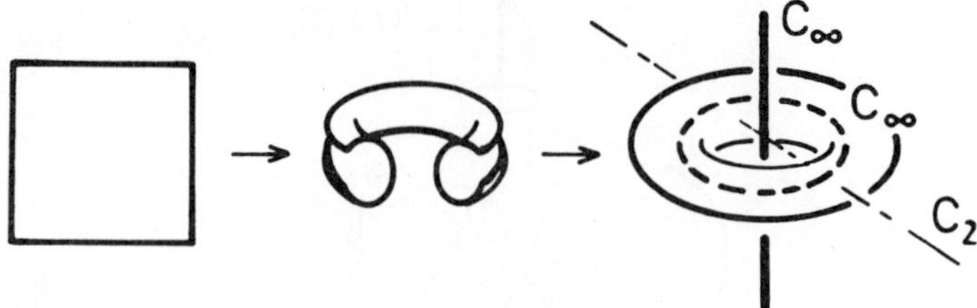

fig. 29

The spherical torus T_2 can be built by identification of a square sheet in S_3, as this is done in a curved space the sheet suffers no distorsion, the torus represented here is a stereographic projection in R_3.

III-The $S_2 \star R_1$ cylindrical space.

First consider the usual cylinder $S_1 \star R_1$. It can be embedded in an Euclidian 3-D space. It is a developable surface with a zero Gaussian curvature and can be considered as an infinite flat strip with two opposite sides identified by rolling. Geodesic lines on the flat strip are straight lines, then by rolling these lines can give circle S_1, when they are orthogonal to the sides of the strip, or they can remain straight lines, when they are parallel to the sides of the strip, or helicoïdal lines in the other cases. There are therefore closed and infinite geodesics.

From an intrinsic point of view, area measurements indicate a zero Gaussian curvature. The well known system of cylindrical coordinates is the appropriate system to descibe this space.

Similarly to the case of the usual cylinder $S_1 \star R_1$ which is embedded in R_3, the cylindrical 3-D space $S_2 \star R_1$ can be embedded in the 4-D Euclidean space. It has several kinds of geodesics: great circles of S_2, straight lines orthogonal to the great circles and also helicoidal lines. From an intrinsic point of view , measurements in the space itself allow to determine an intrinsic constant curvature which is positive. Nevertheless the curvature of this space is not homogeneous as there are different kinds of geodesics. This is discussed further with the introduction of the intrinsic curvature.

Appropriate cylindrical coordinates are obtained by adding a z parameter to the 2-D spherical coordinates. The space can be described as a set of cylinders whose generators are the circles of one family of parallel circles on S_2, including the equator and the two poles, as shown in fig. 30. Among them the cylinder which admits the equator as a generator separates the space in two equivalent sub-spaces.

IV- Intrinsic curvature

The intrinsic curvature of a space has to be obtained by measurements in the space itself without any hypothesis on the space of higher dimensionality in which

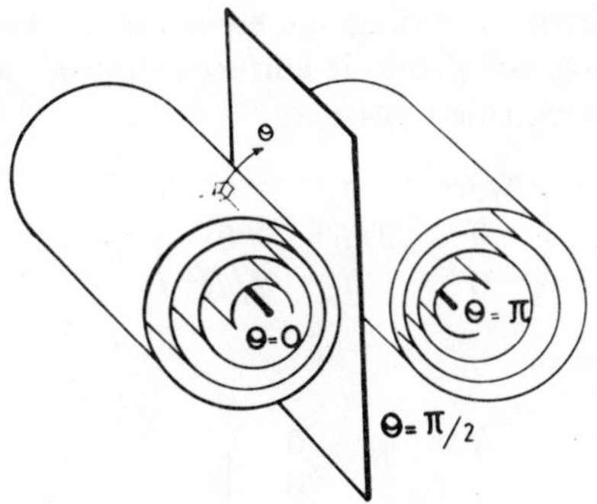

fig. 30
mapping in R_3 of the family of cylinders in $S_2 \star R_1$ with $\theta = $ constant,
the great cylinder corresponds to $\theta = \pi/2$.

it is embedded[41]. In the case of a 2-D surface this can be done by determining the area of a small disk of radius ρ measured along a geodesic. This area is related to the intrinsic curvature, which is also Gaussian curvature in this case, by $s = \pi\rho^2(1 - \rho^2\kappa/12)$, a formula which can be easily tested on a sphere with $\kappa = 1/R^2$. A measure of the length of a small circle also could be used to define the curvature; a derivation of the last relation gives: $l = ds/d\rho$ or $l = 2\pi\rho(1 - \rho^2\kappa/6)$.

In the case of a curved 3-D space the volume content of a small ball differs from that of a ball with the same radius in the Euclidean space. This allows to define a scalar quantity κ characteristic of the intrinsic curvature of the space. E. Cartan uses the measure of the volume of the small ball $V = 4/3\pi\rho^3(1 - \kappa\rho^2/15)$, or that of its area $S = 4\pi\rho^2(1 - \kappa\rho^2/9)$. This definition is an extention of the Gaussian curvature to 3-D, but as it is a scalar quantity it contains no informations on directional properties of the 3D curvature at a given point. S_3 or $S_2 \star R_1$ have the same curvature, with an adequate choice of the two radii, but their curvatures have different directional properties. This is made clear if one considers geodesic surfaces (surfaces whose geodesics are also geodesics of the embedding space) in the 3-D spaces. The need for a more precise characteristic of the curvature appears clearly. In the case of S_3 all these geodesic surfaces through a point are great spheres with a Gaussian curvature $\kappa = 1/R^2$. In the case of $S_2 \star R_1$, through a point there is a geodesic surface which is a 2-sphere, and others which are cylinders (formed by the product of a great circle of S_2 and R_1). In the last case the Gaussian curvature is zero.

So it is of a great interest to have a way to caracterise this anisotropy. The Ricci tensor, which is a contraction of the Riemann tensor [42] has this property. Using appropriate coordinates (spherical for S_3 and cylindrical for $S_2 \star R_1$) the Ricci

41) E. Cartan, Leçons sur la Géométrie des espaces de Riemann, Gauthier-Villars (1963)
42) C. W. Misner, K. S. Thorne and J. A. Wheeler, Gravitation, Freeman and co.

tensor is a diagonal 3x3 matrix. the diagonal elements are two time the Gaussian curvature of the three orthogonal geodesic surfaces through a same point.

In the case of the space S_3 this tensor is:

$$\begin{pmatrix} 2/R^2 & 0 & 0 \\ 0 & 2/R^2 & 0 \\ 0 & 0 & 2/R^2 \end{pmatrix}$$

and in the case of $S_2 \star R_1$ it is:

$$\begin{pmatrix} 0 & 0 & 0 \\ 0 & 0 & 0 \\ 0 & 0 & 2/R^2 \end{pmatrix}$$

The trace of the Ricci tensor is the scalar curvature $\Re$, the real intrinsic curvature, which is two time the the κ curvature used by E. Cartan: $\Re = 6/R^2$ for S_3 and $\Re = 2/R^2$ for $S_2 \star R_1$

V- Fibrations

A space can be considered as a fibre bundle if there is a sub-space (the fibre) which can be reproduced by a displacement so that any point of the space is on a fibre and only one. For exemple the Euclidean space R_3 can be considered as a fibre bundle of straight lines, all perpendicular to the same plane.

If fibres are 1-D lines, which is the case for all examples presented here , it is possible to determine a point on a fibre by one parameter, then it remains two parameters to characterise the fibre itself. So there is a 2-D space in which a point characterises a fibre. This space is called the base. In the simple example of R_3 space the 2-D base space is just the plane orthogonal to the fibres; in this case the base is a sub-space of the whole space, but this property is not general and it can happen that the base is not embedded in the fibre bundle space.

V-1. Fibration of $S_2 \star R_1$

When a space is a product between two spaces it could easily be considered as a fibre bundle. If fibres are straight lines in $S_2 \star R_1$ then the base is the sphere S_2. It is also possible to have helicoidal fibres; in this case the base S_2 is not orthogonal to the fibres (fig. 31).

V-2. Fibration of S_3

Successive toric layers appear naturally if S_3 is described using toroidal coordinates: a torus is defined by a constant ϕ parameter.

Each of these tori could be considered as a 2-D space equivalent to a rectangle with opposite sides identified two by two (or a square in the particular case of the spherical torus). All diagonals of these rectangles have the same length and give a great circle of S_3 after the identification of the sides which close this line. So it is possible to draw on a torus all a family of great circles, which appear as parallel lines on the equivalent rectangle. All these great circles drawn on a whole family of toric layers defined by their two common axes form a fibre bundle. This is the Hopf's fibration of S_3 (fig. 32).

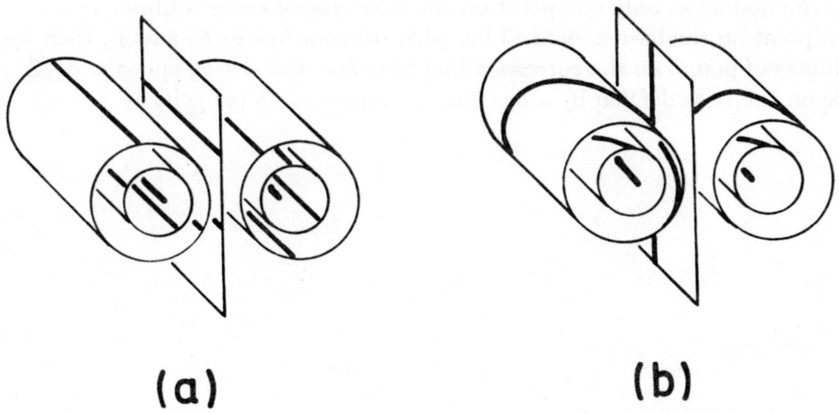

fig. 31

Fibration of $S_3 \star R_1$ by straight lines (a) and helices (b).

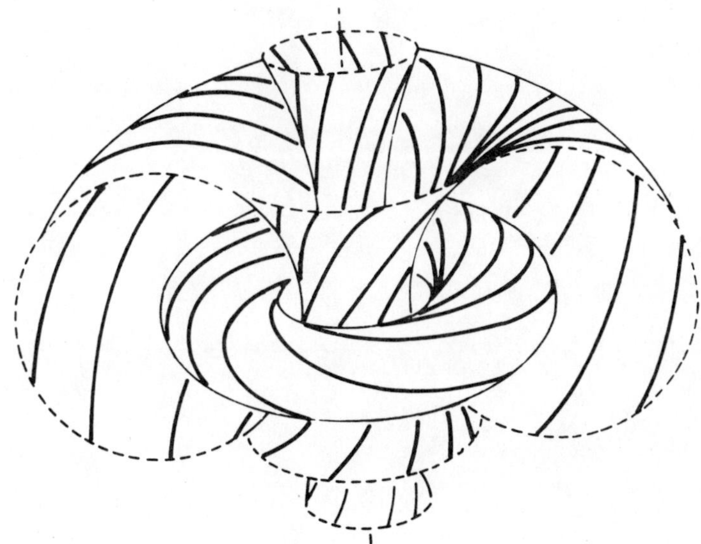

fig. 32

Hopf's fibration of S_3, the fibres are great circles which can be drawn
on the family of tori (see fig. 29).

The base is a 2-sphere, but this sphere is not embedded in S_3. If it was, it
would have two common points with a fibre, as a circle cut a sphere in two points.

This is impossible as only one point on the base characterises a fibre.

If a point on the base is defined by spherical coordinates θ_o and ϕ_o then toric coordinates of points on the corresponding fibre are: $\theta = \omega + \theta_o$ and $\phi = \phi_o/2$. So a circle on the base defined by a constant ϕ_o represents a torus in S_3.

Chapter III
WHICH UNIVERSE FOR BLUE PHASE

Elizabeth DUBOIS-VIOLETTE and BRIGITTE PANSU

WHICH UNIVERSE FOR BLUE PHASES ?

E. Dubois-Violette, B. Pansu

Laboratoire de Physique des Solides, Bât. 510, Université de Paris-Sud

91405 Orsay, France

1 - INTRODUCTION

Blue Phases appear in binary mixtures of nematic and cholesteric liquid crystals in a small range of temperature close to the isotropic transition[1][2][3][4]. Except the Blue Fog, their mean feature is their crystalline habit [5]. Observation of the shape of the crystallites (facets) as well as the study of the scattered light indicate cubic structures. Theoretical analysis suggests an interpretation of these structures in terms of arrays of defect lines (disclinations). In that sense, they could appear either as ordered or disordered phases, as soon as the notion of order itself should be specified. As we shall see later, a local constraint of minimum energy implies short range order with double twist. But this local order cannot be extended over large distances ; it cannot lead to long range order, the system is frustrated. The presence of disclination lines reveals this frustration. The occurrence of frustration is not specific of Blue Phase liquid crystals, it also appears in many other physical systems as various as triangular antiferromagnetic Ising models, biological materials... It is not within the scope of this paper to give a catalogue of all frustrated systems. What we want to point out is that this frustration can sometimes be interpreted in terms of geometrical features and therefore is linked to some topological or metrical properties of the space $\mathcal{R}^3$. We shall focus our attention to this geometrical aspect. The important point is that when the frustration is of a geometrical nature it can released by considering other manifolds than $\mathcal{R}^3$, i.e. spaces with different geometrical properties. The local order leading to frustration in $\mathcal{R}^3$ can therefore be extended to another space leading to long range ordered and the ideal perfect order phase can exist.

Dimension and intrinsic curvature of the space will be of paramount importance. We shall illustrate the notion of geometrical frustration by some examples and show how this frustration can be released by considering either space of higher dimension (noted D here after) or curved space or both. In these ideal new spaces the systems are described as perfect ordered phases. But the next step is to come back to the true physical space either by projecting or decurving the space. These processes introduce some imperfections or defects characteristic of the nature of the order of the ideal structures. Let us begin with the case where a perfect ordered description is obtained when the object is considered as living in a space of higher dimension.

A striking illustration of the utility of changing the space is given by quasi crystals. Up to recent years, positional long range order was linked to perfect periodicity and traduced by the existence of a Bravais lattice. This classical notion of crystalline order is revealed by the appearance of well separated intense Bragg spots in the diffraction pattern[7]. In 1984, the discovery of fivefold symmetry on the diffraction pattern of metallic alloys has upset this concept of long range ordering, since the observed symmetry is incompatible with a periodic structure. In fact in these materials Bragg peaks appear in a large number (dense set). In terms of mathematics this distinguishes the Fourier transform of a quasi-periodic structure from that of a periodic one. Both of them are constituted by a set of countable delta functions, the difference is that the set is not dense for a periodic structure but dense for a quasi-periodic one.

Quasicrystals can be described in terms of a perfect crystal (in the sense of a periodic structure) but in a higher dimension space. A nice description of diffraction patterns with fivefold symmetry is obtained by considering for example a perfect ordered crystal in a 6D Euclidean space where the icosahedral symmetry is allowed. In that 6D space the crystal is a perfect periodic structure. But the quasicrystal leaves in our 3D physical space and so do diffraction patterns. The real quasicrystal is described as resulting from a projection in a 3D space of some portion of the perfect 6D crystal. The projection is performed in an appropriate direction in order to preserve the fivefold symmetry. In that procedure what is lost or more precisely which kind of imperfection is introduced ? The process of cut and projection[7] of the hypercubic lattice transforms a periodic distribution of points in a 6D space into a non periodic one in a 3D space which gives a dense countable set of delta function peaks in a scattering experiment.

The example of quasicrystals emphasizes the role played by the dimension of the space. The notion of order is recovered by changing the dimension of the manifold but by keeping its Euclidean nature. In the following examples frustration will be released by changing metric properties of the underlying space. For example, a perfect order impossible in the 2D Euclidean space (plane) will exist on a 2D sphere or pseudosphere (curved spaces).

A simple example at 2D of the inadequacy between short range order and long range order (frustration) is given by the tesselation of the Euclidean space ($=$ plane). Assume that we want to realize a $\{p, q\}$ regular tesselation where $\{p, q\}$ in the Schläfli notation[8] means a tesselation with at each vertex q regular polygons with p sides. It is easy to show that only a $\{4,4\}$ tesselation or the two duals $\{3,6\}$ and $\{6,3\}$ tesselations are possible. This corresponds to the regular tiling of the plane with squares, triangles and hexagons (Bravais lattices). But what about a $\{5,3\}$ tesselation ? The geometrical frustration is evident since around one vertex three pentagones can be set without overlap but there exists a positive angular deficit (fig. 1a) which prevents the complete tiling of the plane. This deficit is : $\Delta\varphi = 2\pi - 3\left(\frac{3\pi}{5}\right) = \frac{\pi}{10}$. It is also very simple to see how, in that case, geometrical frustration is relieved by considering the same tiling but in a curved space.

a)   b)

Fig. 1 :   *a) A perfect tiling of the plane with regular pentagons cannot exist. There is an angular deficit.*
b) Tesselation of the sphere S^2 with twelve pentagons.

The dodecahedron (fig. 1b) corresponds to the regular tiling of the 2D sphere S^2 by twelve pentagons. Then the local constraint (three pentagones at each vertex) does not lead to long range order on $\mathcal{R}^3$ but on the contrary leads to a regular tiling (long range order) on S^2 which is a manifold with constant positive curvature. If we now want to realize a $\{6,4\}$ tesselation it corresponds to a negative deficit angle : $\Delta\varphi = 2\pi - 4\left(\frac{2\pi}{3}\right) = -\frac{2\pi}{3}$. Here again this tesselation is only possible on a surface with constant negative curvature since the angular deficit is negative. In reality this is possible on the pseudosphere (hyperbolic plane) [9] which is a surface with a constant negative curvature (fig. 4a).

A frustration of the same nature is also encountered in 3D. Local constraint leading to a criterium of more dense packing does exist for example in amorphous systems [10] where tetrahedra or icosahedra packing are required. Here again, as in the above example at two dimensions, this kind of tiling cannot be realized in a perfect manner in $\mathcal{R}^3$. To be convinced just remark that the dihedral angle of a tetrahedron 70°32 ' does not fit with 360° ; there will always remain some angular deficit in any arrangement of tetrahedra. Perfect structures are obtained in that case on S^3 as described in the paper of Mosseri and Sadoc in that book. They also account for the structure of Frank-Kasper phases [11] which are crystalline and as Blue Phases present an organization "major skeleton" of disclination lines. Here again this reveals the geometrical frustration resulting from the impossibility of

136

filling the space periodically with tetrahedral close packed arrangement (tetrahedra or icosahedra). The frustration is relieved in a curved space where a regular close packed structure is obtained, by piling 5 tetrahedra meeting at every edge (polytope $\{3,3,5\}$). This ideal structure only exists in a curved space and to come back to a physical description in $\mathcal{R}^3$ we need to decurve the space by introducing disclinations.

Geometrical frustration appearing in various systems results from physical constraints of diverse nature. Another manifestation is observed in bilayers systems. Some perfect structures exist in $\mathcal{R}^3$. In that case, the lamellar phase corresponds to equidistant bilayers of amphiphilic molecules with plane and parallel interfaces (fig. 2 a). Molecules posses a polar head attached to paraffinic chains. In the lamellar phase the surface occupied by the polar-head and the paraffinic chains is the same. A change in physical parameters induce other structures[12] with a more complicated morphology as will be presented by J. Charvolin and Sadoc in the same issue. A nice interpretation is here again given in terms of frustration directly linked to the curvature of the space[13]. Assuming that the polar heads want to occupy a surface smaller than the paraffinic chains this implies a curvature of the interface. In the academic example given in fig. 2 we see for example that for a 2D sample the curvature of the midsurface is not compatible with the compactness of the bilayer (fig. 2b). On the contrary, this is perfectly performed on the sphere S^2 (fig. 2c).

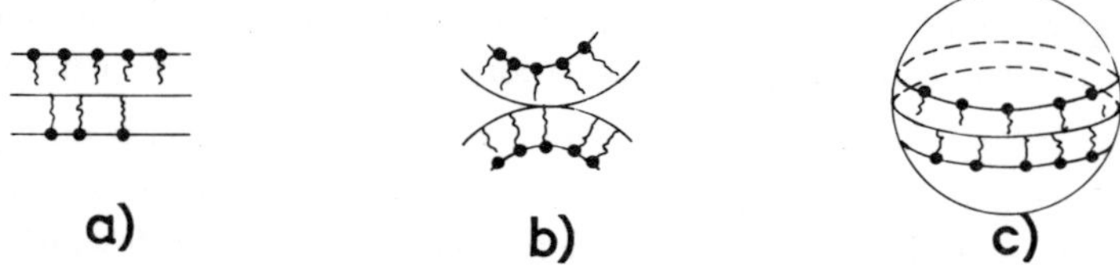

Fig. 2 : *Bilayers structures.*
a) Bilayer with plane interfaces. Surfaces occupied by polar heads and paraffinic chains are the same.
b) The surface occupied by polar head is smaller than the surface occupied by paraffinic chain. This induces a curvature of the interface and frustration (the bilayer is no more compact).
c) On the sphere it is possible to have curvature of the interface and to keep the compactness of the bilayer.

The last feature we now want to introduce is the nature, a little more intricate, of the frustration existing in the Blue Phases. Blue Phases may be described as cholesteric phases with a uniaxial order parameter [13][14] : the director **n** (where $\pm$ **n** are equivalent) orientated along the molecules.

Cholesteric liquid crystals correspond to phases where a twist of the molecules exists in only one direction of the space (fig. 10). A perfect ordered phase is obtained for a director field such as :

$$\begin{cases} n^x(x) &= 0, \\ n^y(x) &= \cos\tilde{p}_o x, \\ n^z(x) &= \sin\tilde{p}_o x, \end{cases} \tag{1}$$

where $p = 2\pi/\tilde{p}_o$ defines the pitch of the helical structure. Twist only occurs in one direction (x) of the space.

In Blue Phases the local constraint of minimum energy is assumed to be double twist[14][16], i. e. twist occurs in more than one direction of the space. It can be written in the Euclidean frame as :

$$\partial_k n^j + \tilde{p}_o \; \epsilon_{kji} \; n^i \; \delta^{jp} = 0 \tag{2}$$

where ϵ_{kji} is the completely antisymmetric permutation symbol, δ^{jp} the Kronecker symbol and an implicit summation is performed on identical indices. Assuming an initial orientation $n_o^z = 1$ for the molecules, this condition reads :

$$\begin{aligned} \frac{\delta n^y}{\delta x} &= -\tilde{p}_o, \\ \frac{\delta n^x}{\delta y} &= +\tilde{p}_o, \\ \delta n^z &= 0. \end{aligned} \tag{3}$$

Contrary to expression (1) which corresponds to long range order in cholesteric, condition (2) cannot lead to long range order in Blue Phases as can be seen in fig. 3. The local rule (2) implies geometrical frustration on the orientation of the molecules. The problem is the following : let us consider the orientation of the molecules $n_o(O)$ at some reference point O and satisfy the local double twist (in the x and y direction) condition, do we generate a unique orientation of the molecules at some other point P ?

We see easily that moving first in the x direction up to point A induces a twist characterized by a component $\delta n^y(x)$ leading to the orientation n_A at point A. Then motion from A to P implies no more twist and molecules are orientated along n_A at point P. On the other circuit we move from O to P through the point B (fig. 3). Along OB the twist of molecules induces a component $\delta n^x(y)$ with an orientation n_B at point B. Along the path BP the molecules remain

138

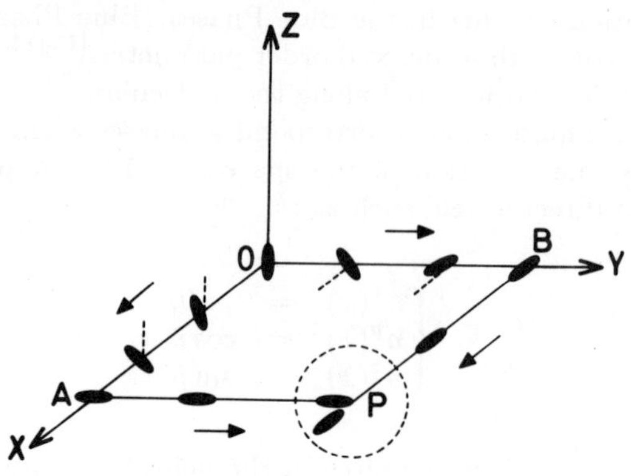

Fig. 3 : *Following the double twist rules from O to P along two different pathes leads to an undefined orientation of the director at point P : there is frustration.*

orientated along n_B. The situation is frustrated since the orientation at point P is not uniquely defined i.e. following the double twist rule step by step does not define a director field. The double twist condition is a rule of transport for a vector, in mathematical terms it is a connection. The frustration corresponds to the fact that this rule of transport or connection does not define a vector field. In mathematical terms, as we shall see later, this results from the fact that the connection associated to the double twist rule has curvature. As shown in section 3, considering the director field in a curved space as the hypersphere S^3 instead of $\mathcal{R}^3$ will suppress the frustration. This implies that the double twist rule, or the associated connection will define a vector field on S^3. We shall explicit the precise nature of the geometrical frustration in Blue Phases and how it is released on S^3. In the following section 2 we present some notions of differential geometry useful for the description of main features of non Euclidean spaces.

The description of the Blue Phase requires the notion of vectors and of a rule of transport for vectors on a manifold. The curvature of connections and the different curvatures of any manifold are concepts of importance to understand sections 4 and 5. We shall now introduce all these general ideas before evoking the Blue Phase structure on S^3.

2 - GEOMETRICAL LANDSCAPE

Let us consider different geometrical objects such as shown for example in fig. 4. Fig. 4a which looks like a trumpet represents a part of the pseudosphere (hyperbolic plane). Fig. 4c is a drawing of the classical S^2 sphere. Fig. 4b represents the Euclidean plane and Fig. 4d a physical torus such as a doughnut. They all are 2D manifolds. But from a topological point of view they are different. A potatoe and a sphere for example are topologically equivalent. The sphere and the torus, even if they are both compact, are not : the torus gets a hole and the sphere does not. We shall squeeze the topological description and focus our attention to some precise geometrical properties. Parts of the description of these objects will refer to intrinsic properties some other to extrinsic one. The intrinsic point of view corresponds to the one of a small animal moving on the object and ignoring what is around. The extrinsic point of view is the opposite. It is the way we look for example to 2D manifolds which can be embedded in our physical space $\mathcal{R}^3$. A concept such as the normal to a sphere is a typical extrinsic ones since the normal is defined only when we look at the sphere as embedded in another space ($\mathcal{R}^3$ for example).

In many physical problems and more precisely in the case of the Blue Phases one is concerned with defining vectors fields on the manifold. We now want to introduce the concept of tangent space i.e. space of tangent vectors and give a tool to compare vectors at different points Q of the manifold.

Tangent space

From a geometrical point of view a tangent vector $\mathbf{X}_A$ at point A is an intrinsic local quantity which can be defined as the tangent vector to a curve $\mathcal{C}$ on the manifold. More precisely it corresponds to an equivalence class of curves through the point A, (fig. 5), and is independent of the choice of coordinates. In terms of differential geometry a vector is a differential operator (derivative) acting on functions. The tangent vector $\mathbf{X}(x)$ at point $\mathbf{x}(t)$ of a curve parametrized by t is :

$$\mathbf{X}|_x = \frac{d\mathbf{x}(t)}{dt}.$$

If $\left\{x^i\right\}$ is a set of coordinates then :

$$X^i = \frac{dx^i}{dt}. \tag{4}$$

Let us consider the derivative of a function f along the preceding curve $\mathcal{C}$:

$$\left.\frac{df}{dt}\right|_{\mathcal{C}} = \frac{\partial f}{\partial x^i}\frac{\partial x^i}{\partial f} = X^i \frac{\partial}{\partial x^i} f.$$

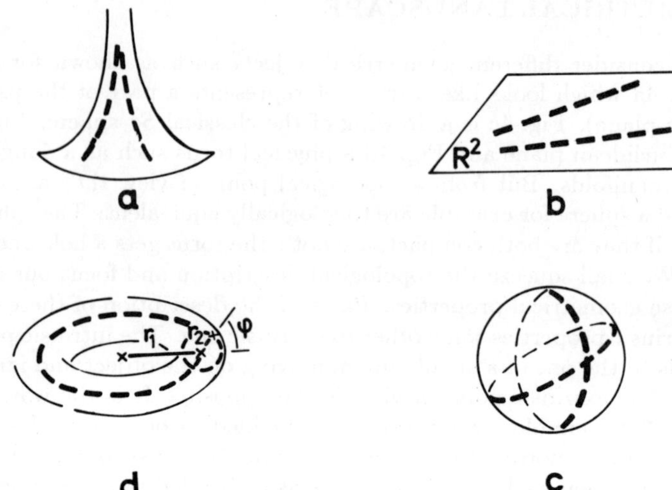

Fig. 4 : *2D surfaces with various intrinsic curvature R :*
a)   Portion of the hyperbolic plane (R= constant < 0) ;
b)   Plane (R = 0) ;
c)   Sphere (R = constant > 0) ;
d)   Doughnut (R is not constant).

Then the vector $\mathbf{X}$ can be seen as a derivative on the functions f in the following way :

$$\mathbf{X}(f) = X^i \frac{\partial}{\partial x^i} f = X^i \mathbf{e}_i(f),$$

where the operator

$$\mathbf{e}_i = \frac{\partial}{\partial x^i}, \tag{5}$$

defines the natural frame $\mathbf{e}_i$, i.e. derivatives with respect to the coordinates. $\mathbf{X}$ is the velocity vector along the curves of coordinates. In a non natural frame :

$$\mathbf{e}_\alpha = W_\alpha^i \mathbf{e}_i,$$

where $W_\alpha^i = \mathbf{e}_\alpha(x^i)$. The vector $\mathbf{X}$ reads :

$$\mathbf{X} = X^\alpha \mathbf{e}_\alpha,$$

but where now $\mathbf{e}_\alpha \neq \frac{\partial}{\partial x^\alpha}$.

Let us emphasize that the tangent vector, defined as the equivalence class of curves, is independent of the choice of coordinates. Only its coordinates depend on the choice of the frame. The space of tangent vectors at point O is the tangent space. We have seen how to define a vector as an intrinsic local quantity, the next question is now how to compare vectors at two different points of the manifold.

Connection - Parallel vector field

Let us remark that, even if the tangent spaces at two different points of the manifold are of the same nature i.e. linear vector spaces, they are distinct vector spaces e.g. we cannot add two vectors at different points. Nevertheless we would like to do some comparison at two different points of the manifold. What is for example a constant vector ? Does it exist a constant vector field ? Before answering we want to introduce a tool which allows the comparison between vectors in the same neighborhood. The mathematical object is a *linear connection* $\nabla_\mathbf{Y}(\mathbf{X})$[17]. It is a local rule of transport for tangent vectors. It defines how a vector $\mathbf{X}(A)$ at a point A is transported when we move in the $\mathbf{Y}$ direction away from point A. We obtain a new vector (transported) $\mathbf{X}'(B)$ at point B as indicated in fig. 5.

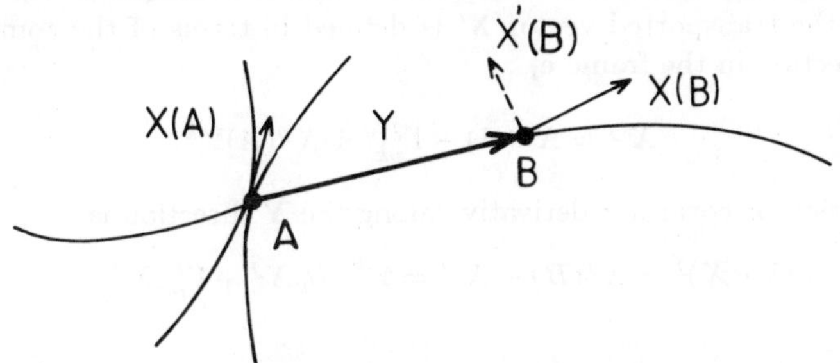

Fig. 5 : *A linear connection gives a rule of transport for a vector. $\mathbf{X}(A)$ is transported along $\mathbf{Y}$ at point B in $\mathbf{X}'(B)$ which may be different from the vector field $\mathbf{X}(B)$.*

Let us emphasize that a connection is not a unique object. We can build different linear connections leading to different transported vectors $\mathbf{X}'_1(B), \mathbf{X}'_2(B)$ at the same point B. We can explicit how a connection is expressed in terms of a local reference frame field. Let us assume we consider the natural frame $\mathbf{e}_i = \frac{\partial}{\partial x_i}$ and write, with use of that frame, the initial vector :

$$\mathbf{X}(A) = X^i \mathbf{e}_i(A),$$

142

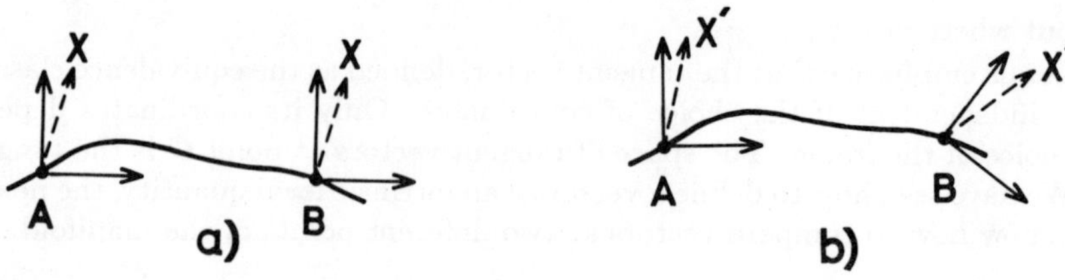

Fig. 6 : *Two different connections in a) and b).*
full line : parallel transported frame.
dashed line : X is a constant vector for connection a). X' is a constant
vector for connection b).

the transported vector :

$$\mathbf{X}'(B) = X'^j \mathbf{e}_j(B),$$

and the direction of transport :

$$\mathbf{Y} = Y^k \mathbf{e}_k(A).$$

In all the following summation on repeated indices is assumed. For a small displacement, the transported vector $\mathbf{X}'$ is defined in terms of the components Γ^i_{jk} of the connection in the frame $\mathbf{e}_i$:

$$X'^j = X^j(A) - \Gamma^j_{pk}(A)X^p(A)Y^k.$$

The connection or covariant derivative along the $\mathbf{Y}$ direction is :

$$(\nabla_Y X)^j = X^j(B) - X'^j = Y^k \left(\partial_k X^j + \Gamma^j_{pk} X^p \right). \tag{6}$$

It is interesting to note that Γ^j_{pk} expresses the fact that the local frame does not necessarily follow the rule of parallel transport and that :

$$\Gamma^j_{pk} = \left(\nabla_{\mathbf{e}_k} \mathbf{e}_p \right)^j. \tag{7}$$

Although the linear connection is an intrinsic quantity let us point out that the local component Γ^i_{jk} of the connection does not transform as a good tensor by a change of coordinates.

We are now able to introduce the notion of a *constant vector* with use of a connection (eq. 8). A constant vector $\mathbf{X}$, or parallel transported by the connection, satisfies the condition :

$$\nabla_{\mathbf{Y}}(\mathbf{X}) = 0. \tag{8}$$

It has constant components in a parallel transported frame which of course depends on the choice of the connection (fig. 6).

The next question of importance is : with this notion of constant vector, is it possible to find a constant vector field ? In other words does a parallel transported vector field exist for any connection ? It depends in fact on the connection. The definition (eq. 8) of a connection depends explicitly on the direction in which we move on the manifold. As a result the parallel transport from one point A to another point B of the manifold depends on the path used to go from point A to point B as indicated in fig. 7.

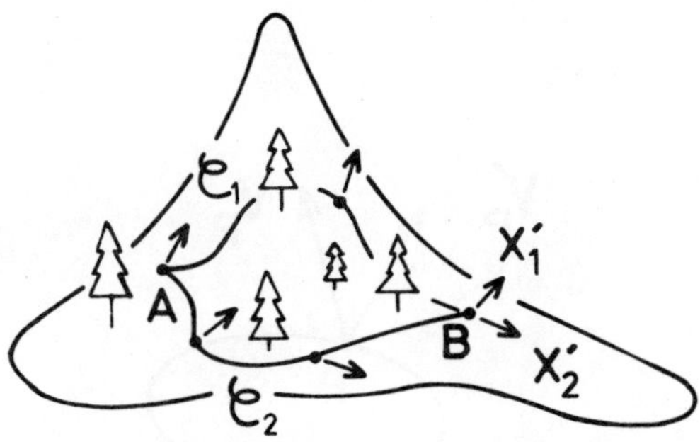

Fig. 7 : *Connection with curvature. The vector parallel transported depends on the path. We obtain X'_1 along C_1 and $X'_2 \neq X'_1$ along curve C_2.*

It may occur that the two vectors $\mathbf{X}'_1(B)$ and $\mathbf{X}'_2(B)$ obtained after parallel transport on the two curves C_1 and C_2 differ at point B. In that case a *parallel transported vector field does not exist*. The connection is *said to have curvature*, it implies that transport of a vector $\mathbf{V}_P$ from a point P on a closed curve leads to a new vector $\mathbf{V}'_P$ different from $\mathbf{V}_P$ (fig. 8). A quantitative measurement of this deviation $\delta\mathbf{V} = \mathbf{V}'_P - \mathbf{V}_P$ is obtained from the curvature tensor $R^j_{ik\ell}$ defined as follows[18] :

$$R^j_{ik\ell} = \partial_\ell \Gamma^j_{ik} - \partial_k \Gamma^j_{i\ell} + \Gamma^j_{m\ell}\,\Gamma^m_{ik} - \Gamma^j_{mk}\,\Gamma^m_{i\ell} - C^m_{\ell k}\,\Gamma^j_{im}, \tag{9}$$

where the ℓ component of the commutator C is :

$$C^\ell_{ij} = \left[\mathbf{e}_i, \mathbf{e}_j\right]^\ell, \tag{10}$$

$$C^\ell_{ij} = (\mathbf{e}_i\mathbf{e}_j - \mathbf{e}_j\mathbf{e}_i)^\ell.$$

144

For a natural frame all components vanish since :

$$\left[\mathbf{e}_i, \mathbf{e}_j\right] = \frac{\partial}{\partial x^i}\frac{\partial}{\partial x^j} - \frac{\partial}{\partial x^j}\frac{\partial}{\partial x^i} \equiv 0. \tag{11}$$

The deviation $\delta\mathbf{V}$ of the transported vector along a circuit (fig. 8) characterized by the element $dx^k \wedge dx^\ell$ is :

$$\delta V^i = \int R^i_{jk\ell} V^j_P dx^k \wedge dx^\ell.$$

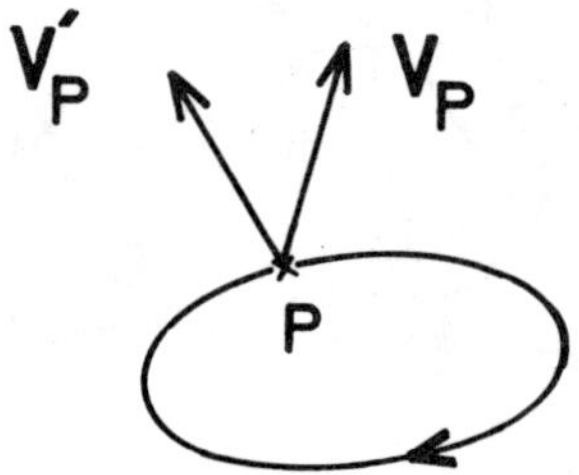

Fig. 8 : *Connection with curvature. A vector V_P transported along a close curve differs from the initial one : $V'_P \neq V_P$.*

If the curvature tensor vanishes on the whole manifold,

$$R^i_{jk\ell} \equiv 0 \quad \Rightarrow \quad V'^i_P \equiv V^i_P \quad \forall P,$$

there exists a parallel vector field for the connection which is said *curvature free*. The curvature of a connection is the obstruction to the existence of a parallel transported frame field. If the connection has no curvature there exists a parallel frame field for that connection.
The next question raised is, given a connection without curvature does it exist a natural frame field associated to coordinate systems which is parallel transported by the connection? Let us introduce the definition of the torsion of a connection :

$$T(\mathbf{V}_1, \mathbf{V}_2) = \nabla_{\mathbf{V}_1}(\mathbf{V}_2) - \nabla_{\mathbf{V}_2}(\mathbf{V}_1) - [\mathbf{V}_1, \mathbf{V}_2],$$

or in terms of local coordinates :

$$T_{ij}^{\ell} = \Gamma_{ij}^{\ell} - \Gamma_{ji}^{\ell} - C_{ij}^{\ell}. \tag{12}$$

For a connection without curvature it results that the torsion is the obstruction to the existence of a natural parallel transported frame field. Let us consider a frame field $\{\mathbf{e}_i\}$ which is constant for the connection :

$$\nabla_{\mathbf{e}_j} \mathbf{e}_i \equiv 0 \ \forall \, i,j,$$

then the torsion reads :

$$T\left(\mathbf{e}_i, \mathbf{e}_j\right) = -\left[\mathbf{e}_i, \mathbf{e}_j\right], \tag{13}$$

and is non zero for a non natural frame field. It is now possible to give the transcription of the double twist rule (eq. 2) in Blue Phases in terms of a connection. The double twist connection in $\mathcal{R}^3$ is defined by the components :

$$\Gamma_{ik}^{j} = -\tilde{p}_o \, \epsilon_{ki\ell} \, \delta^{j\ell}, \tag{14}$$

where δ^{ij} is the Kronecker symbol. The double twist rule is now :

$$\nabla_k \mathbf{n}^j = \partial_k \mathbf{n}^j + \Gamma_{ik}^{j} \mathbf{n}^i = 0. \tag{15}$$

This connection has some curvature ; there does not exist any director field parallel transported for this connection. No perfect (double twisting) Blue Phase can exist in $\mathcal{R}^3$. As we shall see further we can build a double twist connection on any 3D manifold using one particular connection on this manifold. In $\mathcal{R}^3$, for instance, it corresponds to the classical notion of parallelism. This particular connection is called the Levi-Civita connection and is linked to the metric.

Geodesics - Levi-Civita connection

Let us consider two points A and B on a manifold and ask for the shortest way to go from one point to the other one i.e. geodesic lines $\mathbf{x}(t)$. The distance ds between two close points is given in terms of the metric tensor g_{ij} as :

$$ds^2 = g_{ij} dx^i dx^j, \tag{16}$$

where $\left\{x^i(t)\right\}$ is the set of coordinates. Minimization of the length element, $\delta\left\{\int_A^B ds^2\right\} = 0,$ defines the geodesic equation :

$$\frac{\partial^2 x^i}{\partial t^2} + \gamma_{jk}^i \frac{\partial x^j}{\partial t} \frac{\partial x^k}{\partial t} = 0, \tag{17}$$

where the components γ^i_{jk} are :

$$\gamma^i_{jk} = \frac{1}{2}g^{i\ell}\left[\partial_j g_{\ell k} + \partial_k g_{\ell j} - \partial_\ell g_{jk}\right].\tag{18}$$

Straight lines are geodesics for the 2D Euclidean space where $g_{ij} = \delta_{ij}$ and $\gamma^i_{jk} \equiv 0$. In the case of the sphere S^2 geodesics are circles, for the doughnut some of them are also circles (fig. 4).

The γ^i_{jk} given in eq. 18 can be seen as the components of a connection. This is the Levi-Civita connection noted $\tilde{\Gamma}^i_{jk} = \gamma^i_{jk}$. If the tangent vector $\mathbf{X} = \frac{\partial \mathbf{x}}{\partial t}$ to the geodesic line is introduced then eq. 17 becomes :

$$X^k\left[\frac{\partial}{\partial x^k}X^i + \tilde{\Gamma}^i_{jk}X^j\right] = 0,$$

or

$$\tilde{\nabla}_{\mathbf{X}}\left(\mathbf{X}\right) = 0.$$

This shows that the tangent vector to geodesics is parallel transported by the Levi-Civita connection. The Levi-Civita connection also preserves the scalar product $< \mathbf{U} \mid \mathbf{V} >$ during the parallel transport of two vectors $\mathbf{U}$ and $\mathbf{V}$. A vector parallel transported by the Levi-Civita connection from a point A to a point B keeps a constant angle with the geodesic line (joining A to B). In a non natural frame the Christoffel symbol reads :

$$\tilde{\Gamma}^\alpha_{\beta\gamma} = \frac{1}{2}g^{\alpha\mu}\left[\frac{\partial}{\partial x^\gamma}g_{\beta\mu} + \frac{\partial}{\partial x^\beta}g_{\gamma\mu} - \frac{\partial}{\partial x^\mu}g_{\beta\gamma}\right]$$

$$-\frac{1}{2}\left[g^{\alpha\mu}g_{\gamma\delta}\,C^\delta_{\beta\mu} + g^{\alpha\mu}g_{\delta\beta}C^\delta_{\gamma\mu} - C^\alpha_{\beta\gamma}\right].\tag{19}$$

Let us point out that the Levi-Civita connection has no torsion but that its curvature depends on the manifold. Qe see easily that the Levi-Civita connection of $\mathcal{R}^3$ does not get curvature and on the contrary the Levi-Civita connection of the sphere S^3 has curvature. Consider now the simple example of $\mathcal{R}^2$ and S^2 and perform a circuit OAB.

We start on $\mathcal{R}^2$ (fig. 9) with a vector $\mathbf{V}_A$ at point A. We go along the goedesic AB transporting the vector $\mathbf{V}$ with an angle α constant between $\mathbf{V}$ and the geodesic. We arrive with a vector $\mathbf{V}_B$ at B. Along the geodesic BO the tangent vector is parallel transported, we arrive in O with a vector $\mathbf{V}_O$. If now $\mathbf{V}_A$ is transported along the geodesic AO the angle α is conserved and we also

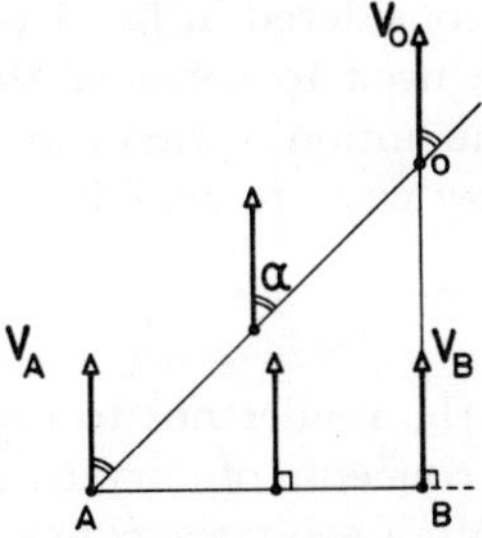

Fig. 9 : *Transport by the Levi-Civita connection in* $\mathcal{R}^2$.

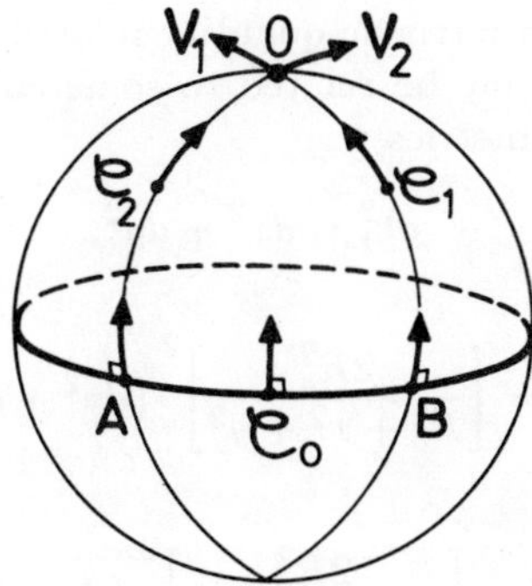

Fig. 10 : *Transport by the Levi-Civita connection on* S^2.

arrive with a vector $\mathbf{V}_O$ at point O. *The transport of the vector does not depend on the path i.e. the connection has no curvature.*

The same procedure on the sphere S^2 is path dependent. Let us consider the circuit ABO along the three geodesics, great circles $\mathcal{C}_o$, $\mathcal{C}_1$, $\mathcal{C}_2$ (fig. 10). $\mathbf{V}_A$ is transported along $\mathcal{C}_o$ with a constant angle and gives $\mathbf{V}_B$ at point B. Then along $\mathcal{C}_1$, $\mathbf{V}$ remains tangent to the circle, we get a vector $\mathbf{V}_1$ tangent to $\mathcal{C}_1$ at point O. Along the other path (AO) the transported vector remains tangent to the circle $\mathcal{C}_2$ and it gives the vector $\mathbf{V}_2$, tangent to $\mathcal{C}_2$ at point O different from the previous one $\mathbf{V}_1$. *This reveals the curvature of the Levi-Civita connection on* S^2 *where no constant vector field can exist.* A similar construction on the hypersphere reveals the curvature of the space.

As a result we have seen how with the metric it was possible to find special geometrical lines : the geodesics and to define a specific connection. Looking again at the four manifolds considered in fig. 4 there is one more quantity which distinguishes them. We are used to consider that a plane is flat, a balloon is curved. But under that vague notion of curvature there exists an intrinsic and an extrinsic point of view we now want to exhibit.

Curvature

First we want to advise the reader not to confuse the notion of curvature of a connection with the other concepts of curvature we now want to introduce. As we have seen the curvature of an arbitrary connection has nothing to do with the curvature of the manifold. On a given manifold (flat or curved space) we can build many different connections. Some of them will correspond to connections with curvature, some others to connections without curvature. But among all possible connections there exists a specific connection, in occurrence the Levi-Civita one, which allows to define an intrinsic quantity related to the curvature of the space. To feel how the metric may be related to some curvature, we shall consider the three different following metrics :

$$d\ell_1^2 = dx^2 + dy^2, \tag{20}$$

$$d\ell_2^2 = \left[\frac{2R_o^2}{R_o^2 + x^2 + y^2} \right]^2 \left(dx^2 + dy^2 \right), \tag{21}$$

$$d\ell_3^2 = \left[\frac{2R_o^2}{R_o^2 - x^2 - y^2} \right]^2 \left(dx^2 + dy^2 \right).$$

defined in the disc $x^2 + y^2 < R_o^2$.

The first one $g_{ij}^{(1)} = \delta_{ij}$ corresponds to the well known Euclidean plane. The two others $g_{ij}^{(2)} = \frac{2R_o^2}{R_o^2 + x^2 + y^2} \delta_{ij}$ to the sphere and $g_{ij}^{(3)} = \frac{2R^2}{R_o^2 - x^2 - y^2} \delta_{ij}$ to the hyperbolic plane. To see that $g^{(2)}$ is related to a curved space, let us consider the classical stereographic projection of the sphere S^2 (fig. 11).

A point $P'(\theta, \varphi)$ of the sphere S^2 defined by :

$$x' = R_o \, \sin\theta \, \cos\varphi,$$
$$y' = R_o \, \sin\theta \, \sin\varphi,$$
$$z' = R_o \, \cos\theta.$$

is projected onto the plane π on a point $P(x, y)$. We get : $\tan\frac{\theta}{2} = \frac{r}{R}$ with $x = r\cos\varphi$, $y = r\sin\varphi$. The length element $d\ell_2$ (eq. 21) can be written in terms of φ and θ as : $d\ell_2^2 = R_0^2 \left[(d\theta)^2 + \sin^2\theta(d\varphi)^2 \right]$ where we recognize the familiar

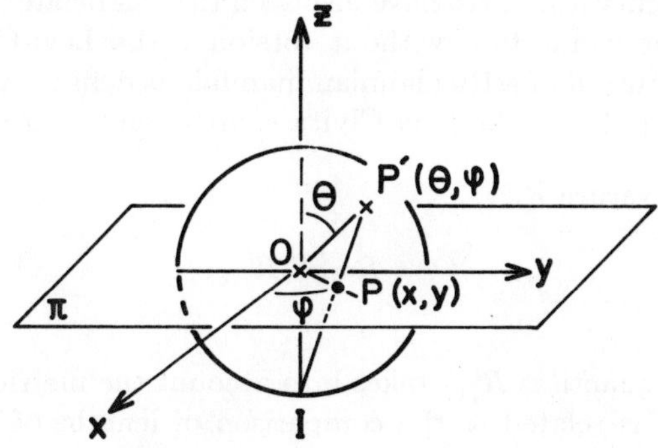

Fig. 11 : *Stereographic projection of the sphere S^2 on $\mathcal{R}^2$. P' is projected onto P.*

metric of the sphere. On that simple example we remark how a change of the metric is associated to a change of curvature since changing the metric $g_{ij}^{(1)}$ to the metric $g_{ij}^{(2)}$ corresponds to the change of the Euclidean plane (flat) to the sphere (constant positive curvature). In the same way the metric $g_{ij}^{(3)}$ pertains to the hyperbolic plane. Considering the stereographic projection, a point $P_1(\tau,\varphi)$ of the pseudosphere defined by :

$$x_1 = R_o \ \sinh\tau \ \cos\varphi,$$
$$y_1 = R_o \ \sinh\tau \ \sin\varphi,$$
$$z_1 = R_o \ \cosh\tau.$$

is projected onto the plane π on a point $P(x,y)$. We get $\frac{r}{R} = \tan\frac{\tau}{2}$, $x = r\cos\varphi$, $y = \sin\varphi$ and obtain : $d\ell_3^2 = R_0^2\left((d\tau)^2 + sh^2\tau(d\varphi)^2\right)$ which is the familiar metric of the pseudosphere. Let us point out that points on the pseudosphere satisfy the relation :

$$-z_1^2 + y_1^2 + x_1^2 = -R_0^2.$$

The pseudosphere is not embedded in $\mathcal{R}^3$ but in a 3D space with the metric :

$$dl_3^2 = -dz_1^2 + dy_1^2 + dx_1^2.$$

In the three exemples given just above, we have seen how the different metrics $g^{(1)}, g^{(2)}, g^{(3)}$ were associated to manifolds of different nature, namely to manifolds with different curvature. We have also seen that associated to a given metric there exists a unique connection without torsion : the Levi-Civita connection. The intrinsic curvature R of a Riemannian manifold is defined with use of the curvature tensor $R^i_{jk\ell}$ (eq. 10) of the Levi-Civita connection (eq. 19).

The scalar curvature R,

$$R = g^{jk} R^i_{jik},$$ (22)

is an intrinsic quantity. $R^i_{jk\ell}$ takes into account the metric and its derivatives. In other words it is related to the comparison of lengths of different trajectories in the same neighbourhood. See for instance ref.[19].

A very nice geometrical interpretation of the intrinsic property of the scalar curvature was given by Gauss for a 2D surface. He explicited[20] the relation between the integral curvature of a triangle constituted by arcs of geodesics and an angular deficit (fig. 12). If ABC is the geodesic triangle and $\hat{A}, \hat{B}, \hat{C}$ the angles between the geodesics we get :

$$\int_{triangle\,ABC} \frac{R}{2} dS = \hat{A} + \hat{B} + \hat{C} - \pi.$$ (23)

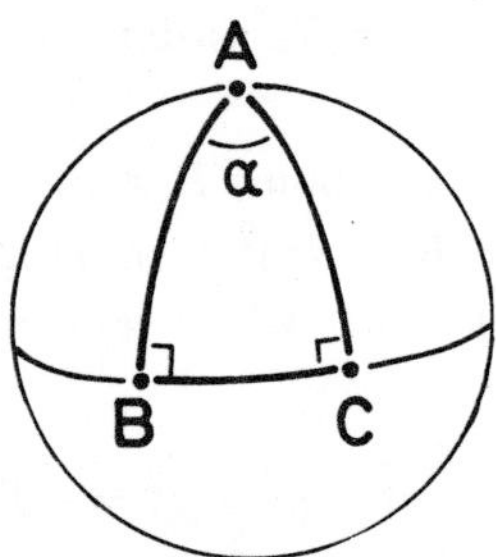

Fig. 12 : *Geodesic triangle. The angle α is proportional to the surface area of the triangle.*

The scalar curvature R can be computed with use of eqs. 10 and 19 for the four manifolds we have previously considered. The 2D plane corresponds to $R = 0$

and we recover that the sum of the three angles of a triangle is π. We find for the S^2 sphere a constant positive scalar curvature $R = \frac{2}{R_o^2}$ where R_o is the radius of the sphere. We are familiar to geometrical features of the sphere and if we consider the geodesic triangle ABC of fig. 12 where $\hat{B} = \hat{C} = \pi/2$, we find that the angle $\hat{A}$ is proportional to the surface area of the triangle. For the half sphere the integration : $\int \frac{R}{2} dS = \frac{1}{R_0^2} 2\pi R_0^2 = 2\pi$ fits with $\hat{A} = 2\pi$.

In the same vein the pseudosphere corresponds to a negative constant scalar curvature $R = -\frac{2}{R_0^2}$ and the angular deficit in a geodesic triangle is now negative. The doughnut torus has an intrinsic curvature which varies on the surface. Let us define the torus as shown in fig. 4d, then the metric is given by :

$$d\ell^2 = r_2^2 (d\varphi)^2 + (r_1 + r_2 \cos\varphi)^2 (d\theta)^2,$$
$$g_{\varphi\varphi} = r_2^2, \quad g_{\theta\theta} = (r_1 + r_2 \cos\varphi)^2.$$

After computation we find : $R = \frac{2\cos\varphi}{r_2(r_1 + r_2\cos\varphi)}$ which shows that for $\varphi = 0$ the curvature is positive and minimum. For $\varphi = \pi$, it is negative and maximum (inner part). On the contrary on the two circles $\varphi = \pm\,\pi/2$ the curvature is zero.

We have only given examples of intrinsic curvature for 2D manifolds. This is a general notion valid for any manifold $\mathcal{E}_m$ of dimension m. But now if we consider the manifold $\mathcal{E}_m$ as embedded in another manifold $\mathcal{E}_p$ other quantities which are no more intrinsic, but depending on the embedding, can be introduced[18].
We shall focus our attention to the case of codimension 1, i.e. $m = p - 1$ where the tangent space $T(\mathcal{E}_m)$ is of dimension m and the normal space $N(\mathcal{E}_m)$ is of dimension 1. Let assume that coordinates of the manifold $\mathcal{E}_m$ are labelled with greek indices y^α ($\alpha = 1....m$) and those of $\mathcal{E}_p$ with latin indices $x^i (i = 1, p)$. The arc length ds associated with a displacement from point P to point $P + \delta P$ on $\mathcal{E}_m$ is :

$$ds^2 = g_{hj} dx^h dx^j, \tag{24}$$

where g_{ij} is the metric of $\mathcal{E}_p$. If $x^j = x^j(y^\alpha)$ and $W_\alpha^j = \frac{\partial x^j}{\partial y^\alpha}$ we get :

$$ds^2 = g_{hj} W_\alpha^h W_\beta^j dy^\alpha dy^\beta = g_{\alpha\beta} dy^\alpha dy^\beta,$$

$g_{\alpha\beta}$ is the metric induced on $\mathcal{E}_m$ by the metric of $\mathcal{E}_p$. The matrix with elements $g_{\alpha\beta}$ defines the *first fundamental form* F_1.

The embedding of the space allows the determination of one normal $\mathbf{n}$ at point P. When we move on $\mathcal{E}_m$ from P to the point $P + \delta P$ there is a change in the

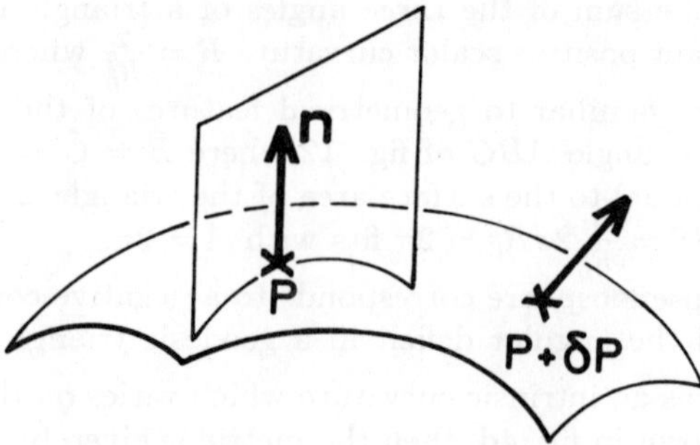

Fig. 13 : *Change in the normal* **n** *when moving from point P to point $P + \delta P$.*

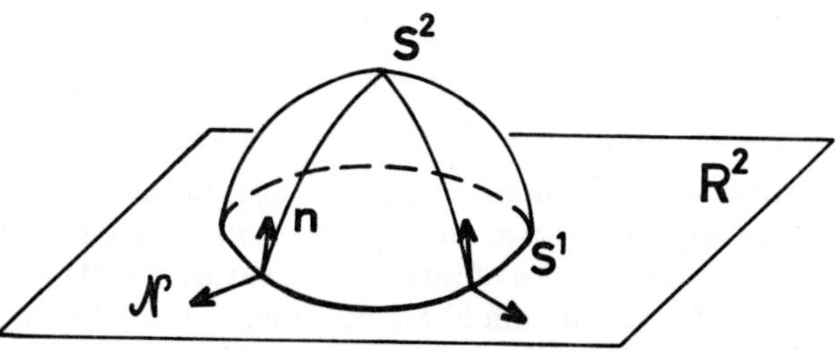

Fig. 14 : $\mathcal{N}$ *is the normal to the circle S^1 embedded in $\mathcal{R}^2$.*
n *is the normal to the circle S^1 embedded in S^2.*

normal **n** (fig. 13). In order to feel how the notion of normal depends on the embedding we can consider the simple example given in fig. 14 where we see easily that the normals to a circle S^1 as embedded in $\mathcal{R}^2$ differ from the normals to S^1 as embedded in S^2.

The j-th component of the variation of **n** when moving in the β direction defines the coefficient $K_{\alpha\beta}(\mathbf{n})$ of the *second fundamental form* F_2 (matrix with elements $K_{\alpha\beta}$).

$$\theta_\beta^j(\mathbf{n}) = \left(\tilde{\nabla}_\beta^p \mathbf{n}\right)^j = \left\langle \left.\frac{\partial}{\partial x^j}\right| \tilde{\nabla}_\beta^p \mathbf{n}\right\rangle,$$

where $\tilde{\nabla}^p$ is the Levi-Civita connection of $\mathcal{E}_p$:

$$K_{\alpha\beta}(\mathbf{n}) = -g_{kj} W_\alpha^k \theta_\beta^j = g_{kj} n^k \tilde{\nabla}_\beta^p \left(W_\alpha^j\right).$$

Diagonalization of $K_{\alpha\beta}$ with respect to the metric gives the principal curvatures ρ_m (eigenvalues) in the m principal directions.

$$\left(K_{\alpha\beta} - \frac{1}{\rho}g_{\alpha\beta}\right)u^\beta = 0.$$

If we introduce :

$$K^\alpha_\beta = g^{\alpha\delta}K_{\delta\beta}, \tag{25}$$

eq. 25 also reads :

$$\left(K^\alpha_\beta - \frac{1}{\rho}\delta^\alpha_\beta\right)u^\beta = 0.$$

Fundamental invariants are symmetric functions of curvatures $\frac{1}{\rho_1}, \frac{1}{\rho_2}..., \frac{1}{\rho_m}$ and product of principal curvatures. For example :

$$S_1 = \frac{1}{\rho_1} + \frac{1}{\rho_2} + \frac{1}{\rho_m},$$

$$S_2 = \frac{1}{\rho_1\rho_2} + \frac{1}{\rho_2\rho_3} + ...$$

$$\cdot$$
$$\cdot$$
$$\cdot$$

$$S_m = \frac{1}{\rho_1\rho_2...\rho_m}.$$

We can easily express the two extreme invariants in terms of the second fundamental form. The mean curvature corresponds to the first invariant :

$$C_M = \sum_i \frac{1}{\rho_i} = g^{\alpha\beta}K_{\alpha\beta} = Trace\left(K^\alpha_\beta\right),$$

and the product of curvature defines the curvature C_T :

$$C_T = \prod_i \frac{1}{\rho_i} = (-1)^{n-1}det\left(K^\alpha_\beta\right).$$

Let us emphasize that C_M and C_T are *not intrinsic quantities*, they both depend on the embedding. There exists a relation expressing the curvature tensor $R^i_{jk\ell}$

of the manifold $\mathcal{E}_m$ in terms of the curvature tensor $\tilde{R}^{\alpha}_{\beta\gamma\delta}$ of the embedding space $\mathcal{E}_p$:

$$R_{\alpha\beta\gamma\delta} = \tilde{R}_{\alpha\beta\gamma\delta} + K_{\alpha\gamma}K_{\beta\delta} - K_{\alpha\delta}K_{\beta\gamma}, \tag{26}$$

where $\tilde{R}_{\alpha\beta\gamma\delta} = W^i_{\alpha}W^j_{\beta}W^k_{\gamma}W^l_{\delta}\tilde{R}_{ijkl}$. We can compute the scalar curvature R of $\mathcal{E}_m$. We obtain from eq. 26 $R = g^{\alpha\beta}g^{\gamma\delta}R_{\alpha\beta\gamma\delta}$ and :

$$R = \tilde{R} + g^{\alpha\beta}g^{\gamma\delta}\left(K_{\alpha\gamma}K_{\beta\delta} - K_{\alpha\delta}K_{\beta\gamma}\right). \tag{27}$$

Let us emphasize that $\tilde{R} = g^{\alpha\beta}g^{\gamma\delta}\tilde{R}_{\alpha\beta\gamma\delta}$ is not the scalar curvature of the space $\mathcal{E}_p$ but only its restriction to the tangent space of $\mathcal{E}_m$. The corrective term depends on the second fundamental form. R is an intrinsic quantity depending only on the manifold $\mathcal{E}_m$. Both terms $\tilde{R}$ and the corrective term are characteristic of the embedding of $\mathcal{E}_m$ in $\mathcal{E}_p$. We now want to do some remarks on the usual curvatures pertinent in physical problems and concerning the case of 2D surfaces imbedded in a 3D space.

2D Surface embedded in a 3D space

A cut of the surface $\mathcal{E}_2$ by a plane containing the normal performs a normal section at point P of the surface (fig. 13) and generates a curve with curvature $\frac{1}{\rho}$. Considering all normal sections enables to find the two principal directions corresponding to the extreme values $\frac{1}{\rho_1}$ and $\frac{1}{\rho_2}$ of the curvature. For 2D surfaces, the mean curvature C_M reduces to :

$$C_M = \frac{1}{\rho_1} + \frac{1}{\rho_2},$$

and the curvature C_T to :

$$C_T = \frac{1}{\rho_1\rho_2}.$$

They are not intrinsic quantities but are of physical importance. The mean curvature for example controls the pressure difference ΔP between the external and internal pressure in a soap bubble. This difference is zero for a film of soap sticked on a support. The mean curvature is then zero and the surface is minimal. There exists a large variety of very nice minimal surfaces depending on the shape of the supports[21]. Another example concerns the shape of red cells[22] which results from a balance between energetic terms depending on both C_M and C_T.

Let us now look to eq. 26 , the corrective term reduces to

$$(traceK)^2 - \left(traceK^2\right) = 2det(K) = \frac{2}{\rho_1\rho_2},$$

which is precisely the curvature C_T and eq. 27 now is :

$$R = \tilde{R} + 2C_T. \tag{28}$$

This formula reveals the non intrinsic nature of the curvature C_T. Indeed the curvature C_T of a surface depends on the curvature of the manifold in which the surface is embedded. In many situations where we look at a *surface as embedded in our physical space* $\mathcal{R}^3$ *without curvature*, C_T is related to the *intrinsic curvature* $\left(C_T = \frac{R}{2}\right)$. Very often physicists are concerned with surfaces embedded in $\mathcal{R}^3$ and call $C_T = \frac{1}{\rho_1 \rho_2}$ the *Gauss curvature* .

To point out the non intrinsic character of the curvature C_T in the general case, look for example at the sphere S^2 with scalar curvature $R = \frac{2}{R_0^2}$ and consider the two different embeddings.

- S^2 embedded in $\mathcal{R}^3$

The curvature tensor of $\mathcal{R}^3$ is $R^i_{jkl} \equiv 0$. $C_T = \frac{1}{R_0^2}$ and eq. 28 simply reads :

$$R = 0 + 2C_T.$$

- S^2 embedded in S^3

We shall give results which can be verified with use of section 4. The sphere S^3 gets a non zero curvature tensor and $\tilde{R} = \frac{2}{R_0^2}$. Computation of C_T gives $C_T = 0$. Eq. 28 is verified but now in the following way :

$$R = \tilde{R} + 0.$$

We shall give in section 4 another illustration of these considerations when we shall look at tori which exist on S^3.

3 - DESCRIPTION OF S^3

We now want to give a geometrical description of S^3 [23]. It is difficult to get a physical perception of S^3 since it is embedded in $\mathcal{R}^4$. We can take a balloon S^2 in our hands but not a sphere S^3. Nevertheless we shall try to proceed by analogy in order to point out the main features of S^3 and use the interplay between analytical and geometrical perception. The geometrical aspect is better emphasized by comparison with S^2. From a mathematical standpoint the analogy is stronger with S^1 since both S^1 and S^3 are isomorphic to groups.

Some geometrical aspects

S^1, S^2, S^3 can be seen as embedded in $\mathcal{R}^2, \mathcal{R}^3, \mathcal{R}^4$ and then defined as points at a constant distance R_o from the origin.

$$S^1 : x_1^2 + x_2^2 = 1,$$
$$S^2 : x_1^2 + x_2^2 + x_3^2 = 1,$$
$$S^3 : x_1^2 + x_2^2 + x_3^2 + x_4^2 = 1.$$

S^1 can be drawn on a piece of paper. S^2 can be built in $\mathcal{R}^3$ with a piece of wood for example. S^3 can only be mentally vizualised. In fig. 15 there is a concrete realization of S^1, a kind of perspective for S^2 and a formal representation for S^3. Antipodals points $+ A$ and $- A$ are :

$$S^1 : A = (+x_1, +x_2), \qquad -A = (-x_1, -x_2),$$
$$S^2 : A = (+x_1, +x_2, +x_3), \qquad -A = (-x_1, -x_2, -x_3),$$
$$S^3 : A = (+x_1, +x_2, +x_3, +x_4), \qquad -A = (-x_1, -x_2, -x_3, -x_4).$$

Points at a constant distance $d_1(A)$ from point A and at a constant distance $d_2(-A)$ from point $- A$ constitute for :

$$S^1 : points\ C,\ D$$
$$S^2 : a\ circle\ S^1$$
$$S^3 : a\ sphere\ S^2$$

If $d_1 = d_2$; points on S^1 are antipodal ; the circle on S^2 is a great circle and the sphere on S^3 is a great sphere.

A circle or great circle S^1 on S^2 can also be seen as the intersection of a 2D plane with S^2. A great circle corresponds to a 2D plane passing through the origin. In the same way, as we shall see now a sphere or a great sphere on S^3 corresponds to the intersection with a 3D plane. A great sphere corresponds to a 3D plane passing through the origin.

Points of S^3 can be described with use of hyperspherical coordinates :

$$\begin{aligned}
x_1 &= R_o \cos\omega, & \omega &\in 0,\ \pi \\
x_2 &= R_o \sin\omega \cos\mu, & \mu &\in 0,\ \pi \\
x_3 &= R_o \sin\omega \sin\mu \cos\varphi, & \varphi &\in 0,\ 2\pi \\
x_4 &= R_o \sin\omega \sin\mu \sin\varphi.
\end{aligned} \qquad (29)$$

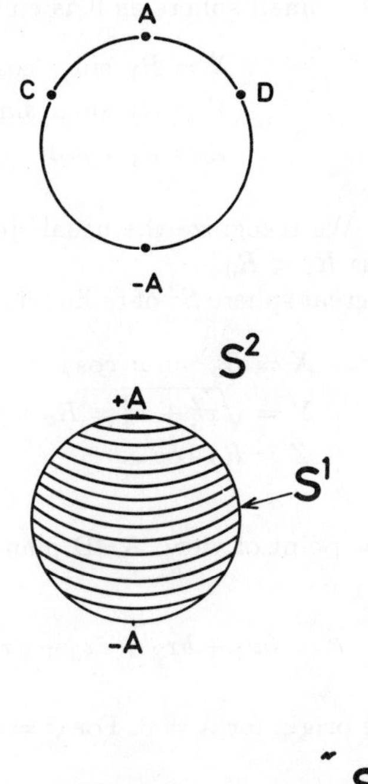

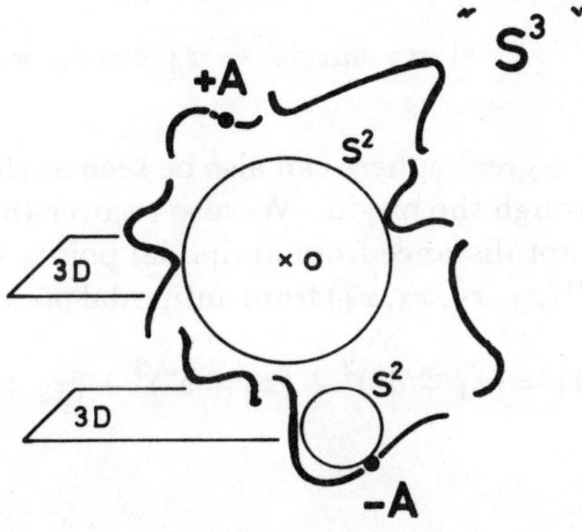

Fig. 15 : *Circle S^1 : Points C and D are at constant distances from antipodal points + A and - A.*
Sphere S^2 : circles are lines at constant distances from antipodal points + A and - A.
Sphere S^3 (symbolic representation) : Spheres S^2 are surfaces at constant distances from antipodal points + A and - A.

- $\omega = \omega_o$ describes a small sphere as it is enlighted by the following change of coordinates :

$$X = R_1 \; \sin\mu \; \cos\varphi,$$
$$Y = R_1 \; \sin\mu \; \sin\varphi,$$
$$Z = R_1 \; \cos\mu.$$

with $R_1 = R_o \sin\omega$. We recognize the usual spherical coordinates (X, Y, Z) of a sphere S^2 with radius $R_1 < R_0$.

- $\varphi = \varphi_o$ defines a great sphere S^2 of radius R_o as seen with the new coordinates :

$$X = R_o \; \sin\omega \; \cos\mu,$$
$$Y = \sqrt{x_3^2 + x_4^2} = R_o \; \sin\omega \; \sin\mu,$$
$$Z = R_o \; \cos\omega.$$

Let us consider a new point of view. A 3D plane generally reads :

$$ax_1 + bx_2 + cx_3 + dx_4 = \lambda,$$

it passes through the origin for $\lambda = 0$. For $\varphi = \varphi_o$ we obtain from eq. 29 :

$$x_3 \; \sin\varphi_o \; - \; x_4 \; \cos\varphi_o \; = \; 0,$$

which shows that a great sphere can also be seen as the intersection of S^3 with a 3D plane passing through the origin. We also recover that a sphere S^2 corresponds to points at a constant distance from antipodal points in the following way : the distance of a point $P(x_1, x_2, x_3, x_4)$ from antipodal points $\pm A = (\pm a^*, \pm b^*, \pm c^*, \pm d^*)$ is :

$$d^2(\pm A) = (x_1 \mp a^*)^2 + (x_2 \mp b^*)^2 + (x_3 \mp c^*)^2 + (x_4 \mp d^*)^2.$$

Noticing that :

$$x_1^2 + x_2^2 + x_3^2 + x_4^2 = R_0^2,$$
$$a^2 + b^2 + c^2 + d^2 = R_0^2,$$

we obtain :

$$d^2(\pm A) = 2R_0^2 \left(1 \mp e^*\right),$$

where $e^* = a^* x_1 + b^* x_1 + c^* x_3 + d^* x_4$. We see that d $=$ constant if $e^* =$ constant. Points P such that $e^* =$ constant are at the intersection of the 3D plane defined by $e^* =$ constant with the sphere S^3 : i.e. it is a sphere S^2. The great sphere corresponds to $e^* = 0$ and $d(+A) = d(-A) = 2R_0^2$. We have reported different manners to describe spheres S^2 on S^3. Special lines such as circle S^1 also exist on S^3. Circles S^1 on S^3 correspond to the intersection of 2D planes with the sphere S^3. A 2D plane is the intersection of two 3D planes :

$$\begin{cases} ax_1 + bx_2 + cx_3 + dx_4 = \lambda_1 \,, \\ a'x_1 + b'x_2 + c'x_3 + d'x_4 = \lambda_2. \end{cases}$$

$\lambda_1 = \lambda_2 = 0$ defines a 2D plane passing through the origin. A first rotation allows to define the first plane by $x_1 = 0$ for example and a second one by $x_2 = 0$. The intersection with the sphere S^3 is obtained from eqs. 29 it corresponds to a circle S^1 of radius R_o :

$$X = x_3 = R_o \cos\varphi,$$
$$Y = x_4 = R_o \sin\varphi.$$

Notice the property of two great circles on S^3 : they are either enlaced or touch themselves at two points.
Up to now we have described special lines and some special surfaces existing on S^3. Before introducing the very specific surfaces (tori) on S^3 we want to report the group structure of S^3 by analogy with the description of S^1.

Group structure of $\mathbf{S^1}$ and $\mathbf{S^3}$

A circle S^1 may be described with use of complex numbers :

$$z = x_1 1 + x_2 i, \qquad x_1, x_2 \in \mathcal{R},$$

with modulus equal to 1 :

$$\|z\| = x_1^2 + x_2^2 = 1 \,.$$

The multiplication law is :

$$\begin{aligned} 1^2 &= 1, \\ i^2 &= -1, \\ 1i &= i1 = i. \end{aligned}$$

- Motion on S^1

Using the isomorphism between S^1 and the group of unit complex numbers we achieve motions on S^1 by left or right multiplication. We move from a point z to a point z' :

$$z \xrightarrow{z_o} z'$$

$$Left \ \ multiplication : z' = z_o z, \tag{30}$$

$$Right \ \ multiplication : z' = z z_o. \tag{31}$$

Since the group is commutative these two motions are equivalent but we introduce these notions to facilitate the forthcoming discussion on the unit quaternion group describing S^3.

- Tangent vector - Left invariant vector field

A tangent vector $\mathbf{X}$ to S^1 at point z is of the following form :

$$\mathbf{X} = zi. \tag{32}$$

At another point z' it also reads :

$$\mathbf{X}' = z'i. \tag{33}$$

Then the field $\mathbf{X} = zi$ is a tangent vector field on S^1. With use of the rule of transport, (eq. 30), we can transport the tangent vector $\mathbf{X}$ (which is a complex number) at a point z in a vector $\mathbf{X}''$ at point z' :

$$Point \qquad z \xrightarrow{z_o} z' = z_o z, \tag{34}$$

$$Tangent \ vector \qquad X \xrightarrow{z_o} X'' = z_o X. \tag{35}$$

It is easy to see that this left transported vector $\mathbf{X}''$ is nothing else than the tangent vector field defined in eq. 32. Indeed with use of eq. 35 and 32, $\mathbf{X}''$ reads :

$$\mathbf{X}'' = z_o \mathbf{X} = z_o z i = z' i, \tag{36}$$

then $\mathbf{X}' \equiv \mathbf{X}''$. The conclusion is that the vector field $\mathbf{X} = zi$ is invariant by the left translation defined in eq. 30. It is important to notice that this tangent vector field (eq. 32) is associated to a constant vector i defined in the tangent space at point $z = 1$. The rule of transport allows to define at everypoint z the tangent vector $\mathbf{X}$ associated with the constant element i. Once again we want to emphasize that this presentation which may seem superfluous and trivial for the circle S^1 is given to facilitate the description of S^3. To conclude on S^1 we have seen that, with use of the group of unit complex numbers, we were able to define a point, a tangent vector, to move to another point of the circle and in the same way to find the tangent vector field (already trivial for S^1) invariant under this motion. We shall now proceed in the same vein for S^3.

S^3 seen as embedded in $\mathcal{R}^4$ is isomorphic to the group of unit quaternions $\mathcal{H}_1$[39]. S^1 was embedded in $\mathcal{R}^2$ and two coordinates x_1, x_2 were necessary to define a point. Here we now work in a 4D space and we need four coordinates x_1, x_2, x_3, x_4 to define a point. A quaternion plays the same role in $\mathcal{R}^4$ than a complex number in $\mathcal{R}^2$. A general quaternion is :

$$q = x_1 1 + x_2 i + x_3 j + x_4 k, \tag{37}$$

also written : $q = (x_1, x_2, x_3, x_4)$. (i,j,k) generalize the notion of complex number. The imaginary part V of q is of dimension 3.

$$V = x_2 i + x_3 j + x_4 k. \tag{38}$$

Product of two quaternions results from the following rules :

$$\begin{aligned}
i^2 &= \quad j^2 = k^2 = -1, \\
ij &= -ji = k, \\
jk &= -kj = i, \\
ki &= -ik = j.
\end{aligned} \tag{39}$$

The sum of two quaternions q (eq. 37) and $q' = y_1 1 + y_2 i + y_3 j + y_4 k$ is :

$$q + q' = (x_1 + y_1) 1 + (x_2 + y_2) i + (x_3 + y_3) j + (x_4 + y_4) k,$$

and the product :

$$qq' = (x_1 y_1 - x_2 y_2 - x_3 y_3 - x_4 y_4) 1 + (x_2 y_1 + x_1 y_2 + x_3 y_4 - x_4 y_3) i$$

$$+ (x_3 y_1 + x_1 y_3 + x_4 y_2 - x_2 y_4) j + (x_1 y_4 + x_4 y_1 + x_2 y_3 - x_3 y_2) k. \tag{40}$$

Resulting from the rules of eq. 39, it is important to notice that $q.q' \neq q'.q$ and that the unit quaternion group is not commutative. Left and right multiplication are different. Since we are concerned with unit quaternions, $\Sigma x_i^2 = 1$, they define points of S^3 assumed here, to simplify the discussion, of radius equal to 1. The *inverse* q^{-1} is :

$$q^{-1} = \frac{x_1 1 - x_2 i - x_3 j - x_4 k}{\Sigma x_i^2} = \quad x_1 1 - x_2 i - x_3 j - x_4 k.$$

The *conjugate* $\bar{q}$ is : $\bar{q} \equiv q^{-1}$ and $\overline{q_1 q_2} = \bar{q}_2 \bar{q}_1$.

- Scalar product and distance

Scalar product is defined with use of the metric in $\mathcal{R}^4$:

$$\langle q \,|\, q' \rangle = x_1 y_1 + x_2 y_2 + x_3 y_3 + x_4 y_4 = \mathcal{R}e\left(\bar{q}.q'\right) = \mathcal{R}e\left(q'.\bar{q}\right),$$

where $\mathcal{R}e$ = real part. The distance r between two points q and q' is :

$$r = \left|q - q'\right|^2 = |q|^2 + \left|q'\right|^2 - 2\mathcal{R}e(q.q'),$$

where $|q|^2 = \sum_i x_i^2$ and $q.q'$ is the product given in eq. 40.

- Motion on S^3

A general motion[24] from a point q to a point q' is generated with use of two quaternions q_1 and q_2 (since the group $\mathcal{H}_1$ is non commutative).

$$q \to q' = q_1 q q_2^{-1}.$$

Special transformations induce :

$$
\begin{aligned}
&Left \;\; screw \quad q' = q q_1^{-1}, &\quad (41)\\
&Right \;\; screw \quad q' = q_1 q, &\quad (42)
\end{aligned}
$$

$$Rotation \qquad q' = q_1 q q_1^{-1}.$$

It leaves the great circle passing through points $q = 1$ and $q = q_1$ invariant.

- Tangent vectors

In the spirit of section 2 we define a tangent vector $\mathbf{X}$ at point q_o as the set of tangent vectors to curves $q(t)$. These curves can be parametrized as :

$$q(t) = q_o(\cos f(t) + A \sin f(t)),$$

where A is a unit imaginary quaternion and $f(t = 0) = 0$. The tangent vector is :
$\mathbf{X} = \left.\dfrac{dq}{dt}\right|_{t=0} = q_o A \left.\dfrac{df}{dt}\right|_{t=0}$. The set of tangent vectors at point q_o is spanned by vectors :

$$\mathbf{X} = q_o V, \tag{43}$$

where V is a pure imaginary quaternion of general form given in eq. 38 (we shall in the following always label V any imaginary quaternion). The tangent space at point q_o is then spanned by the orthonormal frame :

$$\begin{aligned}
\mathbf{e}_i &= q_o i, \\
\mathbf{e}_j &= q_o j, \\
\mathbf{e}_k &= q_o k,
\end{aligned} \tag{44}$$

and is isomorphic to $\mathcal{R}^3$, as the tangent space of S^2 is isomorphic to $\mathcal{R}^2$. Let us remark that, as for S^2, the tangent space of S^3 at point q_o is perpendicular to q_o. Indeed consider the tangent vector $\mathbf{X} = q_o V$ we get :

$$\langle X \,|q_o\rangle = \langle q_o V \,|q_o\rangle,$$

which gives, with use of the scalar product :

$$\begin{aligned}
\langle X \,|q_o\rangle = \mathcal{R}e\left[\left(\overline{q_o V}\right).q_o\right] &= \quad \mathcal{R}e\left[\left(\bar{V}\overline{q_o}\right).q_o\right] \\
&= \quad \mathcal{R}e\left(\bar{V}\right) \equiv 0,
\end{aligned}$$

since V is a pure imaginary quaternion. We now want to show that eq. 41 defines a left invariant vector field.

- Left invariant tangent vector field

Consider the vector field $\mathbf{X}$ defined at every point q, by :

$$\mathbf{X} = q V, \tag{45}$$

with $V = $ constant.
Motion from point q to another point q' by left multiplication :

$$q \quad \xrightarrow{\;q_1\;} \quad q' = q_1 q, \tag{46}$$

also gives a new vector $\mathbf{X}''$ at point q' :

$$\mathbf{X} \xrightarrow{\;q_1\;} \mathbf{X}'' = q_1\mathbf{X}. \tag{47}$$

Eq. 45 defines a vector field i.e. a vector $\mathbf{X}' = q'V$ at point q'. We shall now prove that the left transported vector $\mathbf{X}''$ is nothing else than the vector $\mathbf{X}'$ of the vector field defined in eq. 45. Using eqs. 45 and 46 we can write eq. 47 as :

$$\mathbf{X}'' = q_1\mathbf{X} = q_1 qV = q'V = \mathbf{X}'.$$

It is interesting to remark that the tangent space at the point $q = 1$ of the group is spanned by vectors V, i.e. it is isomorphic to $\mathcal{R}^3$. The tangent vector field given by eq. 45 is then linked to a constant element V of the tangent space at point $q = 1$. There is an isomorphism between the left invariant vector field and the tangent space at the unit element of the group. Equiped with an algebraic structure this corresponds to the *Lie algebra* of the group. In the following we shall refer to V as the Lie algebra element associated with the vector field qV. We shall see in the next section the importance of such a vector field in the description of the Blue Phases. The main interest of this vector field is that it is determined as far as V is fixed. The knowledge of V fixes the determination of a tangent vector at any point q by applying the rule of transport given in eq. 46. Before going on into the detailed structure of the director field describing the Blue Phase, we want to introduce surfaces (tori) specific of S^3.

- Tori on S^3

Let us first write the general equation of a great circle S^1 on S^3 with use of the quaternionic notation :

$$q(t) = q_1 \cos t + q_2 \sin t, \tag{48}$$

where $q_2 \perp q_1$ can be written as : $q_2 = q_1 V$ and :

$$q(t) = q_1(\cos t + V \sin t). \tag{49}$$

S^1 is the intersection of the sphere S^3 with a 2D plane. Let us introduce the two vectors $\mathbf{V}'$ and $\mathbf{V}''$ orthogonal to $\mathbf{V}$:

$$< \mathbf{V}' \mid \mathbf{V} >= 0 \quad < \mathbf{V}'' \mid \mathbf{V} >= 0.$$

The 2D plane corresponds to the intersection of the two 3D planes passing through the origin defined by :

$$\left\langle q_1 V' \,|\, q \right\rangle = \left\langle q_1 V' \,|\,(q_1 \cos t + q_1 V \sin t)) \right\rangle,$$
$$= \mathcal{R}e(\bar{V}'\,(\bar{q}_1 q_1 \cos t + \bar{q}_1 q_1 V \sin t)), \tag{50}$$
$$= 0.$$

since V' is a pure imaginary quaternion and $Re(\bar{V}'V) = 0$ since they are orthogonal vectors. In the same way the equation of the second 3D plane is :

$$\left\langle q_1 V'' \,|\, q \right\rangle = 0. \tag{51}$$

Remark that, with the parametrization given in eq. 48 we recognize the classical property of a circle : $\dfrac{dq}{dt} = qV \perp q$ and $\dfrac{d^2q}{dt^2} = -q$. In the following we shall use the notation :

$$\mathcal{C}_1 = \ (q_1, V), \tag{52}$$

to denote a circle passing through the point q_1. V is the element of the Lie algebra associated with the tangent vector qV at any point q of the circle. It is worth noticing that there always exists a great circle perpendicular to any another one. Consider the circle :

$$\mathcal{C}_2 = (q_2, V) = q_1 \left(\cos t V' + \sin t V'' \right).$$

It is easy to see with use of eqs. 50 and 51 that it is perpendicular to $\mathcal{C}_1$ (eq. 52).

Consider now the simple case with $q_1 = 1$ and the two orthogonal circles :

$$\mathcal{C}_1 = \ (1, i), \tag{53}$$
$$\mathcal{C}_2 = k(1, i). \tag{54}$$

A *torus* is a surface consisting in the set of points of S^3 at a distance d_1 of $\mathcal{C}_1$ and d_2 of $\mathcal{C}_2$. Let us take a new system of coordinates for a point q of S^3.

$$q = \rho_1 \cos \lambda + \rho_1 \sin \lambda i + \rho_2 \cos \mu j + \rho_2 \sin \mu k, \tag{55}$$

with $\rho_1^2 + \rho_2^2 = 1$. The distance d_1 between point q and $\mathcal{C}_1$ is :

$$d_1^2 = Min \left((\cos t - \rho_1 \cos \lambda)^2 + (\sin t - \rho_1 \sin \lambda)^2 + \rho_2^2 \cos^2 \mu + \rho_2^2 \sin^2 \mu \right)$$

where Min = minimum of ( ). Then :

$$d_1^2 = Min\left(2 - 2\rho_1 \cos(\lambda - t)\right),$$
$$d_1^2 = 2\left(1 - \rho_1\right).$$

In the same manner the distance d_2 to the circle C_2 is : $d_2^2 = 2\left(1 - \rho_2\right)$. The tori correspond to points with $\rho_1 = constant$ and $\rho_2 = constant$. The two extreme tori $\rho_1 = 1$, $\rho_2 = 0$ and $\rho_1 = 0$, $\rho_2 = 1$ are the two limit circles C_1 and C_2. To visualize the tori we should see in $\mathcal{R}^4$! But we can get a representation of S^3 on $\mathcal{R}^3$ by a stereographic projection in a way similar to what is done in the stereographic projection of S^2 on $\mathcal{R}^2$. A point $q = (x_1, x_2, x_3, x_4)$ of S^3 is projected as :

$$x = \frac{x_2}{1 - x_1}, \qquad y = \frac{x_3}{1 - x_1}, \qquad z = \frac{x_4}{1 - x_1},$$

where x_1 is the pole of the projection. The projection of one torus defined by $\rho_1 = $ constant in eq. 55 is a torus in $\mathcal{R}^3$. The limit circle C_1 ($\rho_2 = 0$) is projected onto the line $y = z = 0$ and the other limit circle C_2 ($\rho_1 = 0$) on the circle $y^2 + z^2 = \rho_2^2 = 1$. By varying ρ_1 from 0 to 1 a family of tori, successively nested in one another, is generated as shown in fig. 16. This gives a foliation of S^3 similar to the foliation of the sphere S^2 with circles. For the sphere S^2 the foliation is obtained in the following way : we take two antipodal points and consider the circles at a constant distant from each antipodal point. The series of parallel circles allows to cover all the sphere. For S^3 we take two orthogonal limit circles and consider the tori at a constant distance from each limit circle. The series of parallel tori gives a foliation of S^3. As we shall see in the next section this foliation plays an important role in the description of Blue Phase textures. It is interesting to consider the curvature of these tori.

Curvature of the tori on S^3

The intrinsic scalar curvature of all tori is zero. The metric of S^3 in terms of toroidal coordinates introduced in eq. 55 is :

$$g_{\lambda\lambda} = R_o^2 \rho_1^2, \qquad g_{\mu\mu} = R_o^2 \rho_2^2, \qquad g_{\rho\rho} = \frac{R_o^2}{\rho_2^2}, \tag{56}$$

where for convenience we have introduced the radius R_o of the sphere S^3. The induced metric on a torus ($\rho_2 = constant$) is :

$$g_{\lambda\lambda} = R_o^2 \rho_1^2, \qquad g_{\mu\mu} = R_o^2 \rho_2^2,$$

which leads to :

$$ds^2 = R_o^2 \left(\rho_1^2 (d\lambda)^2 + \rho_2^2 (d\mu)^2\right). \tag{57}$$

Fig. 16 : *S^3 is covered by a family of tori nested in one another.*

It results from eq. 10 that every component of the curvature tensor for any torus is zero. Then all the tori have a zero intrinsic scalar curvature. We shall see on the contrary that the mean curvature C_M , which is an extrinsic property, is no more vanishing and depends on parameters linked to a torus. From eq. 58, the first fundamental form reads :

$$F_1 = R_o^2 \begin{pmatrix} \rho_1^2 & 0 \\ 0 & \rho_2^2 \end{pmatrix}.$$

We look for the change $\delta \mathbf{n}$ of the normal $\mathbf{n}$ to one torus seen as embedded in S^3 when we move from point q to point $q + \delta q$ with $\rho_1 = constant$:

$$\delta q = R_o\,(-\rho_1 \sin \lambda \delta \lambda, \rho_1 \cos \lambda \delta \lambda, -\rho_2 \sin \mu \delta \mu, \rho_2 \cos \mu \delta \mu), \tag{58}$$

$$\mathbf{n} = (-\rho_2 \cos \lambda, -\rho_2 \sin \lambda, \rho_1 \cos \mu, \rho_1 \sin \mu)\,,$$

$$\delta \mathbf{n} = (+\rho_2 \sin \lambda \delta \lambda, -\rho_2 \cos \lambda \delta \lambda, -\rho_1 \sin \mu \delta \mu, \rho_1 \cos \mu \delta \mu)\,.$$

From

$$\langle \delta \mathbf{n} \,|\, \delta q \rangle \quad = R_o \rho_1 \rho_2 \left((\delta \mu)^2 - (\delta \lambda)^2\right),$$

the second fundamental form reads :

$$F_2 = R_o \rho_1 \rho_2 \begin{pmatrix} -1 & 0 \\ 0 & 1 \end{pmatrix}.$$

This yields a mean curvature :

$$C_M = \frac{\rho_2^2 - \rho_1^2}{\rho_1 \rho_2} \; \frac{1}{R_o},$$ (59)

which proves that the only torus which is a minimal surface ($C_M = 0$) is the spherical torus $\rho_1 = \rho_2$. It is this spherical torus which is of importance for models of bilayers such as presented by Charvolin and Sadoc in this book. The curvature C_T is :

$$C_T = -\frac{1}{R_o^2}.$$ (60)

Its extrinsic nature is revealed by eq. 28 which depends on the embedding space of the torus. Let us just compute $\tilde{R}$ which only takes into account components of the curvature tensor of S^3 in the tangent space of the torus and here reduces to :

$$\tilde{R} = 2g^{\mu\mu}g^{\lambda\lambda}R_{\mu\lambda\mu\lambda}.$$ (61)

Using the metric of S^3 (eq. 56) and the relation 10 for the curvature tensor we obtain after a straightforward computation :

$$R_{\mu\lambda\mu\lambda} = \rho_1^2 \rho_2^2 R_o^2,$$ (62)

and

$$\tilde{R} = \frac{2}{R_o^2}.$$

Notice here that the intrinsic curvature R_{S^3} of the sphere S^3 is $\frac{6}{R_o^2}$. $\tilde{R}$ only takes into account components of the curvature tensor of S^3 in the tangent space to a torus such as given in eq. 62 and then is different from R_{S^3}. We recover once again how as in eqs. 27, 28 the intrinsic scalar curvature of a torus $R = 0$ is expressed as a balance between two non intrinsic quantities $\tilde{R} = \frac{2}{R_o^2}$ and $2C_T = -\frac{2}{R_0^2}$ depending on the embedding manifold S^3. Let us remark that a torus on S^3 which has no intrinsic curvature is different from the "doughnut" torus introduced previously (with $R \neq 0$).

4 - PERFECT DOUBLE TWISTED S^3 BLUE PHASE

We have seen in the preceding sections that the double twist rule is incompatible with long range order. In terms of the mathematical concepts introduced in section 2, this rule is a connection[24] which simply reads in $\mathcal{R}^3$:

$$\nabla_Y \mathbf{X} = \dot{\nabla}_Y \mathbf{X} + \tilde{p}_o \, \mathbf{X} \wedge \mathbf{Y},$$ (63)

where $\dot{\nabla}$ is the Levi-Civita connection of $\mathcal{R}^3$ which is expressed with use of classical orthonormal coordinates $\mathbf{r} = x\mathbf{i} + y\mathbf{j} + z\mathbf{k}$ as :

$$\dot{\nabla}_\alpha X^\beta = \frac{\partial X^\beta}{\partial x^\alpha},$$

where α stands for x, y or z and

$$(\mathbf{X} \wedge \mathbf{Y})^\beta = \epsilon_{\alpha\gamma\delta}\, g^{\beta\delta} \sqrt{detg}\, X^\alpha Y^\gamma.$$

The double twist connection eq. 63 has a curvature tensor :

$$R^\alpha_{\gamma\rho\delta} = g^{\alpha\beta} R_{\beta\gamma\rho\delta} = \tilde{p}_o^2 \left(\delta^\alpha_\delta \delta_{\gamma\rho} - \delta^\alpha_\rho \delta_{\gamma\delta} \right),$$

which is non zero as first pointed out by Sethna[25][26][27]. This implies that no parallel vector field and therefore no long range (double twist) order can exist. In $\mathcal{R}^3$ the Blue Phase is seen as realizing quasi double twist in some localized parts of the space (tubes) and presenting in between these tubes regions with defects or disclinations[16][28]. The quasi double twist may be described around the Oz axis of one tube as :

$$\mathbf{n} = \sin\left(\tilde{p}_o r\right) \mathbf{e}_\theta + \cos\left(\tilde{p}_o r\right) \mathbf{e}_z,$$

which leads to a twist energy only zero in the limit of $\mathbf{r} \to 0$. See the configuration on fig. 21. Experimental observations suggest cubic structures for most Blue Phases[3][4]. Several models give the interpretation of the different symmetry groups observed in terms of arrangement of tubes of quasi double twist. The tubes (cylinders) are in many structures aligned in preferred directions such as along the axes of the cube. It results an array of disclinations lines (in between the tubes of quasi double twist).

We now want to describe here how the perfect double twist can be realized in S^3. The main idea is to build another double twist connection without curvature by changing the metric of the space. Passing from $\mathcal{R}^3$ to S^3 leads to a change in the Levi-Civita connection. The Levi-Civita connection $\tilde{\nabla}$ of S^3 has a non zero curvature. In terms of the orthonormal non natural frame field $\{\mathbf{e}_i\}$ introduced in eq. 44, this connection is defined by the components :

$$\Gamma^j_{ik} = \left[\tilde{\nabla}_{\mathbf{e}_k}\mathbf{e}_i\right]^j. \tag{64}$$

In the following we shall refer the indices i, j, k to the frame $\{\mathbf{e}_i\}$ of S^3. Using the quaternion $Q = R_o q$ to describe the sphere of radius R_o expression 64 is directly computed by looking at the change of $\mathbf{e}_i$, when we move in the $\mathbf{e}_k$ direction.

170

At point Q, $\mathbf{e}_i$ is :

$$\mathbf{e}_i(Q) = qi = \frac{Q}{R_o}i.$$

At point $Q + \mathbf{e}_k \delta t = Q\left(1 + \frac{k\delta t}{R_o}\right)$ the vector is :

$$\mathbf{e}_i\left(Q + \mathbf{e}_k \delta t\right) = \frac{(Q + \mathbf{e}_k \delta t)}{R_o}i = \frac{Q}{R_o}\left(1 + \frac{k\delta t}{R_o}\right)i. \tag{65}$$

Γ^j_{ik} only takes into account components of eq. 65 in the tangent space i.e.

$$\left(\tilde{\nabla}_{\mathbf{e}_k}\mathbf{e}_i\right)^j = \frac{1}{R_o}\epsilon_{kip}\,\delta^{jp}. \tag{66}$$

This expression could also be computed as shown in ref[24] in a more formal way with use of eq. 19 and the commutator expression for the non natural frame field $\{\mathbf{e}_i\}$:

$$[qi, qj] = \frac{2qk}{R_o},$$

and other relations obtained by circular permutation $i \to j \to k$. We consider a double twist connection on S^3 of the same form as eq. 63 but where the Levi-Civita connection $\dot{\nabla}$ of $\mathcal{R}^3$ is now replaced by the Levi-Civita connection $\tilde{\nabla}$ of S^3.

$$\nabla_Y \mathbf{X} = \tilde{\nabla}_Y \mathbf{X} + \tilde{p}_o \mathbf{X} \wedge \mathbf{Y}.$$

It can now be expressed in the frame $\{\mathbf{e}_i\}$ with use of eq. 66 as :

$$(\nabla_k X)^j = \partial_k X^j - \left(\frac{1}{R_o} - \tilde{p}_o\right)\epsilon_{kip}\,\delta^{jp}X^i. \tag{67}$$

It is interesting to notice that for $R_o = \tilde{p}_o^{-1}$, the last term vanishes. It means that if the radius of the sphere is equal to the pitch of the double twist, the double twist connection has no more curvature. The frame field $\{\mathbf{e}_i\}$ is then a parallel frame field for the double twist connection. A perfect double twisting Blue Phase is obtained by choosing a vector field with constant coordinates in the frame $\{\mathbf{e}_i\}$. Such a field $\mathbf{X}$ is of the general form :

$$\mathbf{X} = qV = q\left(V_1 i + V_2 j + V_3 k\right). \tag{68}$$

Let us point out that V does not depend on the point q ; it means that the knowledge of the Lie algebra element V allows to determine the director $\mathbf{X}$ at any point q of S^3. Recall that a perfect ordered nematic phase is defined in $\mathcal{R}^3$ by a constant director field $\mathbf{n}(r) = a\mathbf{i} + b\mathbf{j} + c\mathbf{k}$ where $\mathbf{i, j, k}$ is the usual orthornomal frame of $\mathcal{R}^3$. Then if this frame is identified to the quaternionic basis i, j, k of the Lie algebra of S^3, we see that the *perfect Blue Phase can be considered as a perfect nematic in the Lie algebra of S^3*. We shall use this property not only to describe a perfect phase but also in section 5 to introduce a disclination line. Let us now give a geometrical description of this director field.

- Director field.

Let us consider to simplify the simple director field :

$$\mathbf{n}(q) = qi. \tag{69}$$

To visualize this director field we need to exhibit such a vector at any point q of the sphere. We have seen that S^3 can be covered by a family of tori whose orientation was fixed by two limit circles. Let us consider the two limit circles introduced in eqs. 53, 54 and the corresponding family of tori. Coordinates of a point q on a torus are defined in eq. 55 we now rewrite as :

$$q(\epsilon, \beta) = q(\epsilon)(\cos \beta + i \sin \beta), \tag{70}$$

where

$$q(\epsilon) = (\rho_1 - \rho_2 k)(\cos \epsilon + i' \sin \epsilon), \tag{71}$$

and

$$i' = 2\rho_1\rho_2 j + \left(\rho_1^2 - \rho_2^2\right) i,$$
$$\epsilon = \frac{(\lambda + \mu)}{2} + \frac{\pi}{4},$$
$$\beta = \frac{(\lambda - \mu)}{2} - \frac{\pi}{4}.$$

We recognize in eq. 70 the equation of a great circle $\mathcal{C}_\beta = (q(\epsilon), i)$ passing through point $q(\epsilon)$ and with a tangent vector $q(\epsilon)i$. Each point $q(\epsilon)$ can also be seen in eq. 71 as on a circle $\mathcal{C}_\epsilon\left(q_A, i'\right)$ passing through point $q_A = \rho_1 - \rho_2 k$ with a tangent vector $q_A i'$.

It means that the director field can be described in the following way. The sphere S^3 is considered as covered by one family of tori successively nested in one another.

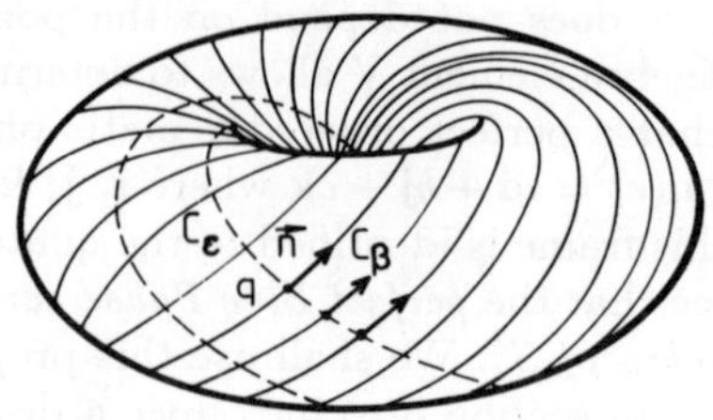

Fig. 17 : *Generation of director field lines along great circles covering each torus.*

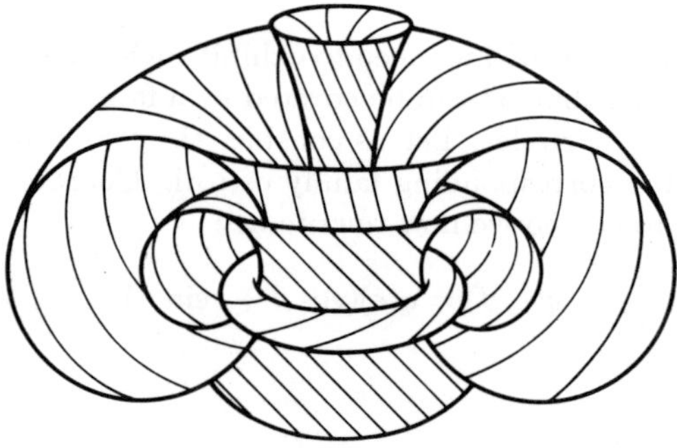

Fig. 18 : *Director field lines lying on the tori and performing double twist.*

On each torus we move on a circle C_ϵ. Director field lines are generated by moving on C_ϵ and considering at each point $q(\epsilon)$ of C_ϵ another great circle C_β passing through that point and with tangent vector $q(\epsilon)$i. The double twist is apparent on the stereographic projection where director field lines are drawn on the family of tori (fig. 18).

In the preceding construction what have we done ? Through each point we have defined a director field line which in fact is a great circle. This corresponds to the Hopf fibration of S^3. To give a simple example of a fiber bundle consider a cylinder which is a fiber bundle with basis S^1 and fiber $\mathcal{R}$. Here for S^3 the fiber is S^1, the choice of the fibration defines the orientation of that great circle on S^3.

All points of S^3 belonging to one great circle project on a single point of S^2. All fibrations corresponding to great circles with tangent vectors of the type qi, qj, qk are equivalent i.e. they belong to the same class of homotopy. They define a perfect ordered Blue Phase with a sign of the twist well defined (here $+ q$). Let us point out that a fibration such that iq corresponds to a twist of opposite sign $(-q)$ but does not belong to the same homotopy class than the first one (qi). Then with use of a Hopf fibration we have been able to build a perfect ordered phase. But are there other possibilities to construct double twisted structures ?

- Textures and Energy

What we call a texture is a configuration without defect. A configuration is defined by a map f from the physical space $\mathcal{P}$ to the space of the order parameter $\mathcal{O}$.

$$\mathcal{P} \quad \xrightarrow{f} \quad \mathcal{O}$$
$$\mathbf{r} \qquad\qquad \mathbf{n} = f(\mathbf{r}).$$

Two different configurations can look either similar or not and therefore we need to explicit the notion of similarity . Two textures defined by the fields $\mathbf{n}_1(\mathbf{r})$ and $\mathbf{n}_2(\mathbf{r})$ are equivalent if there exists a continuous map $\mu(\mathbf{r}, t)$ such that :

$$\mathbf{n}_1(\mathbf{r}) = \mu(\mathbf{r}, t = 0),$$

$$\mathbf{n}_2(\mathbf{r}) = \mu(\mathbf{r}, t = 1).$$

If the two maps $\mathbf{n}_1(\mathbf{r})$ and $\mathbf{n}_2(\mathbf{r})$ are homotopic, it means that one map can be distorted into the other one. From a physical point of view we can imagine a smooth process allowing to transform one configuration into the other one without too much trouble. From a mathematical point of view two textures are equivalent if they belong to the same equivalence class in terms of homotopy[31][32][33]. This theory gives a classification of maps $\mathcal{P} \to \mathcal{O}$ between the two spaces (manifolds) $\mathcal{P}$ and $\mathcal{O}$. It uses the homotopy groups $\Pi_n(X)$ which are the equivalence classes of maps from spheres S_n to a space X (X = $\mathcal{P}$ or $\mathcal{O}$). What we now want to point out is that for nematics , if one no special boundary conditions are imposed, there exists only one class of textures. On the contrary what we shall see is that, for Blue Phases on S^3, several non singular configurations, belonging to different classes of homotopy, are possible. For nematics, as well as for Blue Phases, the order parameter is a director i.e. $+\mathbf{n} \sim -\mathbf{n}$. The space of the order parameter is the projective plane $RP^2 \sim \frac{S^2}{Z_2}$. It can also be seen as points on a sphere with identification of opposite points since $+\mathbf{n}$ and $-\mathbf{n}$ are equivalent. This enlightens the equivalence $RP^2 \sim \frac{S^2}{Z_2}$ where Z_2 is the group with two elements. For a nematic the physical space $\mathcal{P}$ is $\mathcal{R}^3$ and for Blue Phases $\mathcal{P}$ is S^3.

For nematics a configuration is then defined by a map f :

$$\mathcal{R}^3 \quad \xrightarrow{f} \quad RP^2$$
$$\mathbf{r} \qquad\qquad \pm\,\mathbf{n} = f(\mathbf{r}),$$

and for the Blue Phase by a map f :

$$S^3 \quad \xrightarrow{f} \quad RP^2$$
$$q \qquad\qquad \pm\,\mathbf{n} = f(q).$$

For a detailed presentation we send the reader to ref[29]. In a shortway let us say that the space of the order parameter is not simply connected $\Pi_1\left(RP^2\right) = Z$ but in both cases the map f can be lifted to a map h :
for nematics :

$$\mathcal{R}^3 \quad \xrightarrow{h} \quad S^2,$$

or for Blue Phases :

$$S^3 \quad \xrightarrow{h} \quad S^2,$$

since $\Pi_1\left(\mathcal{R}^3\right) = \Pi_1\left(S^3\right) = 0$ and the covering space of RP^2 is S^2. Then, in both cases a texture will be defined by a vector field $\mathbf{n}(\mathbf{r})$ for a nematic or $\mathbf{n}(q)$ for the Blue Phase. For nematics these maps all belong to the same homotopy class : all textures will be equivalent. For exemple one representative texture corresponds to a uniform director field $\mathbf{n}(\mathbf{r}) = $ constant. On the contrary these maps are classified by $\Pi_3\left(S^2\right) = Z$ for the Blue Phases. We have Z classes, labelled by the Hopf number H(f), of non equivalent textures of Blue Phases[29][30]. The texture corresponding to the special fibration of S^3 shown in the preceding paragraph is characteristic of one of the Z homotopy classes. It is a simple configuration corresponding to a director field not too intricate. We now want to explicit energetical considerations in order to point out textures corresponding to a minimum of the energy.

- Nematic

The Frank elastic free energy simply reads[34] :

$$F = \sum_{\alpha,\beta} \int \left(\partial_\alpha n^\beta\right)^2 vol,$$

where *vol* is the volume element and α, β stand for usual orthonormal coordinates. Assuming that all elastic constants are equal and taking into account the constraint $\mathbf{n}^2 = 1$ we obtain the local condition of minimum of F :

$$\Delta\mathbf{n}(\mathbf{r}) = \lambda\mathbf{n}(\mathbf{r}), \tag{72}$$

where Δ is the classical Laplacian operator

$$\Delta = \frac{\partial^2}{\partial x^2} + \frac{\partial^2}{\partial y^2} + \frac{\partial^2}{\partial z^2}.$$

All textures are then homotopic to the uniformly aligned configuration $\mathbf{n(r)} =$ constant which corresponds to an absolute minimum of the energy.

- S^3 Blue Phase

First let us express the elastic free energy $\tilde{\mathcal{F}}$ in $\mathcal{R}^3$ in terms of the double twist connection. At lower order in the molecular order parameter we have :

$$\tilde{\mathcal{F}} = \int \left[K_{11}(div\mathbf{n})^2 + K_{22}(\mathbf{n.curln} + \tilde{p}_o)^2 + K_{33}(\mathbf{n} \wedge \mathbf{curln})^2 \right.$$

$$\left. - K_{22}div(\mathbf{n.curln} + \mathbf{n}div\mathbf{n}) \right] vol.$$

In the isotropic limit $K_{11} = K_{22} = K$ we define the reduced free energy $F = \frac{\tilde{\mathcal{F}}}{K}$. A straighforward computation gives the identity :

$$(\partial_j n^k)^2 = (div\mathbf{n})^2 + (\mathbf{n.curln})^2 + (\mathbf{n} \wedge \mathbf{curln})^2 - div(\mathbf{n.curln} + \mathbf{n}div\mathbf{n}),$$

where i, j, k refer to the classical Euclidean frame of $\mathcal{R}^3$.
The reduced free energy :

$$F = \int \left[(\partial_j n^k)^2 + 2\tilde{p}_o(\mathbf{n.curln}) + \tilde{p}_o^2 \right] vol,$$

can be tranformed, with use of the two relations :

$$\mathbf{n.curln} = -n^i \epsilon_{kij} \partial_k n^j,$$

and

$$2\tilde{p}_0^2 = \tilde{p}_0^2 (\epsilon_{kij} n^j)^2,$$

as :

$$F = \int \left[(\partial_j n^k - \tilde{p}_o \epsilon_{jik} n^i)^2 - \tilde{p}_o^2 \right] vol.$$

Except the constant term we have :

$$F = \int (\dot{\nabla}_i n^j)^2 vol.$$

In the same way the free energy on S^3 is expressed in terms of the double twist connection :

$$F = \int \left(\nabla_i n^j \right) \left(\nabla_k n^\ell \right) g^{ik} g_{j\ell} vol,$$

176

where now i, j, k refer to the frame $\{e_i\}$ of eq. 44. The local condition is still of the type of eq. 72 but where Δ is the Laplace-Beltrami operator defined on S^3 as :

$$\Delta n^\ell = \sum e_i e_i \left(n^\ell\right) - \left\langle \tilde{\nabla}_{ei}\left(e_i\right) \mid \mathbf{grad}\left(n^\ell\right)\right\rangle .$$

The last term vanishes for the frame $\{e_i\}$ which is directed along geodesics of S^3 and we simply obtain :

$$\Delta n^\ell = \sum_i e_i e_i \left(n^\ell\right) . \tag{73}$$

An explicit expression of Δ in terms of coordinates λ, μ, ρ_2 defined in eq. 55 is obtained in the following manner. We need to compute e_i, e_j, e_k in terms of the vectors :

$$\mathbf{Q}_\lambda = \frac{\partial}{\partial \lambda},$$

$$\mathbf{Q}_\mu = \frac{\partial}{\partial \mu},$$

$$\mathbf{Q}_\rho = \frac{\partial}{\partial \rho_2}.$$

From expression 58 we get easily :

$$\mathbf{Q}_\lambda = -\rho_1 \sin \lambda 1 + \rho_1 \cos \lambda i,$$

$$\mathbf{Q}_\mu = -\rho_2 \sin \mu j + \rho_2 \cos \mu k,$$

$$\mathbf{Q}_\rho = -\frac{\rho_2}{\rho_1} \cos \lambda 1 + -\frac{\rho_2}{\rho_1} \sin \lambda i + \cos \mu j + \sin \mu k.$$

Assuming the following forms :

$$\mathbf{Q}_\lambda = q\mathbf{E}_\lambda, \ \mathbf{Q}_\mu = q\mathbf{E}_\mu, \ \mathbf{Q}_\rho = q\mathbf{E}_\rho.$$

We compute $q\mathbf{E}_\lambda = q^{-1}\mathbf{Q}_\lambda ...$ which, as a result, may be written as :

$$\mathbf{E}_\lambda = Ai + Bj + Ck,$$

and so on for the other vectors $\mathbf{E}_\mu$ and $\mathbf{E}_\rho$. From the above expression we deduce the relation :

$$\mathbf{Q}_\lambda = A(qi) + B(qj) + C(qk),$$

or :

$$\mathbf{Q}_\lambda = Ae_i + Be_j + Ce_k.$$

After a straightforward computation we obtain $\mathbf{e}_i, \mathbf{e}_j, \mathbf{e}_k$ in terms of $\mathbf{Q}_\lambda, \mathbf{Q}_\mu, \mathbf{Q}_\rho$:

$$\mathbf{e}_i = \cos(\lambda - \mu)(\frac{\rho_2}{\rho_1}\frac{\partial}{\partial\lambda} + \frac{\rho_1}{\rho_2}\frac{\partial}{\partial\mu}) - \rho_1\sin(\lambda - \mu)\frac{\partial}{\partial\rho_2},$$

$$\mathbf{e}_j = \sin(\lambda - \mu)(\frac{\rho_2}{\rho_1}\frac{\partial}{\partial\lambda} + \frac{\rho_1}{\rho_2}\frac{\partial}{\partial\mu}) + \rho_1\cos(\lambda - \mu)\frac{\partial}{\partial\rho_2},$$

$$\mathbf{e}_k = \frac{\partial}{\partial\lambda} - \frac{\partial}{\partial\mu}.$$

Explicit expression of eq. 73 is :

$$\Delta = \rho_1^2\frac{\partial^2}{\partial\rho_2^2} + (\frac{\rho_1^2}{\rho_2} - 2\rho_2)\frac{\partial}{\partial\rho_2} + \frac{1}{\rho_1^2}\frac{\partial^2}{\partial\lambda^2} + \frac{1}{\rho_2^2}\frac{\partial^2}{\partial\mu^2}.$$

It is trivial to verify that, the textures we considered previously, defined by the director field $\mathbf{n} = qi$, corresponds to an absolute minimum of the energy. Metastable textures satisfying eq. 72 and 73 correspond to harmonic maps. As pointed out in ref[29] it is difficult to exhibit harmonic maps in each of the Z homotopy classes of textures classified by the Hopf number H(f). The texture we have considered before with a director $\mathbf{n} = qi$ corresponds to an harmonic map belonging to the homotopy class with Hopf number H(f) = 0. In that case director field lines are great circles lying on the surface of the tori. One texture in the class with H(f) = 1 gets director field lines such that $\mathbf{n} = iq$ which also lye on the surface of the tori. In other classes, textures exhibit director field lines with a much more complicated configuration[29]. Director field lines are no more lying on the toroidal surfaces, they are rolling around great circles of the tori.

We now want to return to the fundamental texture defined by the director field $\mathbf{n} = qi$ and in order to compare with the physical situation in $\mathcal{R}^3$ we need to relieve the curvature of S^3. This is a process which can be done by introducing many disclination lines. We shall now give an explicit construction of a disclination line.

5 - DISCLINATION IN THE S^3 BLUE PHASE.

We shall proceed by analogy with the nematic phase in $\mathcal{R}^3$. From a geometrical standpoint the introduction of a disclination line in the perfect double twisted textures on S^3 can be described by a process similar to the Volterra one generating a defect in a nematic phase. The analogy is justified by the fact that the perfect S^3 Blue Phase can be seen as a nematic one in the Lie algebra of S^3. In that spirit the analytical description is stricto sensu the introduction of a disclination line as for a nematic defect but in the Lie algebra of S^3. Doing that process we obtain the distorted configuration by applying the general rule of transport (eq. 47) of any tangent vector. But contrary to the perfect state which was described in a simple way by the Hopf foliation the distorted state will not have such a global and simple geometrical interpretation. Before going on into the details let us first justify why it only exists one type of defects in S^3.

The disclination line in S^3 will be a great circle of S^3. Configurations with defects are classified as textures, in the preceding section by the homotopy classes of the maps from the physical space $\mathcal{P}$ to the space of the order parameter $\mathcal{O}$. In presence of a defect the physical space is for a nematic with a straight disclination line :

$$\mathcal{P} = \mathcal{R}^3 - 1 \; straight \; line,$$

for the S^3 Blue Phase with a great circle disclination line :

$$\mathcal{P} = S^3 - 1 \; great \; circle.$$

Then for a nematic $\mathcal{P} \sim$ a circle S^1 and defects are classified by the maps :

$$S^1 \quad \xrightarrow{\;f\;} \quad RP^2 \sim \frac{S^2}{Z_2}.$$

The homotopy classes of these maps are classified by $\Pi_1 \left(\frac{S^2}{Z_2} \right) = Z_2$ and there only exists one class of stable linear defects. For the S^3 Blue Phase, $\mathcal{P}$ is homeomorphic to $\mathcal{R}^3 - 1 \; straight \; line$. This homeomorphism appears simply with the view of the stereographic projection. If the pole of the projection is taken on the great circle, S^3 is projected on $\mathcal{R}^3$ and the great circle on a straight line.

As a conclusion loops of defects on S^3 are also classified by $\Pi_1 \left(RP^2 \right)$. Here again only one class of stable defects does exist. It is worth noticing that configurations with a disclination line in Blue Phases with right or left twist belong to the same homotopy class which is different for textures without defect. We shall illustrate the topology of these defects by considering a special case such as a disclination labelled by S = - 1/2.

Disclination line S = - 1/2 in a nematic.

First we want to explicit the terminology of classification of defects. The index S = - 1/2 means that doing a closed circuit around the disclination induces a jump $\Delta\varphi$ in the vector orientation of $\mathbf{n}$ and $S = \frac{\Delta\varphi}{2\pi}$. Notice that a jump of $\Delta\varphi = - \pi$ on $\mathbf{n}$ gives a well defined director orientation since $+\mathbf{n}$ and $-\mathbf{n}$ are equivalent see fig. 19.

The Volterra process [35] consists in introducing in a perfect ordered state some new materials (half a space) in the following way. Consider the perfect ordered

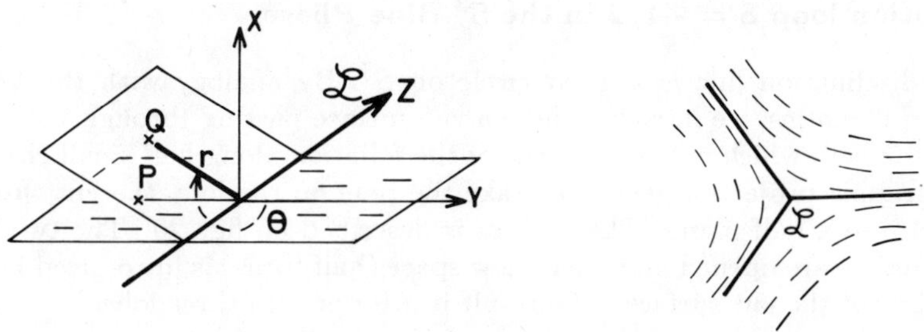

Fig. 19 : *Volterra process generating a disclination $S = -1/2$ in a nematic in $\mathcal{R}^3$.*

state $\mathbf{n}(r) = constant = \mathbf{i}$ as shown in fig. 19. To introduce some new material we consider the disclination line $\mathcal{L}$ (along Oz) and a surface containing this line along which we cut the sample (half plane yOz with $y < 0$). We introduce half space of perfect nematic in the void created by separating the two lips of the cut surface. Then we let the sample relax, we obtain a configuration with a three fold symmetry since in the process 1 space is transformed into $1 + 1/2 = 3/2$ space. In this operation a point P of the space moves to a new point Q with coordinates :

$$P \quad\longrightarrow\quad Q \tag{74}$$
$$(r_o, \theta_o, z) \qquad (r, \theta, z) \quad = (r_o, \theta = \tfrac{2}{3}\theta_o, z).$$

The origin of θ is taken along the Oy axis as indicated in fig. 19. This corresponds to a rotation $\varphi = \theta - \theta_o = -\frac{\theta_o}{\frac{3}{2}} = -\frac{\theta}{2}$ around the z axis. The new configuration is obtained from the initial one by rotating both each point of the space and the associated director. The new director field $\mathbf{n}'(\mathbf{r})$ is :

$$\mathbf{n}'(\mathbf{r}) = \cos\left(\frac{\theta}{2}\right)\mathbf{i} - \sin\left(\frac{\theta}{2}\right)\mathbf{j}, \tag{75}$$

or in polar coordinates :

$$\mathbf{n}'(\mathbf{r}) = -\sin\left(\frac{3\theta}{2}\right)\mathbf{e}_r - \cos\left(\frac{3\theta}{2}\right)\mathbf{e}_\theta, \tag{76}$$

which indicates that the new field corresponds in polar coordinates to the change $\theta \to \frac{3\theta}{2}$.

Disclination loop S = - 1/2 in the S^3 Blue Phase

The disclination line is a great circle on S^3. By analogy with the Volterra process in nematics, we introduce here a cut surface passing through the line ; it is a great sphere which cuts each torus of the foliation along half small circles. In a stereographic projection where we take the pole on the line, the cut surface is projected into a half plane. The process is described on fig. 20. The two lips of the cut surface are opened and a half new space (half torus) is introduced between the two lips of the cut surface. The result is a torus with three lobes.

To get an analytical description of this operation we introduce for example the disclination along the circle (1, k) and new coordinates (r_2, θ, α) revealing the angular symmetry around the line :

$$q(r_2, \theta, \alpha) = (r_1 \cos \alpha, -r_2 \sin \theta, r_2 \cos \theta, r_1 sin\alpha), \tag{77}$$

with $r_1 > 0, r_2 > 0$ and $r_1^2 + r_2^2 = R_0^2$.

$d_2 = \sqrt{2(1 - r_2)}$ is the distance between a point q and the disclination line (circle (1, k)). (r_2, θ, α) are similar to cylindrical coordinates (r, θ, z) around the line in $\mathcal{R}^3$. θ is still the polar angle around the line, it also corresponds to the polar angle around the z axis in the stereographic projection (projection of the disclination line). Surfaces $\theta = $ constant are great spheres in S^3 containing the dislination line whose stereographic projection are planes. The cut surface corresponds to $\theta = \pi$. Surfaces $\alpha = $ constant are great spheres perpendicular to the line. α plays the same role as z in $\mathcal{R}^3$, the spheres the same role as the plane perpendicular to the line. We have seen for a nematic that the introduction of a line corresponds to the two equivalent transformations : simultaneous rotation of angle φ of a point $\mathbf{r}$ and of the director (eqs. 74 and 75) or in polar angle $\theta \rightarrow \frac{3\theta}{2}$ (eq. 76).
In a similar way we want to induce the rotation around the line :

$$(r_2, \theta, \alpha) \qquad \xrightarrow{\mathcal{R}} \qquad (r_2, \theta + \varphi, \alpha),$$

with $\varphi = $ - $\theta/2$. It can be obtained with use of the following quaternionic rotation $\mathcal{R}$ which works as well on the point q as on tangent vectors $\mathbf{X}$:

$$\begin{aligned} q &\xrightarrow{\mathcal{R}} q' = q_1 q q_1^{-1}, \\ X &\qquad X' = q_1 X q_1^{-1}, \end{aligned} \tag{78}$$

$$\text{with} \quad q_1 = \cos\left(\frac{\varphi}{2}\right) + \sin\left(\frac{\varphi}{2}\right) k.$$

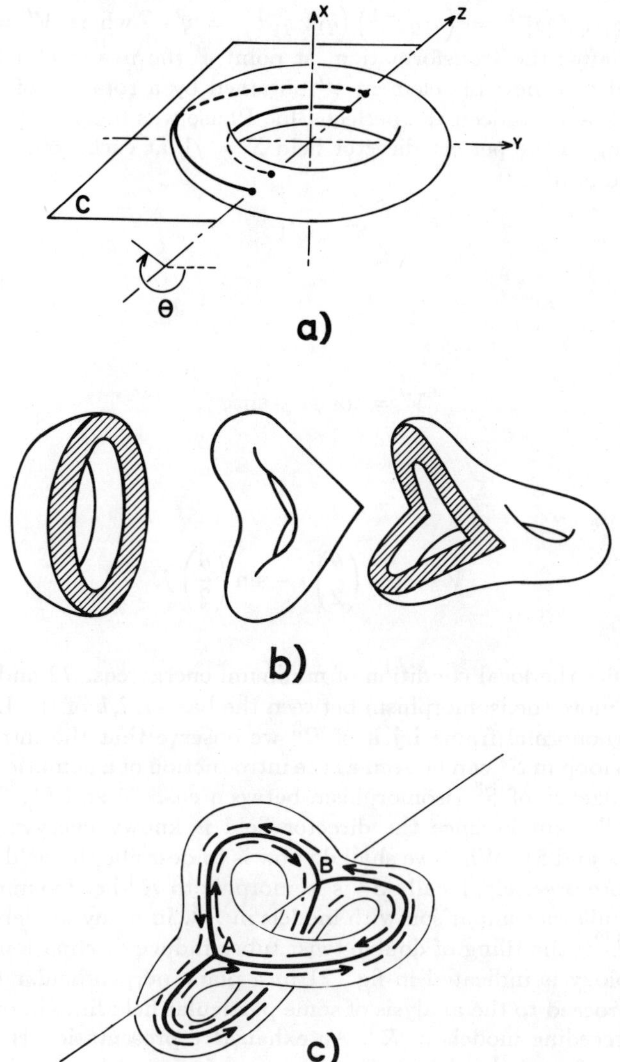

Fig. 20 : *Creation of a disclination $S = -1/2$ in a Blue Phase on S^3.*
   a) A cut surface C on each torus is introduced ;
   b) A half new space is introduced between the two lips of the cut surface ;
   c) Disclinated torus with new director field lines.

First notice it leaves the circle $(1, k)$ invariant and second the vector qV is transformed as $q_1(qV)q_1^{-1} = \left(q_1 q q_1^{-1}\right)\left(q_1 V q_1^{-1}\right) = q'V'$ where $V' = q_1 V q_1^{-1}$. It means that after the transformation, at point q' the new vector field $\mathbf{X}' = q'V'$ is associated to a new Lie element V' obtained by a rotation of angle φ from V around the k axis. Indeed the perfect Blue Phase was defined by the Lie element $V = i$ leading to the perfect director field $\mathrm{X} = qV$ at each point q. Now the new director field is X' :

$$\mathbf{X}' = q'V', \tag{79}$$

with

$$V' = \cos\varphi i + \sin\varphi j, \tag{80}$$

or

$$V' = \cos\left(\frac{\theta}{2}\right) i - \sin\left(\frac{\theta}{2}\right) j, \tag{81}$$

which satisfies the local condition of minimum energy eqs. 72 and 73.

Using once more the isomorphism between the basis i, j, k of the Lie algebra of S^3 and the orthonormal frame $\mathbf{i}, \mathbf{j}, \mathbf{k}$ of $\mathcal{R}^3$ we observe that the introduction of the disclination loop in S^3 can be seen as the introduction of a nematic disclination line in the Lie algebra of S^3 (isomorphism between eqs. 75 and 81. This description is analytically simple since the director field is known everywhere with use of equations 79 and 81. What we shall do now is to describe the field in some specific regions. More precisely, locally S^3 is isomorphic to $\mathcal{R}^3$ i.e. looking in the tangent space of S^3 allows comparison with models in $\mathcal{R}^3$. In many models as for example given in ref.[27] the tiling of double twist tubes induce disclination lines S = - 1/2 whose topology is indicated in fig. 21 in a plane perpendicular to the line. We shall now proceed to the analysis of some particular field lines in order to compare with the preceding models in $\mathcal{R}^3$. An exhaustive presentation is given in ref.[19]. The introduction of the disclination line transforms each torus into a three lobe torus whose cross section is modified in that process. Before introducing the defect the director field line was a circle which is then transformed into three half circles cuting the line at points A and B as indicated in fig. 20c. A director field line close to these three half circles intertwins twice the disclination line sweeping two times, in opposite sense at each of the three lobes.

We now look at the director configuration in a section perpendicular to the line i.e. on spheres $\alpha = $ constant.

- Double twist near the disclination line.

To compare the distorted configuration on S^3 with the one with models such as in fig. 21, we look for the field on surfaces, spheres $\alpha = $ constant, perpendicular to the line. Each sphere is swept by a family of great circles perpendicular to the disclination line. This is analogous in $\mathcal{R}^3$ to a plane swept by a family of straight lines perpendicular to the line.

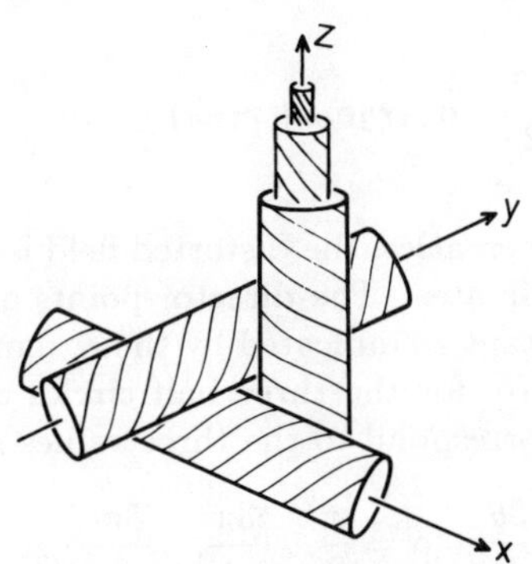

Fig. 21 : *Cylinders of double twist aligned along the three directions Ox, Oy and Oz.*

To enlighten the description we introduce a new orthonormal frame field relative to the "cylindrical coordinates" (r_2, θ, α) :

$$\mathbf{e}_{r_2} = r_1 \frac{\partial}{\partial r_2},$$

$$\mathbf{e}_\theta = \frac{1}{r_2} \frac{\partial}{\partial \theta},$$

$$\mathbf{e}_\alpha = \frac{1}{r_1} \frac{\partial}{\partial \alpha}.$$

This new frame field can simply be expressed as a function of the previous one $(qi,\ qj,\ qk)$ as :

$$\mathbf{e}_\alpha = -r_2 \cos(\theta - \alpha)qi - r_2 \sin(\theta - \alpha)qj + r_1 qk,$$

$$\mathbf{e}_\theta = -r_1 \cos(\theta - \alpha)qi - r_1 \sin(\theta - \alpha)qj + r_2 qk,$$

$$\mathbf{e}_{r_2} = -\sin(\theta - \alpha)qi + \cos(\theta - \alpha)qj,$$

184

or

$$qi = -\cos(\theta - \alpha)\,(r_2\mathbf{e}_\alpha + r_1\mathbf{e}_\theta) - \sin(\theta - \alpha)\mathbf{e}_{r_2},$$
$$qj = -\sin(\theta - \alpha)\,(r_2\mathbf{e}_\alpha + r_1\mathbf{e}_\theta) - \cos(\theta - \alpha)\mathbf{e}_{r_2},$$
$$qk = (r_1\mathbf{e}_\alpha - r_2\mathbf{e}_\theta).$$

The distorted field $qi' = qi\cos(\frac{\theta}{2}) - qj\sin(\frac{\theta}{2})$ now reads :

$$qi' = -\cos(\frac{3\theta}{2} - \alpha)\,(r_2\mathbf{e}_\alpha + r_1\mathbf{e}_\theta) - \sin(\frac{3\theta}{2} - \alpha)\mathbf{e}_{r_2}.$$

First notice that, as for nematics, the distorted field is obtained by the change $\theta \to 3\theta/2$ in the polar coordinates. The director points out of the surface of the spheres defined by $\alpha = $ constant as indicated by the α component of the distorted field. This component is zero, for the three half circles of fig. 20c lying on the surface of the torus which correspond to the three values of θ :

$$\frac{3\theta}{2} + \alpha = \frac{\pi}{2}, \quad \frac{3\pi}{2}, \quad \frac{5\pi}{2}.$$

A better comparison with the local topology in $\mathcal{R}^3$ is obtained by introducing new coordinates enlightening the double twist :

$$r_1 = \cos\tilde{p}_o r,$$
$$r_2 = \sin\tilde{p}_o r,$$
$$\alpha = \tilde{p}_o z,$$

then

$$n_r = \sin\left(\frac{3\theta}{2} - \tilde{p}_o z\right),$$
$$n_\theta = \cos\left(\frac{3\theta}{2} - \tilde{p}_o z\right)\cos\tilde{p}_o r,$$
$$n_z = \cos\left(\frac{3\theta}{2} - \tilde{p}_o z\right)\sin\tilde{p}_o r.$$

Fig. 23 shows the stereographic projection of the sphere $\alpha = \pi$. We recognize the projections D_1, D_2, D_3 of the three half circles corresponding to director field lines lying on the surface of the tori. These lines are in correspondence with the three lines D_1, D_2, D_3 of fig. 22 pointing along the director orientation at the points where the three cylinders of double twist touch themselves.

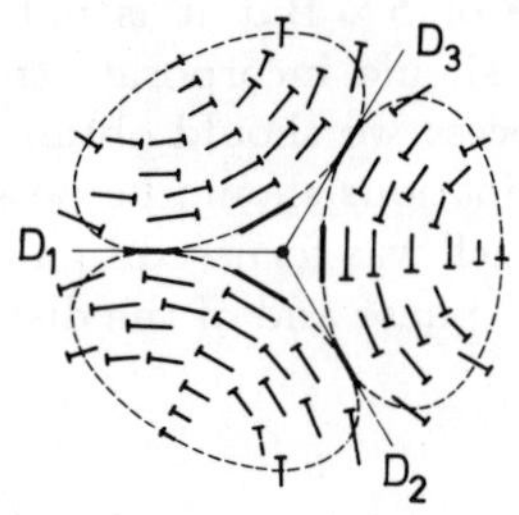

Fig. 22 : *Cut of fig. 21 perpendicular to the disclination line (direction 111) with nails representing the molecules. The size of the nails depends on the component of the molecules in the plane. - (or .) corresponds to molecules in (or perpendicular to) the plane.*

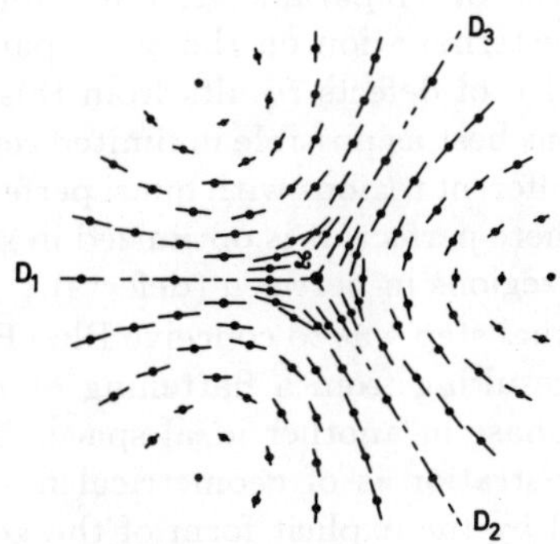

Fig. 23 : *Stereographic projection of a sphere perpendicular to the disclination line $\mathcal{L}$. We see the double twist of the molecules.*
- (or .) corresponds to molecules in (or perpendicular to) the plane.

Our description gives the complete director field configuration. In the model[16], illustrated in fig. 21 this configuration is only determined inside the double twist tubes. Nevertheless it is interesting to note that in the region where the director field is well defined it is very similar to the corresponding local configuration in S^3. From the energetical point of view we also obtain similar results. The energy of one disclination of order S is $F = 4\pi^2 S^2 Log(r_c)$ where r_c is the cutoff introduced by the core defect. This result is similar to the one found by Sethna[14].

As a final remark let us point out that the introduction of one disclination line decreases the curvature of S^3. But it is not a complete process to relieve the all curvature of S^3. We should incorporate step by step other disclinations. At the very end of that process we should obtain $\mathcal{R}^3$. Comparison with models incorporating arrays of disclinations should be possible in that limit but it is out of the scope of this paper which was focussed to a complete analytical description of the topology of the perfect phase and of the distorted director field around one disclination line.

6 - CONCLUSION

In this paper we have analyzed in a general geometrical context crystalline Blue Phases ; they are crystalline in the sense of the existence of Bragg reflections implying an order related to some translational invariance. In fact they are arrays of dislocations. The reason why defects are present in a crystalline organization results from the fact that order, imposed by local constraint, cannot be extended at large distance. In other words the local condition of minimum energy leads to a local constraint on the order parameter, but applying step by step such a condition leads to an indertermination on the order parameter at large distance. The structure with a lattice of defects results from this frustration. The system tries to satisfy local order as best as possible in limited regions of the sample. Then the adjustment between different regions with quasi perfect order is performed with defects. The order is nowhere perfect, it is optimized in some finite parts but there is a cost in energy in the regions in between (defects).

An important conceptual step was to conceive Blue Phases and other "ordered phases with defects" as resulting from a flattening or reduction to our physical space of a perfect order phase in another ideal space. This is a natural attempt for systems where the frustration is of geometrical nature. The passage to the ideal space is then guided by the explicit form of the geometrical frustration. In Blue Phases the double twist frustration is linked to the curvature of the double twist connection. The natural approach is to look for an ideal space where we can define a double twist connection without curvature i.e. leading to the construction of a director field performing double twist everywhere. The ideal manifold is, in that case, a 3D surface with constant positive curvature i.e. the sphere S^3. The passage to the real structure implies the flattening of S^3 which is obtained by introducing disclinations. At the end of the process we should obtain $\mathcal{R}^3$ with a well organized set of defects. Compare to models directly built in $\mathcal{R}^3$ the ideal models have several advantages. Contrary to models in $\mathcal{R}^3$ where the perfect order is nowhere achieved, ideal description satisfies both local and long range order. Furthermore we get geometrical and analytical ideal descriptions which is not the case in the real space. Locally the topology of the perfect ordered phase can be compared with the real one. Introduction of one defect is a well established result geometrically and analytically, nevertheless the complete flattening of the space is not yet achieved. A general geometrical description of the director field in cubic structures is still missing.

Precise investigation of cubic structures of Blue Phases has only been performed in Landau theory analysis[37][38]. The predicted scattering peak positions and amplitude are in good agreement with most experimental results. On the contrary, the suggested quasicrystalline structure for the Blue Fog is not yet well established. Blue Fog remains quite mysterious, who knows in what kind of world could we dream of the Blue Fog ?

BIBLIOGRAPHY

[1] Reinitzer F., Monatsch. Chem., $\underline{9}$, 421 (1888)

[2] Lehmann O., Z. Phys. Chem. $\underline{56}$, 750 (1906)

[3] Stegemeyer H., Blümel Th., Hiltrop K., Onusseit H. and Porsch F., Liq. Cryst. $\underline{1}$, n°1, 3 (1986)

[4] Belyakov V. A. and Dmnitrienko V. E., Sov. Phys. Usp. $\underline{28}$, 7 (1985)

[5] Barbet-Massin R., Cladis P. E., Pieranski P., Phys. Rev. A $\underline{30}$, n° 2, 1161 (1984)

[6] Pieranski P., Barbet-Massin R., Phys. Rev. A $\underline{31}$, n°6, 3912 (1985)

[7] International workshop on Aperiodic Crystals, J. Phys. France C3, Editors Gratias D., Michel L.,(1986)

[8] Coxeter H.S.M., "Introduction to Geometry", John Wiley and Sons (1969)

[9] Geometry and the imagination, Hilbert D., Cohn-Vossen S., Chelsea Publishing Company, New York (1952)

[10] " From Crystalline to Amorphous ", Editions de Physique, Godreche C. (1988)

[11] Frank F. C., Kasper J. S., Acta Crystallogr. $\underline{11}$, 184 (1958), $\underline{12}$, 483 (1959)

[12] Charvolin J., J. Phys. France, Colloque C 3, 173, Les Houches (1985)

[13] Charvolin J., Sadoc J. F., J. Phys. France $\underline{48}$, 1559 (1987)

[14] Meiboom S., Sethna J. P., Anderson P. W., Brinkman W. F., Phys. Rev. Lett. $\underline{46}$, 1216 (1981)

[15] Wright D. C. and Mermin N. D., Phys. Rev. A $\underline{31}$, 5, 3498 (1985)

[16] Meiboom S., Sammon M., Brinkman W. F., Phys. Rev. A $\underline{27}$, n°1, 438 (1983)

[17] Choquet-Bruhat Y., De Witt-Morette C., Dillard-Bleick M.," Analysis, Manifolds and Physics" North Holland (1982)

[18] Lovelock D., Rund H.," Tensors, differential forms and variational principles", Wiley and sons (1975)

[19] Dubois-Violette E., Pansu B., " Papers in honor of the 100th anniversary of Liquid Crystal Research", Mol. Cryst. Liq. Cryst. $\underline{165}$ (1988)

[20] Stoker J. J., " Differential Geometry. Pure and applied Mathematics ", Wiley Interscience (1969)

[21] Andersson S., Hyde S. T., Larsson K., Lidin S., Chem. Rev. $\underline{88}$, 221 (1988)

[22] Helfrich W., Course 12 in Physics Defects, Ecole de Physique théorique des Houches, 1980, Balian Ed., North Holland (1981)

[23] Pansu B., Dubois-Violette E., Dandoloff R., J. Phys. France $\underline{48}$, 305 (1987)

[24] Pansu B., Dandoloff R., Dubois-Violette E., J. Phys. France $\underline{48}$, 297 (1987)

[25] Sethna J. P., Wright D. C., Mermin N. D., Phys. Rev. Lett., $\underline{51}$, n°6, 467 (1983)

[26] Sethna J. P., Phys. Rev. Lett., $\underline{51}$, n°24, 2198 (1983)

[27] Sethna J. P., Phys. Rev. B $\underline{31}$, n°10, 6278 (1985)

[28] Meiboom S., Sammon M., Berreman D. W., Phys. Rev. A $\underline{28}$, n°6, 3553 (1983)

[29] Pansu B., Dubois-Violette E., J. Phys. France $\underline{48}$, 1861 (1987)

[30] Dandoloff R., Mosseri R., Europhys. Lett. $\underline{3}$, n°11, 1193 (1987)
[31] Mermin N. D., Rev. Mod. Phys. $\underline{51}$, 591 (1979)
[32] Toulouse G., Kléman M., J. de Physique Lett. $\underline{37}$, 149 (1976)
[33] Trebin H. R., Adv. in Phys. $\underline{31}$, n°3, 195 (1982)
[34] de Gennes P. G., The Physics of Liquid Crystals, Clarendon Press (1974)
[35] Kléman M., Points, Lignes et Parois, Ed. de Physique (1977)
[36] Grebel H., Hornreich R. M., Shtrikman S., Phys. Rev. A $\underline{28}$, n° 2, 1114 (1983)
[37] Rokhsar D. S., Sethna J. P., Phys. Rev. Lett. $\underline{16}$, n°14, 1727 (1986)
[38] Hornreich R. M., Shtrikman S., Phys. Rev. Lett. $\underline{56}$, n°16, 1723 (1986)
[39] Du Val P. Homographies , quaternions and rotations. Oxford Clarendon Press (1964)

Chapter IV
GEOMETRY AND TOPOLOGY OF CELL MEMBRANES

Yves BOULIGAND

GEOMETRY AND TOPOLOGY OF CELL MEMBRANES

Y. Bouligand

Histophysique et Cytophysique
Laboratoire de l'Ecole Pratique des hautes Etudes,
CNRS, 67, rue M.-Günsbourg, 94200 Ivry-sur-Seine (F.)

ABSTRACT

Cells are limited by a membrane which is a fluid bilayer of phospholipids to which are associated numerous components, such as cholesterol, polysaccharides, proteins and, among them, many enzymes. Organelles within cells are made for a large part of similar bilayers including phospholipids and various molecules. The cell membrane forms architectures closely related to those observed in liquid crystalline phases given by water-lipid systems (purified amphiphilic molecules in presence of water and oily components).

The cell is divided into a series of compartments with definite topological relations, which are rehandled more or less profoundly in diverse circumstances as endocytosis, exocytosis, mitosis etc. There are several geometric arrangements of membrane sets : parallel membranes, hexagonal packing of tubes, cubic systems made of tubes joining either three by three, or four by four, or six by six. There are other arrangements less directly related to liquid crysyalline structures (annulate lamellae, tubes and lamellae with nematic symmetries, randomly joining tubes).

Comparisons of structures in cellular membranes and in water-lipid systems reveal important differences. If geometries are often similar, water percentage and scales are distinct and bilayers observed *in vitro* present a symmetry which is broken in cell membrane bilayers. The curvature effects observed in water-lipid systems mainly come from a density difference between polar heads and corresponding paraffinic chains within a monolayer, whereas, in biological membranes, the asymmetry lies between the two monolayers and their associated molecules. Both systems produce saddle-shaped bilayers arranging into cubic lattices separating two aqueous compartments. In water-lipid systems, the coupling at an interface of two different areas seems to predominate, whereas in biological membranes, mechanisms are different and probably originate from geometric properties of proteins included within bilayers.

194

I-CELLS AND THEIR MEMBRANES

Animals and plants are made of *cells,* which are fundamental units observed in all living organisms, but it is worthy to note that *extracellular materials* are highly developed in several tissues, mainly in the animal organization [8]. For instance, various types of fibrous lattices ar built out of cells and insure the coherence of tissues, particularly in the skeleton. Calcium phosphates or other minerals are secreted in some of these extracellular matrices, namely in bones and teeth [8, 82]. Different body fluids fill extracellular compartments and in particular the blood plasma and the lymph [82]. Cell membranes separate cells from other cells, but also from the extracellular medium, which corresponds either to the environment, external to the organism, or to the 'milieu intérieur' defined by Claude Bernard [5]. In many tissues, the extracellular medium is reduced to extremely thin films extending between the two apposed cell membranes [8].

Cells are the smallest units of life and often are autonomous (protozoa, cells in culture). Each cell is limited by a thin film, called plasma membrane, separating two aqueous domains : the cytoplasm within the cell and the extracellular medium. This membrane is a bilayer of phospholipids with various associated components : cholesterol, proteins and polysaccharides.

Phospholipids are easily extracted from biological materials, with high contents of membranes, such as soybean oil, egg yolk, red cells from the blood, white matter of the central nervous system, etc. Very different phospholipids are found in biological membranes : phosphatidylcholine for instance, which is the lecithin *sensu stricto* and is represented in Fig.1.

$$(CH_3)_3N^+\text{-}CH_2\text{-}CH_2\text{-}O\text{-}P\text{-}O\text{-}CH_2$$

$$CHOCOR_1$$

$$CH_2OCOR_2$$

Figure 1. Formula of phosphatidylcholine; R_1 and R_2 : paraffinic chains.

The other examples of phospholipids common in biological membranes are phosphatidylethanolamine, phosphatidylserine, sphingomyelin etc. These phospholipids are glycerol esters which associate strongly hydrophobic paraffinic chains (saturated or not), corresponding to fatty acids (palmitic or oleic for instance) and polar heads, with a phosphoric group and a base. Such molecules are said to be *amphiphilic* (or amphipathic), since they are made of parts which would normally segregate into distinct phases, if they were not held together. Detailed accounts about membrane chemistry are numerous [19,53].

The basic organization common to cells consists of a nucleus and a cytoplasm limited by a thin membrane [33]. The nucleus itself is enveloped in a double membrane, fenestrated with pores and containing the chromatin, a complex of desoxyribonucleic acid (DNA) and basic proteins (Fig.2). The double helical DNA molecules are extremely long and form the genetic material.

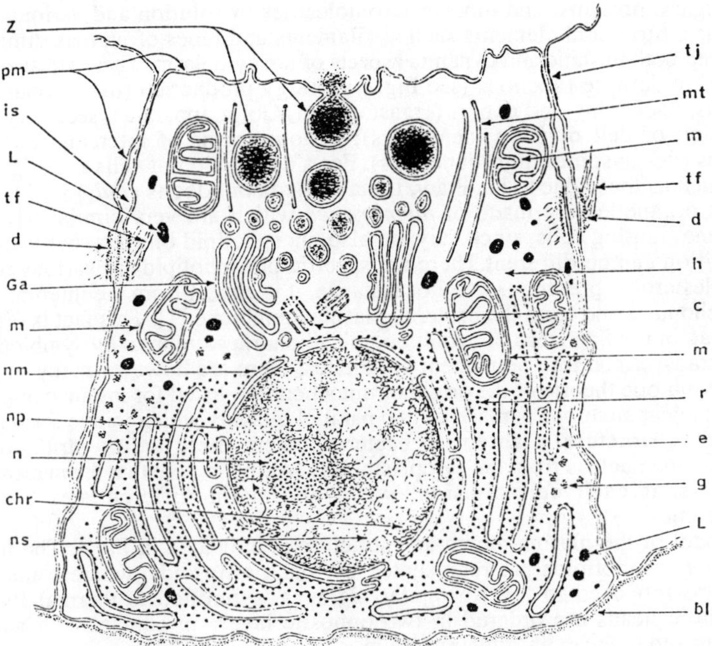

Figure 2. Schematic representation of a pancreatic cell involved in secretion of digestive enzymes. This cell type is close to the ideal generalized cell, with all fundamental organelles. These cells produce amylases for the hydrolysis of starch and other polysaccharides, lipases cleaving lipids into glycerol and fatty acids and proteases: carboxypeptidase breaking peptide bonds of small polypeptides, trypsin and chemotrypsin which cut heavier proteins, but are first secreted in an inactive form, to preserve cell structures and are further activated in the small intestine. All the represented machinery is involved in enzyme synthesis (reproduced from Bouligand, 1978). bl : basement lamina; c : centrioles; chr : chromatin; d : desmosome; e : rough endoplamic reticulum = ergastoplasm (note the presence of numerous ribosomes attached to the surface of this reticulum; ribosomes are complex systems of macromolecules involved in the synthesis of proteins); g : glycogen; Ga : Golgi apparatus; h : hyaloplasm = fraction of the cytoplasm extending between the organelles, being an isotropic sol in general and sometime a gel; is: intercellular space; L : lipidic droplet; m : mitochondrion with its internal partitions or cristae; mt: microtubule; n : nucleolus, a dense region of the nucleus which is rich in ribonucleic acid, and has its own DNA forming a fibrous network (the granular fraction is made of precursors of ribosomes, such as those visible at the surface of ergastoplasm); nm : nuclear membrane (double); np : nuclear pore; ns : nuclear sap (medium free between chromatin and nucleolus); pm : plasma membrane (bilayer); r: ribosome; tf : tonofilament attached to desmosome and forming a cytoskeleton with microtubules; tj: tight junction (electrophysiological role); Z : secretion of an enzyme complex: a zymogen filled vesicle.

The *cytoplasm* surrounding the nucleus consists mainly of an aqueous medium with salts, sugars, proteins, and other macromolecules in solution and various suspended organelles. Structural elements such as filaments and tubes of various dimensions are implicated both in static and dynamic aspects of the cytoplasm. There are also *organelles* involved in definite functions (see Fig.2) : energy production (mitochondria), protein synthesis, packaging and export (ergastoplasm, Golgi apparatus, secretory vacuoles), orientation of cell division (centrioles); various forms of intercell attachments or junctions (desmosomes, tight-junctions). For a description of cells and organelles, see books such as those due to Du Praw, Lima de Faria or Alberts *et al.* [31, 51,1].

Most organelles are made of *membranes* which are very similar to the plasma membrane limiting cells, since they are also phospholipid bilayers, but their chemical composition can be different.There are specific phospholipids in various proportions and cholesterol is present or not. For instance, it is absent from the internal membrane of mitochondria and from the bacterial plasma membrane [1]. This fact is often used to argue that, in the first steps of evolution, mitochondria were initially symbiotic bacteria, further transformed into permanent organelles; there are however many objections to such a symbiotic theorie of the origin of mitochondria, listed for instance in [28].

The nuclear envelope is made of two parallel bilayers which join to form the nuclear pores. This structure is absent in bacteria, where there are no structures sharply separating the nucleus from the cytoplasm. It is worthy to note that the nuclear envelope disappears at each cell division, when chromatin differentiates into distinct chromosomes.

A model of the plasma membrane structure is indicated in Fig.3. The hydrophilic heads of phospholipids are represented by black ellipsoids, whereas the paraffinic chains are supposed to form irregular zig-zags, submitted to thermal fluctuations. Hydrophilic heads are ordered in two opposite layers in contact with water in the cytoplasm and in the extracellular medium.

The biological bilayers are strongly *asymmetrical* in contrast with those simply produced by addition of water to lecithin. Phospholipids have different concentrations in the two layers of membranes and, for instance, in red cells, phosphatidylserine and phosphatidylethanolamine are more concentrated in the layer facing the cytoplasm. "Flip-flop" or diffusion of phospholipids from one layer to the other, through the fatty median level, appears to be slow. The asymmetric distribution of phospholipids is controlled by a mechanism involving enzymes called "flipases", which are proteins scanning the whole thickness of the bilayer, and form a more or less cylindrical canal [26,27].

The oriented transport of phospholipids through this canal is coupled with an hydrolysis of adenosinetriphosphate into adenosinediphosphate and inorganic phosphate. These proteins are therefore ATPases and the hydrolysis energy is converted into that required for a flip-flop of a specific phospholipid within the canal. Many other enzymes which have the form of a canal are involved in similar mechanisms and, for instance, unidirectional transport of ions, such as sodium, calcium, protons etc. These enzymes are called pumps and the energy for transport against a concentration gradient comes from the chemical metabolism of cells. The difference of electric potential across biological membranes originates from these oriented transports between the cytoplasm and the surrounding extracellular medium.

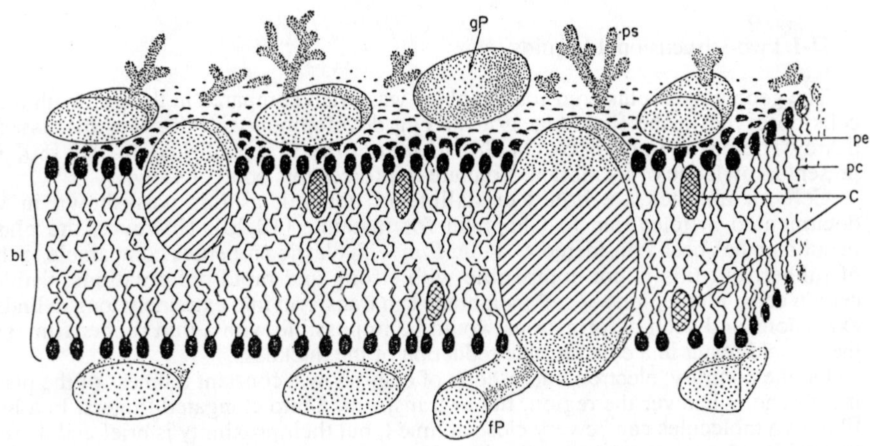

Figure 3. The phospholipidic bilayer (bi) of the plasma membrane (pc : paraffinic chains; pe : polar ends) and the associated molecules (C : cholesterol; fP : cytoplasmic fibrous protein; gP : globular protein and ps: polysaccharide). Certain proteins scan the whole thicknes of the bilayer. The stippled zones represent the polar residues of proteins and the hatched ones the nonpolar aminoacids. This model draws its inspiration from Singer and Nicolson [74], but differs by the absence of hexagonal packing of phospholipids and by the presence of surface proteins as in models of Danielli and Dawson [25] and Robertson [65]. Reproduced from Bouligand [11].

Molecular architectures observed within cells or in the spaces developed between cells are often reminiscent of structures described in liquid crystals. For instance, many biological materials are made of aligned fibrils, the symmetry of the system being that of nematic or cholesteric liquid crystals, and these structures are birefringent but not fluid in general. Examples of non fluid analogues are numerous in skeletal tissues. Nematic and cholesteric structures are also observed in spermatozoid nuclei and in certain types of chromosomes. I published several reviews on these themes [10-14]. Examples of double twist also are frequent and analogues of blue phase were described in the skin of certain worms [34,35,50].

Another domain remains to be reviewed and concerns membranes which form *lamellar, hexagonal or cubic stuctures,* very close to those studied in various water-lipid systems, which also are liquid crystalline.Water-lipid models are very simple compared to the multiphasic systems, with numerous types of macromolecules, encountered in cells and tissues and we will simply consider the geometric analogies between structures of these lyotropic liquid crystals formed with amphiphilic molecules and the arrangements produced by cell membranes.

II-THE FLUIDITY OF CELL MEMBRANES AND THEIR TOPOLOGY

II-1.Two-Dimensional Liquids

The geometry of clustered soap bubbles was often compared in the past with that of cells observed in the first stages of the egg development and this was discussed by d'Arcy Thompson in his book *On Growth and Form* first published in 1917 [80]. Our present knowledge of membrane structure reinforces this point of view.

Cell membranes are liquid cristalline in general, as this was shown in well documented articles due to Chapman for instance [18,20]. However, the plasma membrane has been considered as a thin film with elastic properties comparable to those of rubber, but this is not compatible with strong and rapid deformations visible in certain cells. Flat regions of cell membrane can transform into extremely long cylindrical extensions and such deformations would be impossible with a thin rubber film, since they would result in a considerable reduction of the thickness.

On the contrary, electron microscopy of cells show a constant thikness of the plasma membrane, whatever the region, flat or transformed into elongated fingers. In a liquid film, two molecules can be very close at time t, but their proximity is brief and there are no limiting distances other that the film dimensions themselves. In rubber membranes, the vicinity of two molecules is conserved for long times in general. The distance separating two neighbouring molecules is limited by the elasticity and rupture is unavoidable for longer distances.

The cell limiting membrane, the nuclear envelope and the various internal membranes observed in the cytoplasm are capable of extreme deformations under micromanipulation and this confirms their fluidity as shown in a series of works due to Chambers [16,17].

The liquid character of cell membranes is popularized now among biologists since the publication of an article entitled "the fluid mosaic model of the cell membrane" [74]. This model comes from the interpretation of experiments in immunocytochemistry, which show clearly the diffusion of proteins within the membrane plane itself and it is proposed in this work that cell membrane is an oriented solution of "amphipathic" proteins in a bilayer of phospholipids. Proteins are highly concentrated in areas of the plasma membrane which correspond to junctions between cells. They often form two-dimensional arrays, well vizualized in cryofracture. These lattices are either hexagonal or tetragonal and the fluidity of the membrane is probably abolished in these specialized zones, which correspond to a weak percentage of the total membrane area.

Holes are stable in thin rubber films, even if they deform under the applied tensions, whereas a liquid film disrupts when a hole appears and this is well known for soap films and soap bubbles. Cell membranes have a liquid behaviour similar to that of soap films, with the difference that soap films extend usually in air, whereas biological membranes separate aqueous compartments, as do bilayers produced by purified phospholipids or other amphiphilic components placed in contact with water.

Soap films and biological membranes have no free edge; they always form closed surfaces and this is a rule verified in cell structure, with no exceptions. It is worth remembering a demonstrative experiment which is one century old and is due to Pfeffer, inventor of the first osmometers [63]. He studied the red corpuscles of the blood, which are flattened biconcave vesicles limited by a membrane. These 'cells' have lost their main organelles (nucleus, mitochondria etc.) and are filled with a solution of haemoglobin mainly. Pfeffer tried to tear red cells with glass microneedles, but he only got smaller vesicles, with no apparent leak of hemoglobin in the surrounding plasma [63]. Hemoglobin solutions are simply prepared by exposing red cells to hypotonic saline solutions and further separation of the "ghosts", which correspond to the membrane vesicles after hemolysis.

II-2. The Cell Compartments

The topology of cell membranes is that of closed surfaces. The plasma membrane surrounding a cell has the topology of a sphere, and this is also the case for vesicles within the cytoplasm and the Golgi cisternae. The mitochondria topology is that of two concentric spheres, with a greater area for the inner membrane presenting numerous folds or cristae.

The nuclear envelope is topologically a system of two concentric spheres which are both fenestrated and join toroidally to form nuclear pores. It suffices therefore to enlarge one of the n nuclear pores to visualize the equivalence with a torus having n-1 holes.

A frequent event is the fusion of vesicles with membranes. Separations also occur and, for instance, of vesicle from a membrane after a process of invagination [1,33]. Exocytosis and endocytosis are well known examples of such topological changes and are represented in Fig.4 a and b. Many cells are involved in secretion processes and the acinar pancreatic cell represented in Fig.2 is a classical material.

Proteins are synthesized from specific messenger RNAs, by a kind of ribosome processed translation, and the polypeptide chains enter progressively the lumen of ergastoplasm and are transported to the Golgi region through the canalicular reticulum and vesicles as depicted in Fig.4 c [33]. Proteins are chemically transformed, and associated to various heteropolysaccharides, all along these steps. Some of these vacuoles condense into secretory granules and coalesce with the plasma membrane (exocytosis). It can be observed also that the ergastoplasm is often in continuity with the nuclear envelope (Fig.2). There is also evidence of continuity between ergastoplasmic cisternae, when they are observed in thick section and the ergastoplasm possibly forms a single compartment.

Many cell membranes present specific matching sites to facilitate the fusion with cytoplasmic vesicles and rosettes of protein particles are evidenced by cryofracture at the expected position of coalescence (Fig.4 d)

III-SMECTIC, COLUMNAR AND CUBIC SYMMETRIES

Cell membranes often differentiate into systems presenting symmetries which are close or similar to those of lyotropic liquid crystals and several examples are now discussed.

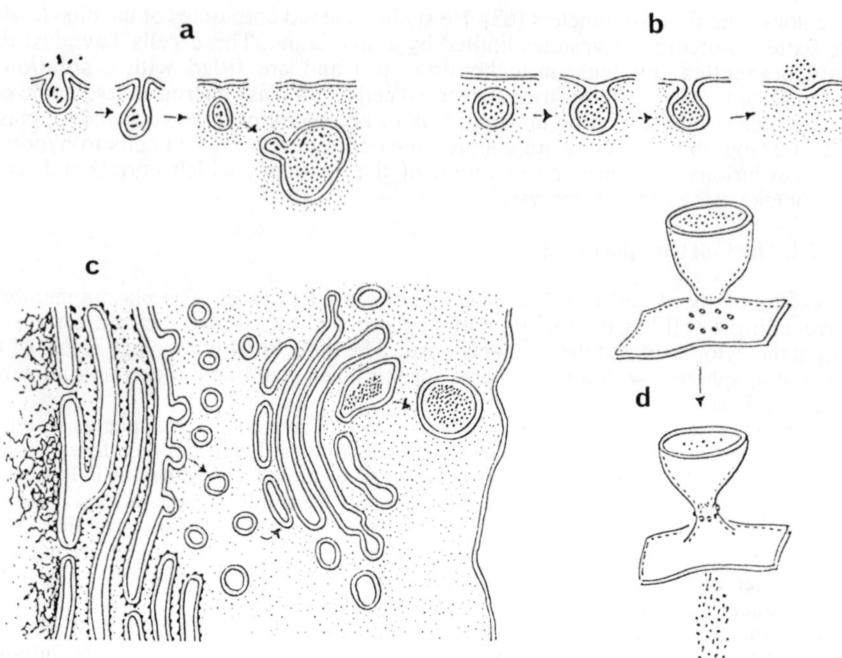

Figure 4. a and b : Schematic representation of exocytosis and endocytosis. The first process is involved in secretory discharge, whereas the second one corresponds to the uptake of external particles and fluids by encirclement with cell membrane and rehandling. c : Relations between the nuclear envelope, the ergastoplasm, the Golgi apparatus and the cell membrane in a secretory cell. The arrows indicate the path followed by the synthesized proteins, which are progressively processed into a secretion product; d : ring of protein particles prepared at a specific site, to facilitate the matching of the plasma membrane with the membrane limiting a mucocyst, a secretory organelle of the ciliated protozoa *Tetrahymena*. A simple hypothesis is that the protein particles decrease locally the splay elastic constant of the bilayer (model drawn after micrographs due to Satir [68].

III-1. Parallel Membranes

Membranes are often superimposed and lie more or less parallel according to the stacking density. An instance of loose parallel packing is seen in figure 2, with the ergastoplasm, but this figure is schematic. The alignment of ergastoplasmic membranes is drawn after a micrograph of a thin section in Fig.5a [33]. The spacing is rather regular despite the presence of various materials such as ribosomes and probably cytoskeletal filaments. Parallel bilayers with considerable aqueous intervals were observed by Nageotte [59] in preparations of brain lecithin in water and the interval separating bilayers depends on water concentration [69].

The distribution of cristae in mitochondria is another example of loose parallel packing as shown in Fig. 5b. In the case of the Golgi apparatus, the parallelism of membranes is sharp in the central region, whereas it is loose at the periphery, where it transforms into a system of branched canaliculi (Fig.5c).

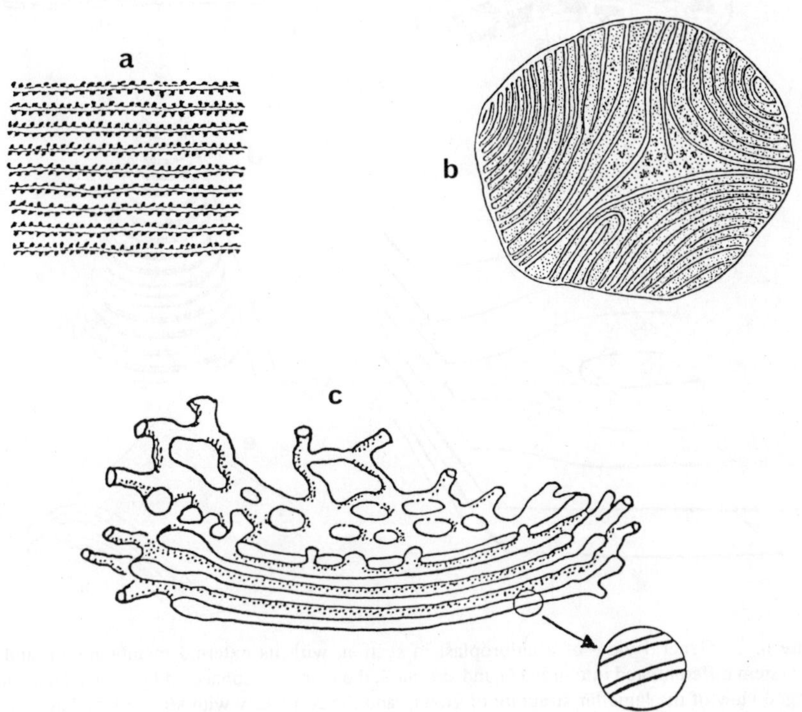

Figure 5. Examples of membrane alignment within the cytoplasm and its organelles. a : Cross section of ergastoplasm within a pancratic cell, redrawn after a micrograph in [33]; b : section of a mitochondria with parallel internal folds or cristae; edge dislocation and -πdisclination in a mitochondria of a brown adipose cell from the bat, *Myotis lucifugus* after a micrograh in [33]; c : structure of the Golgi apparatus of Sertoli cells of the rat. Redrawn after Rambourg *et al.* [64].

Membranes form regular parallel stacks (Fig.6) in various types of photoreceptors such as chloroplasts (plant cell organelles involved in photosynthesis) and cones or rods (from visual cells).

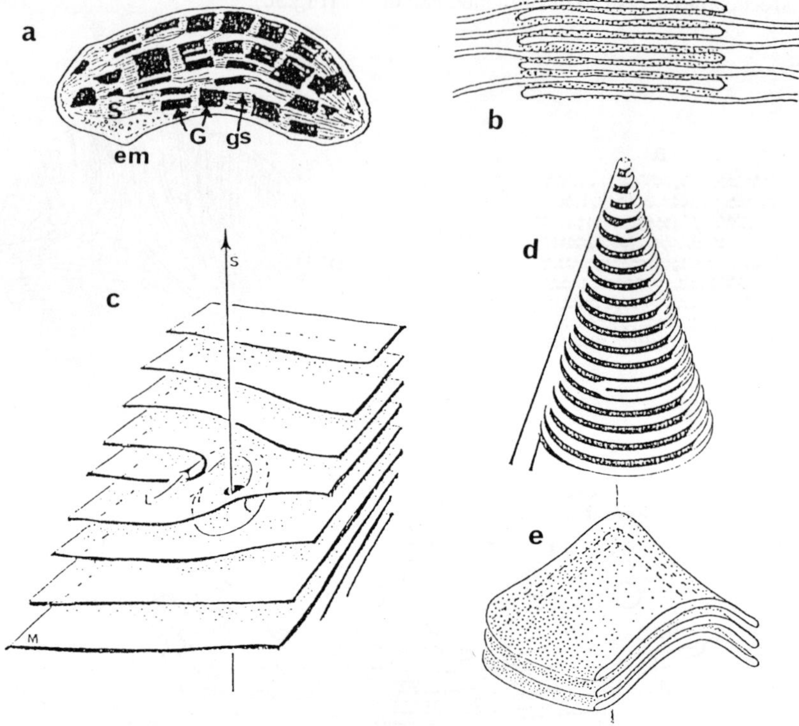

Figure 6. a : General view of a chloroplast in section, with its external membrane em and the internal system differentiated into grana G and stroma S; the grana are connected by stroma lamellae gs. b : enlarged view of the lamellar structure of grana, and the continuity with stroma lamellae [46]. c : screw dislocation connecting the stroma lamellae. d : edge dislocation corresponding to the cleavage of a photoreptor disk in a cone outer segments of the tadpole retina, after [32]; for a complete representation of visual cells, see [1]. e : conic deformations of nested disks in a rod outer segment of the human retina. Representation suggested from micrographs in [45,48].

Chloroplasts are limited by an external membrane, which is topologically equivalent to a sphere, whereas the internal membrane limits a compartment named *stroma*.. The internal membrane folds and forms a complicated system containing the chlorophylls and other important pigments (Fig 6a). These membranes are densely assembled into *grana* and loosely arranged in the stroma between the grana, as shown in Fig.6b.

Several studies indicate that membrane pairs are connected by screw-dislocations [61] within the stroma (Fig.6c) and it appears that the internal membrane system is continuous and separates two compartments, a topological situation comparable to that of mitochondria. It has been often supposed that chloroplasts are modified mitochondria or the converse or that these organelles have a common origin, progressively transformed symbiotic bacteria.

Visual cells present in the vertebrate retina are called rods or cones, owing to the characteristic shape of their outer segment, the photoreceptor, which is a cylindrical or a conical stack of discs or flattened vesicles [1]. New discs form by the nucleation of edge dislocations [32] in this system, which is a smectic analogue (Fig.6d). Nested conical deformations are frequent in rods [45,48] and this structure reproduces locally that of a focal conics, as in smectics (Fig.6e). There is a possible role of these defects which is to concentrate light and to facilitate the photochemical reactions to be initiated.

Another analogue of smectic liquid crystals is the myelin sheath wrapped around the axons of white nerves of the central nervous system. The myelin speeds up the propagation of action potentials along axons and is the result of an extensive development of the membrane of adjacent 'Schwann cells' which form a spiralized coating as can be seen from Fig.7. Myelin is thus a stacking of plasma membranes.

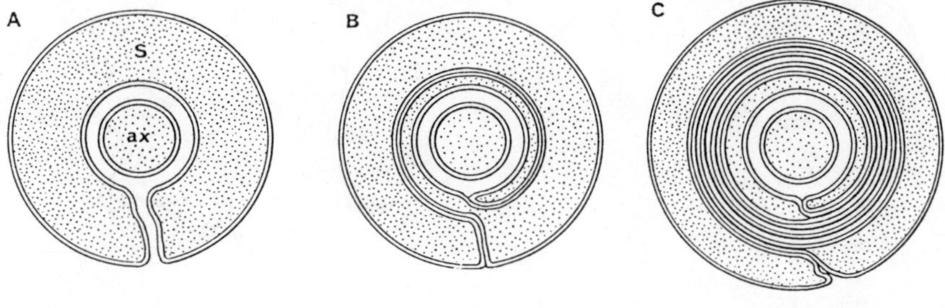

Figure 7. Cross sections showing the origin of myelin around an axon (ax). Axons are cylindrical extensions of nerve cell involved in the transmission of impulses. Axons are often surrounded by glial cells called Schwann cells : S. The plasma membrane of Schwann cells develops extensively and forms a spiralized coating , as shown in the three steps represented in A, B, C; redrawn after Geren and Robertson [36,65].

Axons are polarized systems, since there is a definite direction from the cell to the axon end, which is generally that of propagating action potentials. However, the two possible orientations of myelin spirals are observed in micrographs, for axons belonging to the same nerve[8]. The myelin sheath is perfect and there are neither edge nor screw-dislocations. Focal conics are absent and there are no disclinations.

Phospholipid-rich smectic liquid crystals are produced by specialized cells of the alveolar epithelium of the lung, to form a fluid secretion which is the pulmonary surfactant. The smectic secretory granules observed before secretion within the cells (pneumocytes II) are called multilamellar bodies and their examination in thin section shows clearly the presence of dislocations, focal conics and disclinations $+\pi$ or $-\pi$, the three main types of defects encountered in smectic liquids [75,76,86,87].

III-2. Hexagonal and Square Arrays

Shape transformations are often important in cell membranes, as discussed above. Strong deformations are frequent in red cells, for instance, and particularly the passage from the 'discocyte' to the 'echinocyte' (Fig.8), when a biconcave disk transforms into a crenated corpuscle, without any variation of volume and cell area [6,15]. The lateral projections of red cells are obtained *in vitro* by the addition of molecules able to intercalate asymmetrically within the bilayer, what modifies the spontaneous curvature.

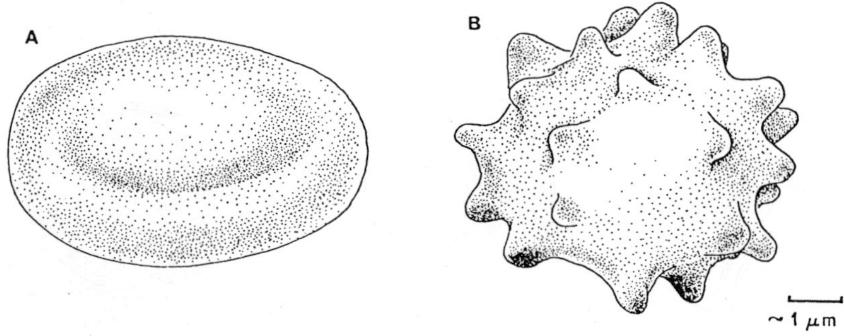

Figure 8. In diverse conditions, the normal biconcave shape of red cells or discocytes, represented in A, can transform into a spiculated form called echinocyte, as shown in B. After [6] and [15].

Several models were proposed to explain the polymorphism of red cells and are based on fluidity and asymmetry of cell membranes [73].The fingers which differentiate in echinocytes resemble microvilli, which will be considered now. In the intestinal epithelium, most cells are specialized for absorption and present a 'brush border' made of closely packed cylindrical microvilli of equal diameter and length, as represented in fig.9. The packing is generally hexagonal. The bilayer structure of membranes limiting these fine cytoplasmic digitations is well evidenced in cross section [33]. Microvilli are specializations of the cell surface, which stongly increase the absorptive area. Each microvillus contains a bundle of parallel actin filaments, whereas thin branching polysaccharides form an external mat developed mainly at the tip [33]. Some other characters underline the asymmetry of the membrane and, for instance, the endocytosis often observed at the basis of microvilli. There are examples of microvilli containing delicate extensions of the endoplasmic reticulum and each microvillus corresponds to a set of two coaxial cylinders, each one made of a bilayer.

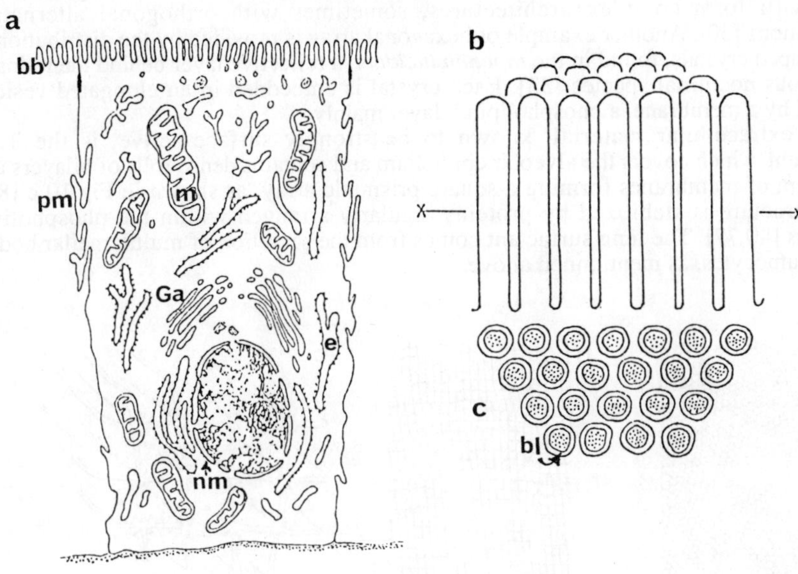

Figure 9. a. Schematic drawing of an enterocyte, captions as in fig.2. Note that cell membranes are represented by a single line. Their bilayer character is not recalled here, as in fig.2. The cell membrane is differentiated into a brush-border, along the lumen of the intestine. b. A perspective of the brush border is represented and its hexagonal array appears in cross-section at the level x--x, as shown in c. ac : actin filaments in cross section; bb : brush border; bl : bilayer limiting microvilli; mv : microvilli.

In several groups of invertebrates, the brush border of the midgut epithelium produces a fine lattice of chitin fibrils associated with proteins, called 'peritrophic membrane'. This fibrous lattice is assembled at the tip of microvilli and separates from the epithelium, to form a continuous wrapping of the gut content [62]. The secretion possibly occurs at the basis of microvilli [58]. The peritrophic membrane can be mounted onto a carbon-coated grid for electron microscopy and shows the hexagonal or the quadratic symmetries of the array of microvilli (Fig.10 a and b). The epidermis of most invertebrates also develops a system of microvilli which probably plays an important role in ordering the secretion of filaments which associate to form a protective external layer, named cuticle [9]. The mechanism is similar to that involved in the formation of peritrophic membranes, but cuticles are thicker in general [50].

We now indicate briefly some comparable examples of parallel packing of microvilli or tubes. Brush borders are common in epithelia, not only in the digestive tract, but also in the renal convoluted tubule and in many other organs [33]. Hexagonal arrays of microvilli are observed in photoreceptor cells of the compound eye of arthropods. These microvilli form complex architectures, sometimes with orthogonal alternating orientations [30]. Another example of hexagonal array is provided by the distribution of rod-shaped crystals, found in the *tapetum lucidum,* a reflective layer behind the retina of numerous nocturnal species [33]. Each crystal is embedded in an elongated vesicle, limited by a membrane, a phospholipid bilayer mainly.

An extracellular material, known to be strongly surface-active, is the 'lung surfactant' which covers the alveolar epithelium and contains dense rolls of bilayers and a system of membranes forming a square prismatic array, as shown in Fig.10 c [83]. This structure is stabilized by proteins regularly arranged within the phospholipid bilayers [40,77]. The lung surfactant comes from the secretion of multilamellar bodies by pneumocytes, as mentionned above.

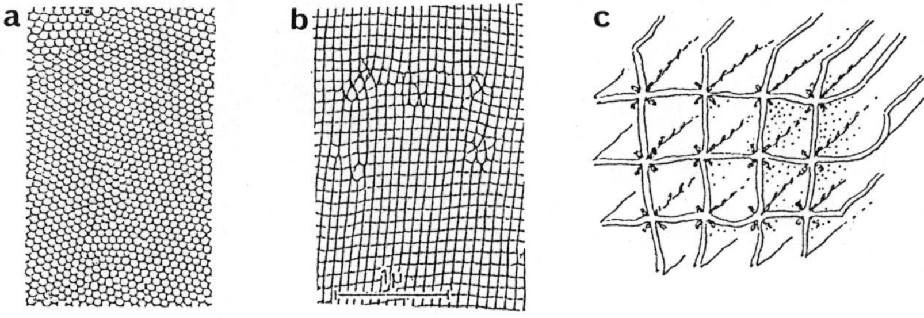

Figure 10. a. The peritrophic membrane and its hexagonal lattice in *Orictes nasicornis,* Coleoptera [62]. b. Example of square grid in the peritrophic membrane of *Perlodes intricata,* Plecoptera [62]. c. Perspective view of the prismatic myelin in the lung surfactant of a fetal rat. The proteins P are supposed to stabilaze the square pattern [40,83].

Mitochondrial cristae are often tubular and can form a more or less regular hexagonal packing (Fig.11 a). There are also examples of cristae which are prismatic, with a triangular section, and are aligned hexagonally (Fig.11 b).

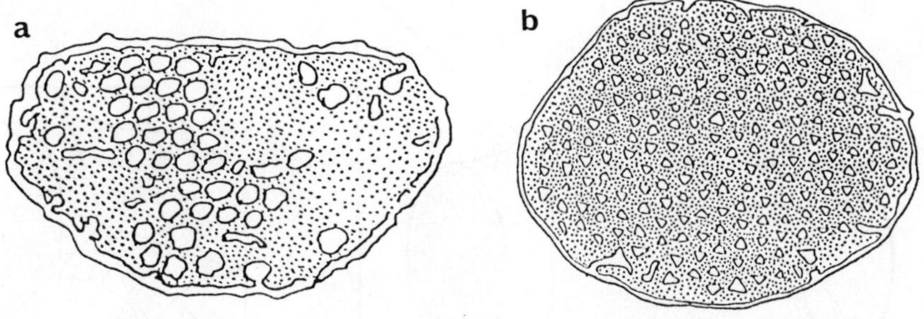

Figure 11. Two examples of mitochondria with tubular or prismatic cristae. a. Mitochondrion from testicular Leydig cells of the ground squirrel; the hexagonal packing is seen in an oblique section. Drawn after a micrograph due to J. Pudney [33]. b. Mitochondrion from an astrocyte, in the hamster brain. Whereas the hexagonal lattice is regular. Drawn after a micrograph [7,33].

III-3. Cubic Membrane Systems

There are examples of cell membranes resembling periodic minimal surfaces, with cubic symmetries, often studied by mathematicians and crystallographers [43,44,57,70,71,79]. In such systems, the bilayer is continuous and separates two sub-spaces filled mainly with water. Cubic structures of this type are observed within cytoplasmic organelles such as mitochondria [33,47], chloroplasts [37,38,56], vesicles [41] etc. These surfaces are considered elsewhere in this book, particularly at the chapter devoted to 'stratified fluids' by Charvolin and Sadoc. Amphiphilic molecules assemble into bilayers submitted to geometrical constraints, and form periodical arrangements either in one or in two or in three dimensions. Such systems have the symmetries of smectic, columnar and cubic liquid crystals. Among the inventoried infinite periodical minimal surfaces (IPMS), which are free of self-intersection, there are only three, called respectively P, F and G, which are expected to occur in the cubic phases of lyotropic liquid crystals [22]. Let us consider the P surface, the simplest one to visualize, whose unit cell is represented in figure 12a. The regular packing of such units is centered cubic, since the space is divided into two parts, which are superimposable by a translation along a half long diagonal of the cube. The mean curvature of minimal surfaces is: $1/R_1 - 1/R_2 = 0$ and the two curvature radii are therefore equal and opposite ($R_1 = R_2$). Each cell is composed of eight subunits where the minimal surface is limited by a series of circular arcs of radius $R_1 = R_2 = R$, forming a curvilinear hexagon, as shown in fig.12b. The lines of curvature are drawn and present a star singularity at the intersection with the threefold symmetry axis.

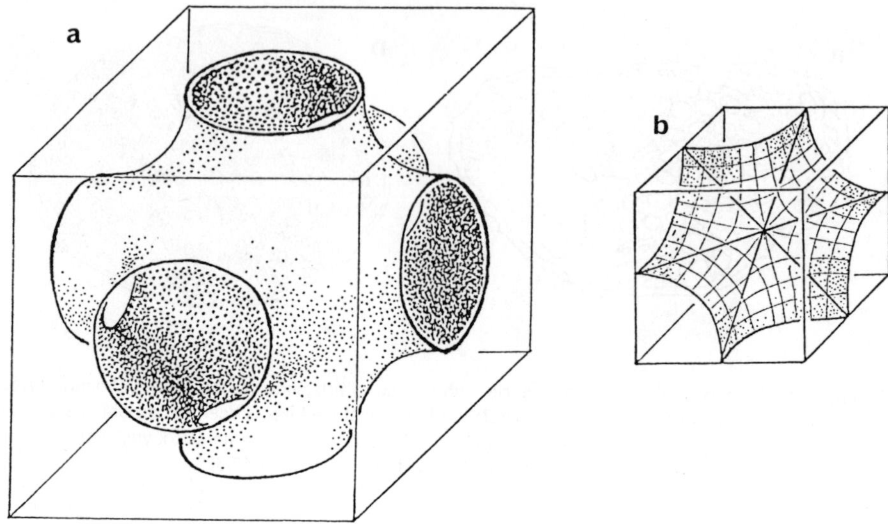

Figure 12. The P surface . a. A unit cell. b. A subunit of the cell with its curvature lines.

A membrane system very close to the P surface is observed in etiolated plant cell chloroplasts and is called 'prolamellar body' [46]. The set of grana described above, in figs 6 a and b, is then replaced by a cubic lattice, which gives various patterns in thin section, depending on the cutting plane orientation. These patterns were carefully studied by Günning [38,39] and the corresponding structure is represented in Fig.13.

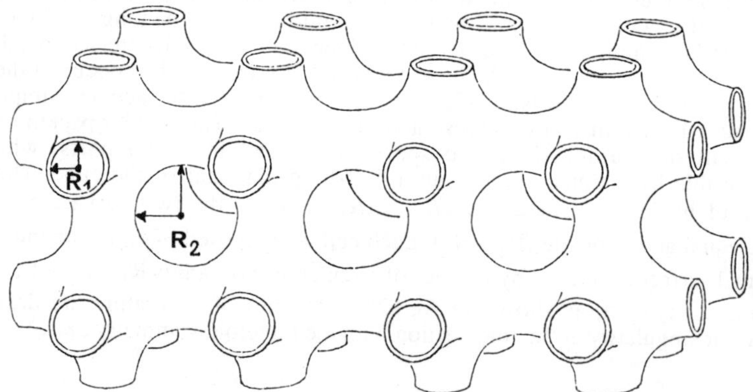

Figure 13. Perspective of the membrane surface in a chloroplast prolamellar body. Two different radii R$_1$ and R$_2$ are observed in the symmetry planes. Redrawn after Günning [38].

The mean curvature $1/R_1+1/R_2$ differs from zero in this membrane network and is probably constant in a first approximation. This suggests the existence of a difference Δp in the osmotic pressure between the two compartments. The structure is simple cubic and offers a new illustration of the cell membrane asymmetry. In these systems, the membrane folds along a surface which is not far from being parallel to a P surface. Let be M a point of P and μ the oriented normal to P at point M. We define m on μ, such as Mm = ΔR. We call p the surface obtained with m, when M describes P, ΔR being constant and less than R, the radius of P in the mirror planes. The two curvature radii of p are R+ΔR and R-ΔR, in these planes, and the mean curvature is constant, being approximately $2\Delta R/R^2$. The p surface is said to be parallel to P and differs from P surface by the absence of centered cubic symmetry.

The set of fourfold axes of the P surface forms a system of two intertwined labyrinths, which are equivalent in the centered cubic symmetry, even if they are represented in black and in white to well distinguish them in Fig.14a. These fourfold axes are conserved in the p surface and accordingly the two intertwined labyrinths, but they cease here to be equivalent.

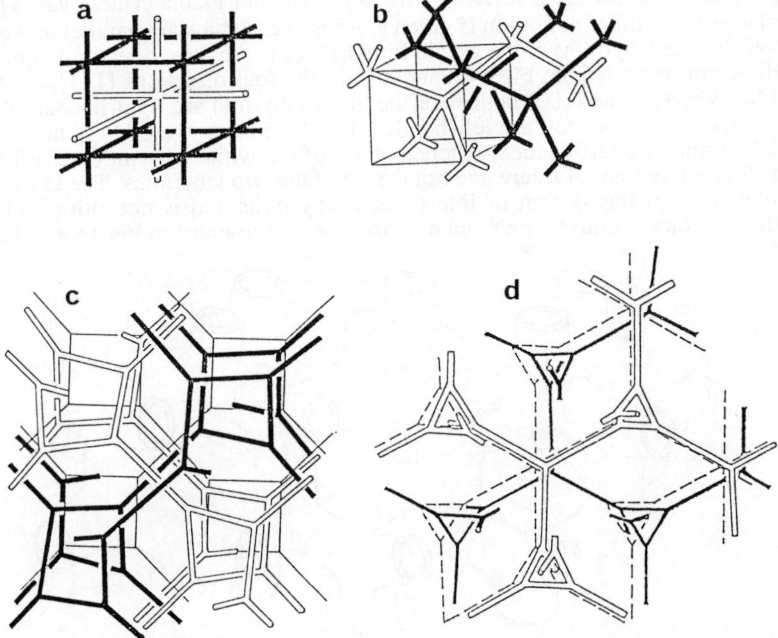

Figure 14. Intertwined labyrinths of the three IPMS, the P labyrinths represented in a, the F ones in b, corresponding to the diamond lattice, and the G labyrinths observed either along the [100] direction (c) or along the [111] direction (d). Reproduced from [54,55,78].

210

The membranes of prolamellar bodies in chloroplasts do not differ significantly from p surfaces. The asymmetry of membrane is underlined by the presence of ribosome-like particles in one of the two compartments, that which is in continuity with the stroma [38]. Surfaces p are encountered in some mitochondria of tumours affecting striated muscle cells [41]. Another instance of p surface is given by calcite plates or spicules which form the skeleton of echinoderms, these well known marine invertebrates, such as sea-urchins, sea-stars, sea-cucumbers etc., with a remarkable pentagonal symmetry of the whole organism. Most skeletal elements of these animals are monocristals of a magnesium rich calcite, each cristal being not limited, as expected, by planar faces but by complex curved surfaces, and there are examples of p surfaces, which probably correspond to the geometry of internal cell membranes [29,60].

The F surface is the centered cubic IPMS separating two intertwined and equal compartments, whose labyrinths correspond to the diamond lattice (Fig.14b). Whereas the P labyrinths associate equal segments which join six by six, the F labyrinths associate them four by four. Similarly the G surface or *gyroid* separates two equal compartments whose labyrinths associate segments which join three by three. Two different aspects of these G-labyrinths are presented in figures 14 c and d, according to the viewing direction [100] or [111]. We define f surfaces as being parallel to a F surface, as we did for the p surfaces, with a ΔR smaller than a critical value to avoid singularities. A similar definition is given to g surfaces, which are parallel to the gyroid.

Note that the labyrinths of the gyroid form helices in the three directions equivalent to [100], as can be seen from Fig.14c, and also in the four directions [111], as shown in Fig.14d. Whereas, the [100] helices of the black labyrinth are right-handed, the white ones in the same direction are left-handed. On the contrary, the [111] helices of the black labyrinth are left-handed, whereas those of the white labyrinth are left-handed. Each labyrinth is a chiral figure and not the set of the two labyrinths. The G surface has the symmetry of the system of intertwined labyrinths and is not chiral, whereas a parallel g surface is chiral. The f and g surfaces are represented in Fig.15 and 16.

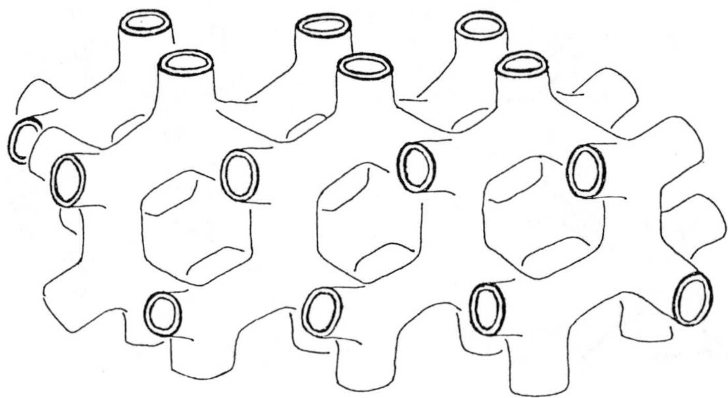

Figure 15. Perspective of a 'f surface'.

Symmetry planes are present in the f surface. On the contrary, there are neither mirror planes, nor symmetry centres in the g surface. We think that examples of f surfaces exist in mitochondria of certain specialized cells in seminiferous tubules in some vertebrates [33,47]. The occurence of g surfaces (Fig.16) is less obvious, but a possible example is provided with the endoplasmic reticulum which form 'photogenic grains' in luminous cells of certain invertebrates [2-4]. Since these membrane crystals are chiral, the question as to whether the emitted light presents a circularly polarized component might be asked.

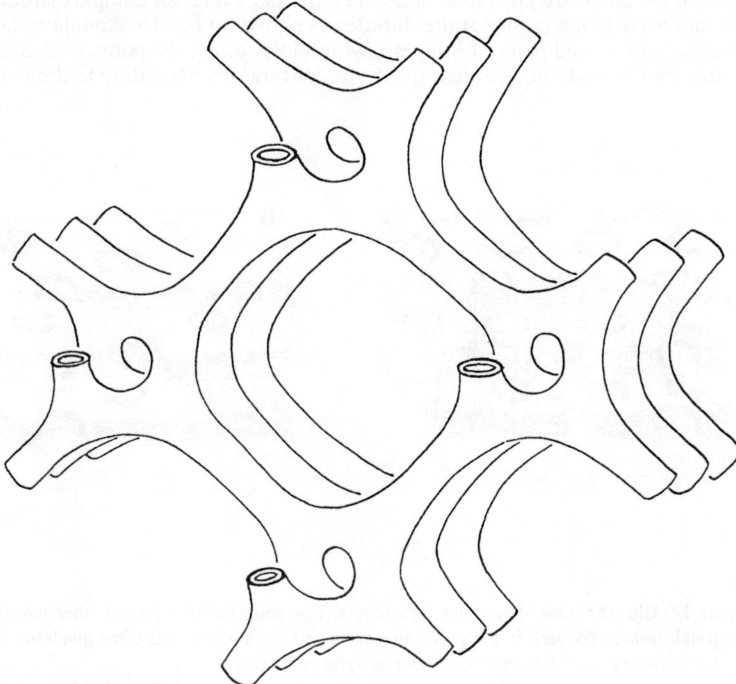

Figure 16. Perspective of a 'g surface'.

IV-OTHER SYSTEMS

Many different architectures are produced by cell membranes, showing crystal symmetries or liquid crystal symmetries, but there are many other anisotropic or isotropic systems, with extremely differentiated morphologies. We briefly review a few examples.

212

IV-1. Annulate Lamellae

These organelles were first observed in cytoplasm of sea-urchin eggs, as birefringent stuctures, and the ultrastructural studies showed that they probably originate from the nuclear envelope [review in 33].These lamellar systems are present in many other germ cells, namely in vertebrate oocytes and in spermatogonia, lying often within the nucleus itself. Annulate lamellae are also found in Sertoli cells of seminiferous epithelia in the human testis, and in the myocardium of the chick embryo. The characteristic pores of the nuclear envelope are present in annulate lamellae with their complex structure, a kind of cylinder associating eight parallel tubules as shown in Fig.16. Annulates lamellae are made of superimposed pairs of bilayers, which join along the pores whose distribution is various (hexagonal, more or less quadratic, or random, according to the material).

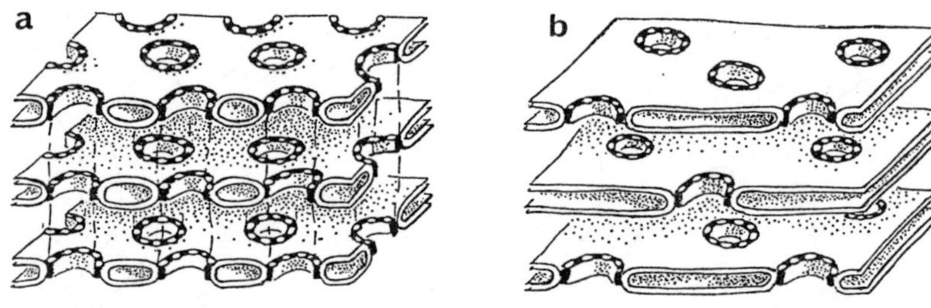

Figure 17. Ultrastructure of annulate lamellae. a. The pores are facing each other and the symmetry is hexagonal (human oocyte). b. the pores alternates and the symmetry is often quadratic (Sertoli cells in man for instance). Reconstructed after micrographs published in [33].

IV-2. Tubes and Lamellae with Nematic Symmetries

Cytoplasmic membranes form either flattened sacs or tubes. The usual examples of superimposed sacs are the Gogi apparatus and the rough endoplasmic reticulum or ergastoplasm (Fig.2). The term 'rough' comes from the presence of small particles which are ribosomes and are as sprinkled at the surface of membranes. These ribosomes are involved in protein synthesis. Another form of endoplasmic reticulum is said 'smooth', since there are no particles at the surface. This smooth reticulum forms tubes, which arrange irregularly, bifurcate here and there and anastomose, as indicated by the term 'reticulum'. In numerous cases, the tubes show a preferential orientation and the whole symmetry is therefore that of a nematic liquid crystal (Fig.18a). This symmetry is also observed in sets of parallel thin axons, limited by cell membranes in close aposition (Fig.18b). On the contrary, there are examples of bilayers forming flattened sacs, which are irregularly superimposed (Fig.18c). Anastomoses are often observed. The whole symmetry is nematic, since layers do not show any periodic distribution. The uniaxial nematic symmetry of these two systems have opposite signs. The existence of membrane sets with a global biaxial nematic symmetry is not excluded but, to this date, we do not know any examples. The origin of nematic symmetries is not clear in general. The preferential alignment along a direction can be due to an elongation or to any other cell deformation, and is not necessarily the result of a spontaneous alignment.

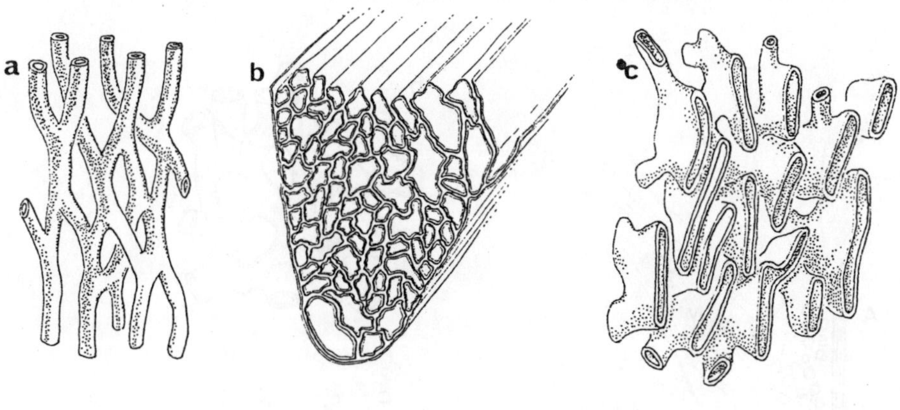

Fig.18. Membrane systems with a global nematic symmetry. a. Smooth endoplasmic reticulum, with a preferential orientation. b. Bilayers limiting parallel axons. Drawn after a micrograph [33]. c. Irregularly superimposed and anastomosed membrane sacs.

IV-3. Isotropic systems

Examples are numerous of cytoplasmic domains filled by isometric vesicles, limited by a bilayered membrane. This is the case in synapses, this specialized area maintaining a close contact between an axon terminal and another neuron (Fig.19a). Such aggregates of spherical vesicles are frequent in other cell types.

There are examples of membrane systems resembling prolamellar bodies described above, but instead of being cubic, they are amorphous and show no periodic structures, without any preferential alignment. For instance, charasomes are organelles found in a family of fresh-water algae, whose role is still uncertain [52]. The membrane separates two intertwined aqueous domains, one being more developed than the complementary one, what underlines the membrane asymmetry considered above (Fig.19b).

Prolamellar bodies often show pentagonal patterns, instead of hexagonal, as they ordinarily appear in the plane (111). These pentagons do not arrange into 'quasi-crystals' and generally associate irregularly, the whole system being isotropic (Fig18c). According to Günning [39], certain regular lattice correspond to a dense packing of slightly deformed pentagonal dodecahedra, associated with 'disclined dodecahedra', the term adopted by Charvolin and Sadoc, chapter 5 and [23]. See also [84]. Such lattices, if they exist, have a high density of defects and are globally isotropic. We are in a situation often encountered in amorphous materials, highly structured at the molecular level, with short range order.

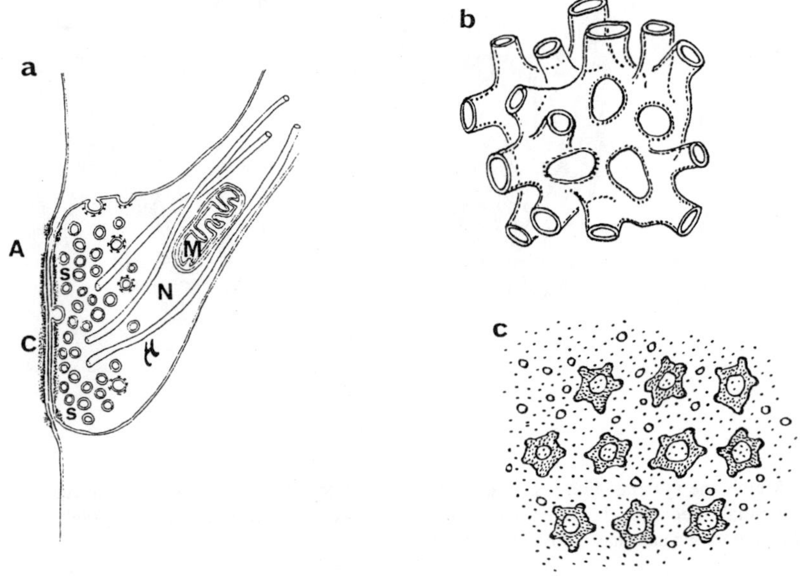

Figure 19. Isotropic membrane systems in cells. a. Synapse between axon A and neuron N; s: sysnaptic vesicles; C: synaptic cleft, where the bilayer is stabilized by an hexagonal or trigonal array of proteins, not represented here; μ: microtubules; M: mitochondria. b. Charasome, a randomly reticulated system of membranous tubes. Drawn after a micrograph [52]. c. Pentagonal patterns observed in prolamellar bodies; drawn after Günning micrographs [38].

V-COMPARISON WITH *IN VITRO* WATER-LIPID SYSTEMS

Amphiphilic molecules are made of one or several paraffinic chains attached to a polar head. Such molecules stongly decrease the surface tension between oil and water, since they assemble into thin films extending along such interfaces. Ternary mixtures are experimented, associating these molecules with water and a component mainly paraffinic, sparingly miscible with water : the decanol, for instance.

Among investigated amphiphilic molecules, let us quote potassium and sodium soaps, saturated or not, and various detersives. Formulae of most experimented amphiphilic molecules are indicated in Fig.20.

a) $CH_3 (CH_2)_{10} CO_2K$

b) $CH_3(CH_2)_9OSO_3Na$

c) 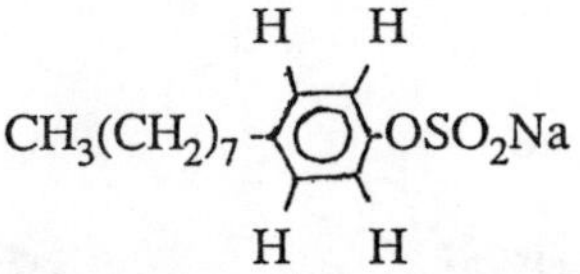 $CH_3(CH_2)_7 -\bigcirc- OSO_2Na$

d) $CH_3 (CH_2)_{11} N(CH_3)Cl$

Figure 20. a. Potassium laurate. b. Sodium decyl sulfate. c. Sodium octyl benzyl sulfonate. d. Chloride of dodecyl trimethyl ammonium. Reproduced from [21].

Phospholipids also can be purified and observed in presence of water and water insoluble components such as cholesterol for instance. In all these examples, the amphiphilic molecules lie generally perpendicular to the interface between water and oil. They form more or less separated units, with definite shapes which are spheres, elongated or flattened ellipsoids, cylinders, ribbons, bilayers and complex entities, as those encountered in cubic phases [24,53,85]. They associate into more or less ordered phases, which are liquid crystalline, even when they form three-dimensional lattices. Rapid diffusion of amphiphilic molecules within these lattices has been observed, even if such phases are extremely viscous [24].

216

V-1. Main Structures in Amphiphilic Mesophases

The structures observed in water-lipid systems are represented in Fig.21 and 22, with the exception of IPMS and related architectures considered in Figs.12 and 14. The basic structures are shown in Fig.21 and the less frequent situations in Fig.22.The symetries are body-centered cubic (Fig.21c, and also 12 and 14), columnar hexagonal (Fig.21d,e and Fig.22b) or rectangular (Fig.22a), smectic (Fig.21f) and nematic (Fig.22c,d).

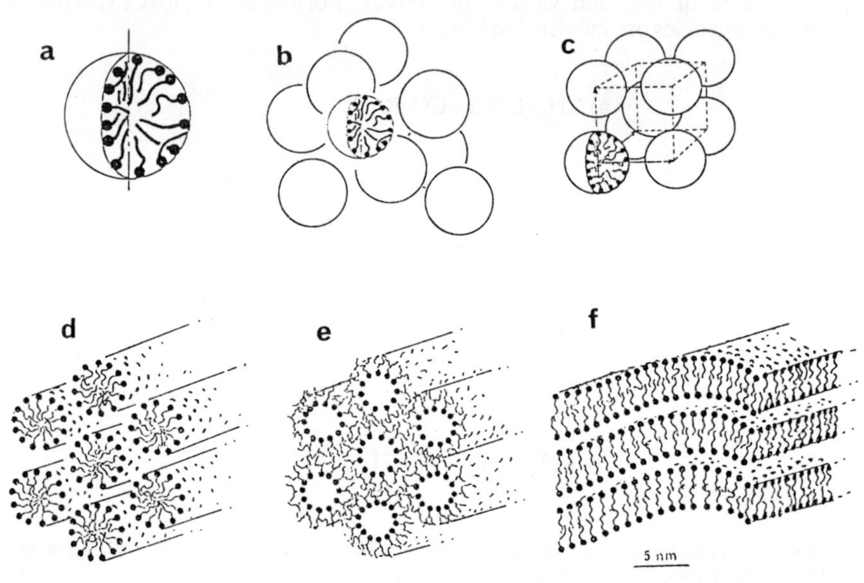

Figure 21. Some of the basic structures observed in water-lipid systems. Figures redrawn after [21,37]. a. Spherical micelle. b. Body centered cubic association of spherical micelles. c and d. Hexagonal association of cylindrical micelles. e. Stacking of parallel bilayers. The core of spherical micelles is paraffinic, but can be polar, with different compositions of the lyotropic preparation. The core of cylindrical micelles is paraffinic in d, whereas it is polar in e.

The figures 21 and 22 mainly concern systems with paraffinic cores embedded in an aqueous environment. The reverse situations also occur, with polar cores in a hydrocarbon environment, as in Fig.21e. The case of superimposed bilayers is particular, since it coincides with its own reverse. These various structural elements and their three-dimensional associations are discussed by Charvolin and Sadoc, particularly about the problem of their origin [22,66].

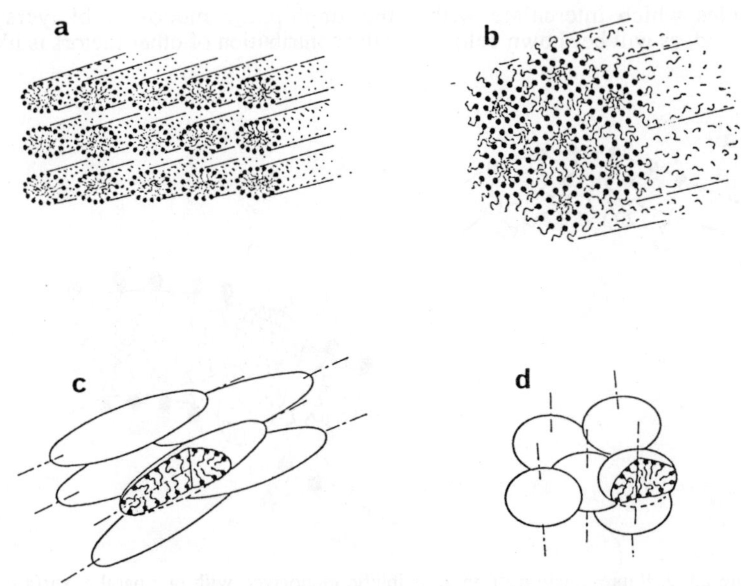

Figure 22. Some other structures encountered in water-lipid systems. Figures redrawn after [21,37]. a. Orthorhombic association of parallel ribbons. Such a system was described by Goodman and Clunie, in the potassium oleate + water system [37]. The original sample contained 45% water at 25°C and was osmium tetroxide fixed. b. Complex hexagonal phase from the potassium oleate + water system. The original sample contained 42% water at 25°C and was osmium tetroxide-fixed [37]. c. Elongated micelles in a uniaxial nematic phase. Actually the length of these micelles is not known. Some of them are possibly indefinite or anastomose as structures represented in Fig.18a. d. Flattened micelles in a uniaxial nematic. Connections between these micelles are not excluded, as in Fig.18c.

A polar head occupies a mean area which differs from that of the corresponding paraffinic chains. According to their concentrations, water molecules intercalate here and there, between polar heads, and increase the area per polar head, whereas oily components added to the system modify the density of paraffinic chains of the amphiphilic molecules. In such conditions, the polar level expands relative to the

218

paraffinic one or the converse and this induces a splay effect at the interface, depending on compositions of the oily and the aqueous phases. Among the useful parameters to consider, there are the density of polar heads and that of the corresponding paraffinic chains, along two parallel surfaces representing the mean levels of these two differentiated regions of the amphiphilic monolayer (Fig.23a). The relative volumes of the oily and of the aqueous phases are also important. The concentrations of certain molecules which intercalate within the amphiphilic mono- or bilayers must be considered, as will be shown below, and the contribution of other factors is likely.

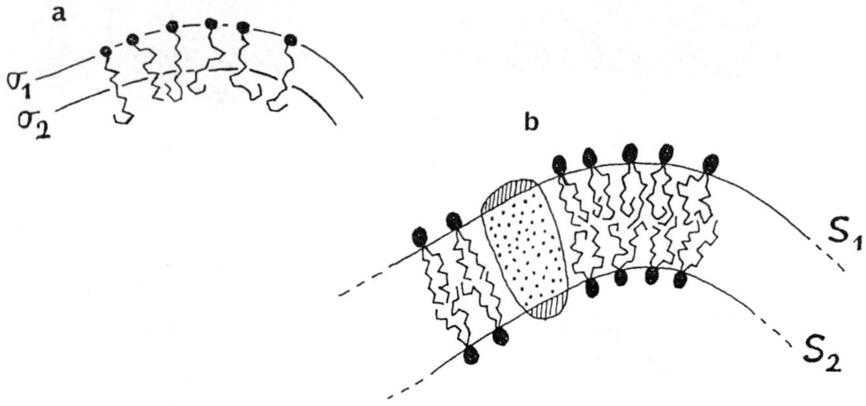

Figure 23. a. Representation of an amphiphilic monolayer, with two parallel surfaces s_1 and s_2 corresponding to the mean levels of the polar heads and the paraffinic chains. b. Representation of a cytomembrane, with two parallel surfaces S_1 and S_2 separating the paraffinic region from that of polar heads.

When polar heads occupy an area which is much more than the area of the corresponding hydrocarbon chains, and when the volume of the oily phase is reduced, at less than 20% for instance, the spherical configuration of Fig.21a is expected and bcc arrays of spheres form in certain cases (Fig.21c), but disaggregate in an excess of water (Fig.21b). If the differences between the relative areas and volumes are less marked, a hexagonal arrangement of cylinders appears (Fig.21d). Lamellar phases are obtained when the surface densities of polar heads and paraffinic chains are comparable.

When the percentage of oil is increased, the reverse structures may appear, according to the opposite order and this is reproduced schematically in Table I, indicating the transitions, but all phases do not necessarily differentiate. Cubic arrays with labyrinths as represented in Fig.14 are not always present but, if they differentiate, their domains are observed between hexagonal and lamellar phases.

This table comes from an idealized diagram, proposed by Winsor and Scriven [72,85] for a binary mixture, prepared with an amphiphilic compound and water. Table 1 is purely theoretical, and indicates the relative positions of phases. The connexity of the water domain and of the oily domain depends mainly on the volume ratio of the two phases. In the lamellar system, the connexity of both phases comes mainly from screw- and edge-dislocations, but also from disclinations and other topological situations.

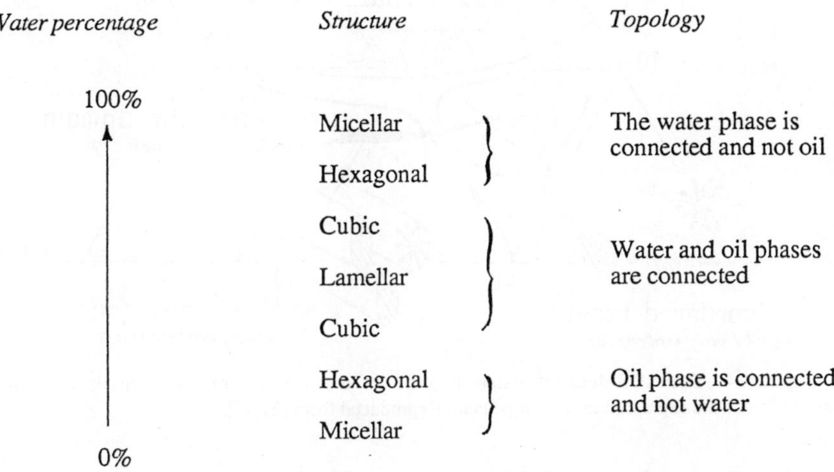

Table 1. Idealized succession of structures in a water-amphiphile binary mixture.

We already indicated the role of factors different from the volume ratio and it is clear that the phase space is not limited to one dimension. Various situations are not considered in table 1, and for instance, that of parallel ribbons in columnar structures, or the various nematic structures, uniaxial as represented in Fig.22 and even biaxial in certain cases. Nematic phases occupy small domains in diagrams, when they exist, but they are generally absent. The position of nematic lyotropic phases is represented in a ternary diagram in Fig.24 [21,42].

Water-lipid systems form many other structures, which are said 'exotic', since they are simply rare, or have not received great attention. We represented such a system in Fig.22b, which is hexagonal, with an oily core and an oily environment, separated by an inverted bilayer. This is the reverse situation of what is realized in microvilli (Fig.9).

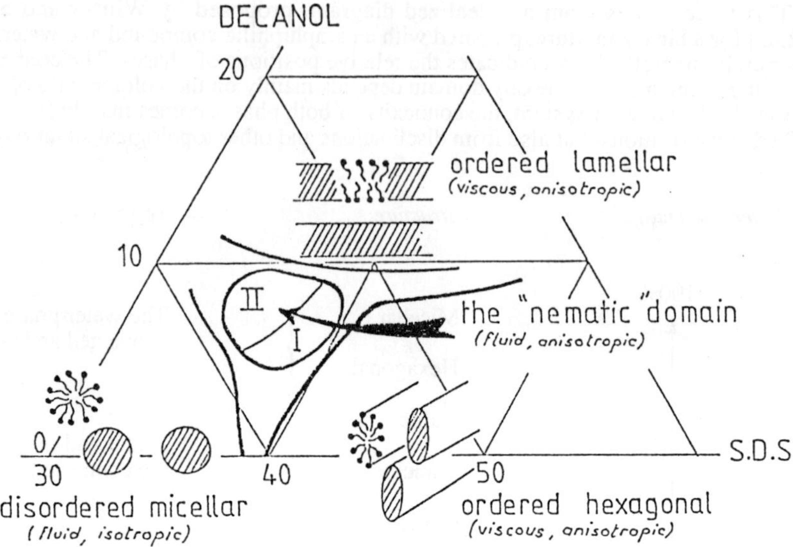

Figure 24. The SDS-decanol-water diagram in the vicinity of the nematic domain. The concentrations are expressed in weight percent. Reproduced from [21,42].

V-2. Differences between Cell Membranes and Water-Lipid Systems

Lamellar, cubic and hexagonal systems are produced by membranes within cell and vesicles also are observed. The analogy with lamellar, hexagonal, cubic and micellar phases is obvious, but differences must be underlined :

1. In cytoplasmic ultrastructures, the water percentage is generally more than 90%.

2. The observed periodicities in membrane systems are larger than the analogous ones in water-lipid systems, by a factor often greater than 10.

3. The core and the environment of vesicles and hexagonally packed tubes are aqueous, whereas, spheres or hexagonal structures have different cores and environment in water-lipid systems.

4. Biological membranes are not symmetrical, whereas bilayers usually considered in water-lipid systems are symmetrical.

5. The shape of biological membranes depends on the ratio of two areas : that of the 'cytoplasmic monolayer' and that of the opposite monolayer, measured for instance along the two parallel surfaces S_1 and S_2 separating the oily structures from the hydrophilic domains (Fig.23b). In water-lipid systems, a different ratio must be taken into account : it is the ratio of densities of polar heads and of corresponding paraffinic chains along two surfaces parallel s_1 and s_2, within a monolayer (Fig.23a). Another ratio is important; it is that of two volumes, the one occupied by hydrophilic structures and the one occupied by hydrophobic structures, as suggested in Table 1.

Curvature of cell membranes is weak in general, with respect to that observed in water-lipid systems. There are however examples of strong curvature, particularly in the course of reassociation of membranes, during exo- and endocytosis. As indicated in Fig.4d, proteins included in the bilayer facilitate the curvature and probably diminish the splay elastic constant.

The large amounts of water in cells eliminate in general the reverse structures. The cores and environments are aqueous and never oily. However, the geometries are similar to those observed in water-lipid-systems. A set of two parallel surfaces is involved in both situations: the surfaces s in water-lipid systems and the surfaces S in biological membranes. The asymmetry between s_1 and s_2 comes from the structure of the amphiphilic compound and of the presence of added hydrophilic or hydrophobic molecules. This asymmetry lies in the structure of the amphiphilic molecule itself, linking two parts which normally segregate. In biological membranes, the asymmetry between S_1 and S_2 is due to the relative proportions of phospholipids in the two monolayers, and those of associated molecules such as cholesterol, polysaccharides, proteins and even, the orientation of proteins which scan the whole thickness of the bilayer. As already pointed out, these proportions of molecules in the two monolayers and their orientation is the result of a set of complex metabolic reactions. The asymmetry of bilayers from which originate the main shapes in cells and organelles is controlled by metabolism, a set of reactions controlled by enzymes, which are part of membranes, whereas in water-lipid models, the asymmetry comes from the structure of the amphiphilic molecule.

VI-SADDLE-SHAPED BILAYERS AND MEMBRANES

Saddle-shaped surfaces are frequent in biological membranes and in water-lipid systems. They are observed at the basis of microvilli or at the end of mitosis, in the narrow zone where two daughter cells separate. They are present in numerous organelles. These saddle-shaped surfaces are mainly developed in cubic water-lipid systems, when the bilayer forms an infinite periodic minimal surface of type P, F or G. Membranes also form such periodical surfaces, but of type p, f or g as defined above. In water-lipid systems, Sadoc and Charvolin [22,66] proposed that the area occupied per polar head can differ from that occupied by the corresponding paraffinic chains and that is enough to create a saddle deformation of bilayers. We think that this factor is probably involved in most situations, but at various degrees and there are cases, namely in biological membranes, where very different factors come into play.

VI-1. Coupling of Layers of Different Area

Let us recall some classical properties of parallel surfaces (Fig.25). Normals to a set of concentric spheres pass through the centre or focus F (Fig.25a). The case of parallel cylindrical surfaces s is easily visualized in considering their section by a plane normal to generators (Fig.25b). In this plane, the cylindrical surfaces are cut along parallel curves and their common normals envelop a curve γ, which in turn is the section of a cylindrical surface c, tangent to all normals common to cylindrical surfaces s. The curvature lines of s are the generators themselves and the sections of s by the normal planes. At a point M, there is a curvature centre C_1, the other one being at infinity. In the general case, at a point M correspond two curvature centres C_1 and C_2 and parallel surfaces lying between them are saddle-shaped, in contrast with those exterior to the segment C_1C_2 (Fig.25c).

The area of the saddle-shaped rectangles passes through a maximum between C_1 and C_2 and this area is nearly zero for rectangles in the vicinity of C_1 and C_2. The rectangle area is stationnary in the vicinity of this maximum. Let us consider now a saddle-shaped bilayer, with its oily median surface corresponding to the maximum of area. The two surrounding polar layers occupy a smaller area. If the thickness of this bilayer is weak, compared to the curvature radii, the area differences are weak and are possibly relaxed by the presence of water, or oily components intercalating between amphiphilic molecules, or also by slight thickness variations of the polar and paraffinic layers. On the contrary, if the curvature radii are comparable to the layer thickness, the effect of different areas at the interface is strong and not easily relaxed.This is represented in Fig.26 where the area occupied by polar heads measured along the surfaces σ_1 is much smaller than tha area occupied by the corresponding paraffinic chains considered at levels σ_2. The median surface Σ of the oily domain corresponds to the maximum area.

In the case of biological membranes forming periodic surfaces p, f or g, with a cubic symmetry, the two radii differ significantly and the oily layer occupies an area intermediate between those of the two polar layers, but is neither smaller nor greater than the areas of the two surrounding polar layers. The principle considered by Sadoc and Charvolin does not predominate in this situation.

Figure 25. Parallel surfaces. a. Spheric surfaces centered in F. b. Cylindrical parallel surfaces with the corresponding cylindrical caustics; C_1 is the curvature centre at finite distance of a surface s at point M. The curvature lines are drawn at point M and the one coinciding with the generator corresponds to a zero curvature. c. The general case with two curvature centres C_1 and C_2 and the two caustics, their tangent planes in C_1 and C_2 lying orthogonally. The parallel surfaces around M have two curvature radii which have the same sign, positive for instance. At the figure bottom, both radii are negative. Between the two curvature centres C_1 and C_2 the radii have opposite signs and the parallel surfaces are saddle-shaped. The represented surfaces are limited by curvilinear rectangles and the area of the saddle-shaped rectangles passes through a maximum at a point lying somewhere between C_1 and C_2. (see next page).

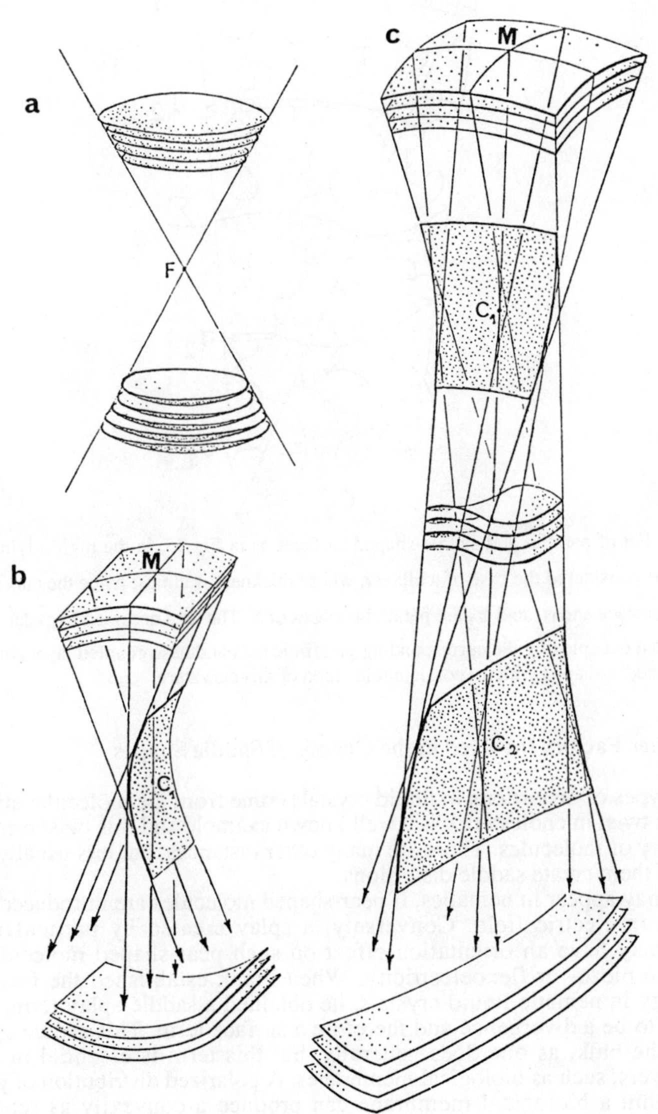

a
F
b
M
C
c
M
C_1
C_2

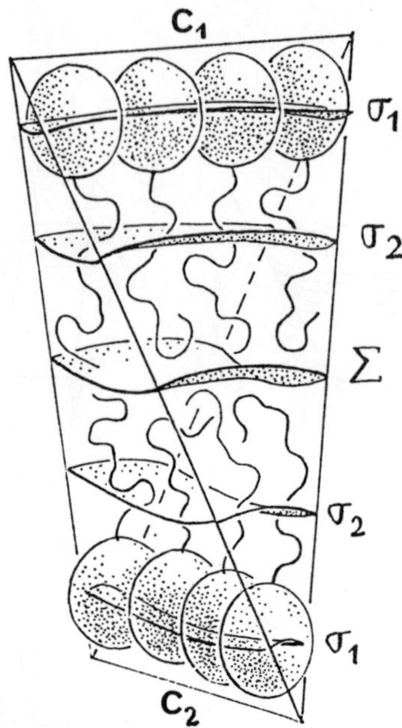

Figure 26. Set of parallel and saddle-shaped surfaces as in Fig.25, in the region lying between C_1 and C_2. Here is considered the case of a bilayer, whose thickness is almost twice the curvature radius. Σ is the median surface and σ_1 and σ_2 are parallel surfaces of Σ. The area occupied by polar heads is much smaller than that occupied by the corresponding paraffinic tails andthese coupled layers of different area form a set of saddle-shaped with no possible relaxation of this curvature.

VI-2. Other Factors Involved in the Genesis of Saddle Shapes

Several types of distortions in liquid crystals come from the molecular structure. The spontaneous twist in cholesterics is a well known example and this twist originates from an asymmetry of molecules. There are many other instances, but less usually considered and some of them create saddle distortions.

A splay may appear in nematics, if pear-shaped molecules are introduced and can be oriented in an electric field. Conversely, a splay created by particular boundary conditions may have an orientation effect on such pear-shaped molecules, leading eventually to piezo- or flexoelectricity. When Frank established the formula of the elastic energy in nematic liquid crystals, he obtained a saddle-splay term, which was shown later to be a divergence and therefore a surface term. This saddle splay can be omitted in the bulk, as one does generally, but this term is essential in the case of isolated bilayers, such as biological membranes. A polarized distribution of pear-shaped proteins within a biological membrane can produce a convexity as represented in Fig.27a. Similarly, proteins in the form of tetrahedron can induce a saddle splay as shown in Fig. 27b. Each tetrahedral molecule is oriented within the bilayer by an adequate distribution of the polar and non polar residues. The saddle splay represented in Figs. 25 and 26 introduces a similar tetrahedral geometry. In Fig.27b, the phospholipids lying in the vicinity of this protein show a local saddle splay and a global effect is obtained with a density of such molecules homogeneously scattered in the bilayer.

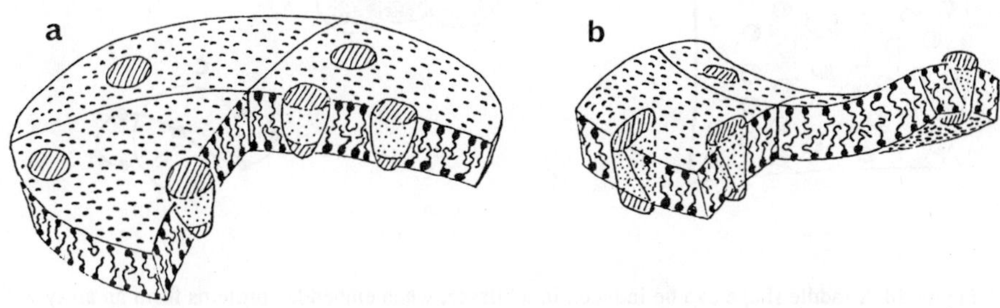

Figure 27. Plausible influence of proteins on the bilayer shape. a. Pear-shaped proteins (P) , with a preferential or constant orientation within the bilayer B, produce a splay of phospholipids and accordingly a definite convexity of the bilayer. b. A saddle splay can be induced by tetrahedral proteins scanning the whole thickness of the bilayer, with an adequate distribution of polar and non-polar aminoacid residues. In both figures a and b, polar regions of proteins are represented by hatched areas.

Let us consider another possible mechanism generating saddle-splay in bilayers. Two kinds of globular proteins are supposed to be included in a bilayer, one small, forming an array, hexagonal for instance, and another one large and less numerous. The presence of the large proteins disturbs the hexagonal array of the small ones and is supposed to introduce a density of heptagons. This is enough to generate a saddle curvature, as shown in Fig.28. The hexagonal array probably reduces considerably the bilayer fluidity. The diffusion of phospholipids within the bilayer is not suppressed but decreased. The hexagonal array is rigid, but not incompatible with fluidity, if a sufficient density of defects is present. Actually, the regular lattice of small proteins is probably not a strict requisite of the saddle curvature. The model associates two structures : proteins of small diameter with six first neighbours and larger ones with seven, this being realized with enough accuracy to produce the saddle effect. A strong accuracy leads to the hexagonal array lying between the disclination points, but possibly the effect is still sensitive without such a high degree of hexagonality. A similar model can be considered with a tetragonal array of small particles within the bilayer and larger particles with five first neighbours. In this case also, the rather bad alignment of particles between disclinations is not icompatible with the saddle effect.

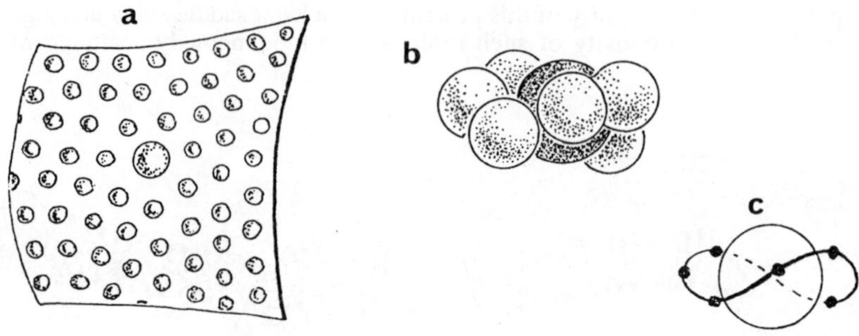

Figure 28 A saddle shape can be induced in a bilayer, when embedded proteins form an array with hexagonal symmetry and local disclinations due for instance to presence of larger proteins associating seven first neighbours instead of six. a. Aspect of the membrane , its hexagonal array and the local disclination. b. Side-view of the large protein surrounded by seven first neighbours. c. Distribution of the centres of small proteins about the large one in a side-view as in b.

Another source of saddle-shapes deserves attention: it is well known that to twist a ribbon creates a helicoid which is a saddle-shaped surface. The twist between neighbouring elongated molecules leads to similar situations and a biological example of that is provided by twisted protein β-sheets [67]. Parallel polypetide chains are stabilized by hydrogen bonds, but the chiral character of aminoacids can introduce a twist of constant handedness. Let us consider an element of a saddle-shaped surface with its two curvature lines as shown in Fig.29a.

The normal vectors to the surface present a twist when they are considered along a curve running obliquely relative to the curvature lines. Note that this property also holds for surfaces with two different cuvatures of same sign. Around each point M of the surface, there are four $\pi/2$ sectors, two of them corresponding to a right-handed twist and two others to a left-handed one (Fig.29a). Parallel curves drawn on this surface, and lying obliquely relative to the curvature lines, present a mutual twist along a curve running perpendicularly to them (Fig.29b). Similarly, if parallel molecules form a thin sheet and have a chiral structure leading to a twist, as in β-sheets, they will arrange along an oblique direction with respect to the curvature lines (Fig.29c). Biological membranes are coated by chiral polymers (polysaccharides and proteins) which could induce a saddle shape but, up to this date, we do not know any example of such an effect in biological bilayers.

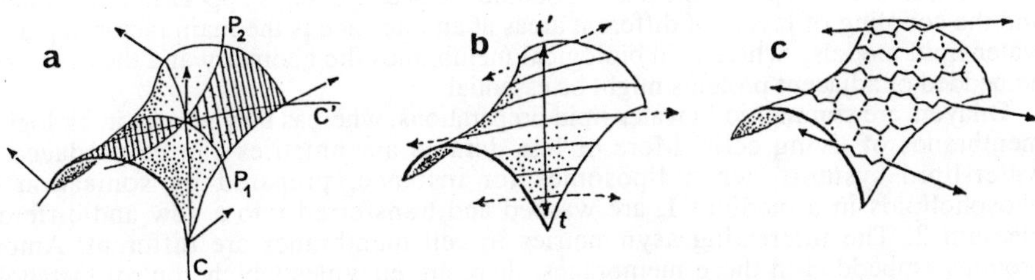

Figure 29. a. Twist of normals to a saddle-shaped surface, along geodetics oriented obliquely relative to the curvature lines. The twist is right-handed in the hatched zone and left-handed in the complementary one. b. The twist of tangent planes along a diagonal direction tt'. c. Saddle-shaped β-sheet formed by a polypeptide. The twist is left-handed. After Salemme [67].

In brief, it appears likely that different factors are involved in the production of saddle shapes in bilayers, and they possibly cooperate, but experimental works on the origin of this morphology are rare. From pure geometric speculations, we think that plausible mechanisms are numerous and four of them were discussed :

1. Coupling of a median layer with lateral layers of smaller area.

2. Presence of tetrahedral proteins with adequate orientation.

3. Presence of proteins forming more or less regular arrays within bilayers, with point disclinations centered about larger proteins.

4. Existence of a twist of preferred chirality in the structure of bilayers or membranes.

The first factor predominates in the production of IPMS by water-lipid systems, but its role is probably secondary in the analogous structures formed by cell membranes, where proteins and chirality may interact with curvature.

VII-CONCLUSIONS

Morphologies of cell membranes are closely related to those observed *in vitro* in liquid crystalline systems prepared with amphiphilic molecules, water and oily components. There are however profound differences (water content, scale, core and environment of micelles, symmetry or broken symmetry of bilayers etc.) The geometric similarity comes from a common scheme, which is that of two parallel and coupled surfaces which occupy different areas. In water-lipid systems these two surfaces correspond to the mean levels of polar heads and paraffinic chains in a monolayer, whereas in biological membranes these two surfaces are those separating the layers of polar heads from the paraffinic chains. Saddle-shaped structures appear in both systems and the coupling of layers of different areas at an interface is the main factor in play in water-lipid models, whereas in biological membranes the geometry and the density of embedded or adjacent proteins might be essential.

Bilayers are symmetric in water-lipid preparations, whereas they are not in biological membranes of living cells. More or less durable asymmetries can be produced in water-lipid systems, when liposomes for instance, prepared by sonication of phospholipids in a medium 1, are washed and transferred into a new and different medium 2. The interesting asymmetries in cell membranes are different. Among proteins embedded in these membranes, there are enzymes which control metabolic parameters and maintain the asymmetry. The future of experiments will be to study water-lipid systems in presence of one or several enzymes and substrates, in conditions producing this bilayer asymmetry, but this is a long term program, however indispensable before considering the origin of complex geometries and topologies as those of mitochondria, Golgi apparatus, nuclear envelope, etc.

Cell structures associate membranes and fibrous components. These latter also are liquid crystalline, or analogues of liquid crystals; they have a nematic or cholesteric geometry, often stabilized and not more fluid [10-14]. These fibrous materials are developed within cells or in extracellular matrices. A great part of the architecture of cells and tissues is based on the coupled distribution of fibrous and membranous structures at the periphery of cells or within cells. New models are expected in the analysis of morphogenesis at the cell scale and possibly some of them will come from the preparation of water-lipid systems in association with liquid crystalline phases of main biopolymers.

REFERENCES

1. Alberts B., Bray D., Lewis J., Raff M., Roberts K., Watson J.D. Molecular Biology of the Cell, Garland Publishing Inc., N.-Y., London (1983).
2. Bassot J. M., J. Cell Biol., <u>31</u>, 135-158 (1966).
3. Bassot J. M., La Recherche, <u>13</u>, 1314-1317 (1982)
4. Bassot J. M. and Nicolas M. T., Experientia, <u>34</u>, 726-728 (1978).
5. Bernard C., Introduction à l'Etude de la Médecine Expérimentale, Paris (1865).
6. Bessis M., Living Blood Cells and Their Ultrastructure, Springer Vlg., Berlin (1973).
7. Blinzinger, K., J. Cell Biol., <u>25</u>, 293 (1965).
8. Bloom W. and Fawcett Don W., A Textbook of Histology, 10th edition, W.B.Saunders (1975).
9. Bouligand Y. J. Microscopie, Paris, <u>3 a</u>, 25-26 (1964).
10. Bouligand Y., Tissue & Cell, 4, 189 (1972).

11. Bouligand Y., in Liquid Crystalline Order in Polymers, A. Blumstein ed., 261-297, Acad. Pr., N.-Y. (1978).
12. Bouligand Y., in Physics of Defects, Balian *et al.* ed., School of Theoretical Physics, Les Houches, 25, 778-811 (1980).
13. Bouligand Y., Solid State Physics, Suppl. 14, 259-294 (1978).
14. Bouligand Y. et Giraud-Guille M.-M., Proc. 1st Intern. Symp. Biol. Invertebrates and Lower Vertebrates Collagens, A. Bairati and R. Garrone eds., 115-134, Plenum (1985).
15. Chailley B., Weed R.I., Leblond P.F. and Maigne J., Nouv. Rev. Fr. Hémat., 13, 71 (1973).
16. Chambers R., Ann. Physiol.Physicochim. Biol., 6, 233 (1930).
17. Chambers R. and Chambers E.L., Exploration into the Nature of the Living Cell, Harvard Univ. Pr., Cambridge Mass. (1961).
18. Chapman D. Ann. N.-Y. Acad. Sci., 137, 745 (1966).
19. Chapman D. in 'Membranes and Ion Transport' E.E. Bittar ed., 1, 24 Wiley, N.-Y. (1970).
20. Chapman D., in 'Plastic Crystals and Liquid Crystals', G.W. Gray and P.W. Winsor eds., E. Horwood publ., Chichester, 1, 288-326 (1974).
21. Charvolin J., J. Chimie Physique , 80, 15 (1983).
22. Charvolin J. and Sadoc J. F., J. Physique, 48, 1559-1569 (1987).
23. Charvolin J. and Sadoc J. F., J. Physique, 49, 521-526 (1988).
24. Charvolin J. and Tardieu A., Solid State Physics, Suppl.14, 209-257 (1978).
25. Danielli J. R. and Dawson H. A. J. Cell Physiol., 5, 495 (1935)
26. Devaux P. F., Biol. Magnetic Reson., 5, 183 (1983).
27. Devaux P. F., Biophys. Biochim. Acta, 822, 63 (1985).
28. Dillon L. S., Ultrastructures, Macromolecules and Evolution, Plenum (1981).
29. Donnay G. and Pawson D.L., Science, 166, 1147-1150 (1969).
30. Doughtie D. G. and Ranga Rao K., Cell Tissue Res., 238, 271-288 (1984).
31. Du Praw E.J., Cell and Molecular Biology, Acad Press, N.-Y. (1968).
32. Eckmiller M.S. J. Cell Biol., 105, 2267-2277(1987).
33. Fawcett, D.W. "The Cell", Second Edition, 862 p., W.B. Saunders Co., Philadelphia (1981).
34. Gaill F. and Bouligand Y. Proc. 1st Intern. Symp. Biol. Invertebrates and Lower Vertebrates Collagens, A. Bairati and R. Garrone eds., 266-274, Plenum (1985).
35. Gaill F. and Bouligand Y., Tissue & Cell, 19, 625-642 (1987).
36. Geren B. B., Exp. Cell Res., 7, 558 (1954).
37. Goodman J. F. and Clunie J.S. in 'Liquid Crystals & Plastic Crystals', G.W. Gray and P.A. Winsor ed., E. Horwood publ., Chichester, 2, 1-23 (1974).
38. Günning B.E.S., Protoplasma, 60, 111-130 (1965).
39. Günning B.E.S. and Steer M. W., Ultrastructure and the Biology of the Plant Cell, E. Arnold (1975).
40. Hassett R.J., Wayne E. and Kuhn C III, J. Ultrastructure Res., 71, 60-67 (1980).
41. Henderson D.W., Papadimitriou J.M., Coleman M., Ultrastructural Appearances of Tumours, 2d Edition, Churchill Livingstone (1986).
42. Hendrikx Y. and Charvolin J. J. Physique, 42, 1427 (1981).
43. Hyde S.T. and Anderson S., Z. Kristallogr., 168, 221-254 (1984).

44. Hyde S.T., Anderson S., Ericsson B. and Larsson K., Z. Kristallogr., 168, 213-219 (1984).
45. Hogan M.J., Alvarado J. A. and Weddell J.E., Histology of the Human Eye: An Atlas and Textbook. Saunders (1971).
46. Jensen W.A. and Park R.B., Cell Ultrastructure, Wadsworth Publ. Co., Inc. (1967).
47. Kalt M., Anat. Rec., 182, 53 (1974).
48. Kunz Y., Ennis S. and Wise C., Cell Tissue Res., 230, 469-486 (1983).
49. Lentz T. L. Cell Fine Structure, Saunders (1971).
50. Lepescheux L., Biol. Cell, 62, 17-31 (1988).
50. Lima-de-Faria A. (ed.), Handbook of Molecular Cytology, North Holland Publ., Amsterdam (1969).
51. Lucas W.J., Brechignac F., Mimura T. and Oross J.W., Protoplasma, 151, 106-114 (1989).
53. Luzzati V. in 'Biological Membranes', D. Chapman ed., Acad. Pr. N.-Y., 71 (1968).
54. Luzzati V. and Spegt P.A., Nature, 215, 701 (1967).
55. Luzzati V., Tardieu A., Gulik-Krzywicki, Rivas E. and Reiss-Husson F., Nature, 220, 458-488 (1968).
56. Mc Lean R. J. and Pessoney G.F., J. Cell Biol., 25, 522-531 (1970).
57. Mackey A. L., Nature, 334, 562 (1988).
58. Martin J. S. and Kirkham J.B., Tissue & Cell, 21, 627-638 (1989).
59. Nageotte J. Actual. Sci. Industr., 431/434 (1936)
60. Nissen H. U., Science, 166, 1150-1152 (1969).
61. Paolillo D. J. Jr., J. Cell Sci., 6, 243-255 (1970)
62. Peters W., Z. Morph. Tiere, 64, 21-58 (1969).
63. Pfeffer W. Osmotic investigations (in German). Von Wilhelm Engelmann (1877). (English translation by G.R. Kepner and E.J. Stadelmann, Van Nostrand Reinhold, N.-Y. (1985).
64. Rambourg A., Clermont Y. and Hermo L., Am. J. Anat., 154, 455-476 (1979).
65. Robertson J. D., Biochem. Soc. Symp., 16, 3 (1959).
66. Sadoc J. F. et Charvolin J., J. Physique, 47, 683 (1986)
67. Salemme F.R. Structural Properties of Protein β-Sheets, Prog. Biophys. Molec. Biol., 42 : 95-133 (1983).
68. Satir B., Symp. Soc. Exper. Biol., 28, 399-418 (1974).
69. Schmitt F. O. and Palmer K. J., Cold Spr. Harb. Symp. Quant. Biol., 8, 94 (1940)
70. Schoen A.H., Infinite Periodic Minimal Surfaces without Self-Intersection, NASA Technical Notes C-98 (1969), D-5541 (1970).
71. Schwarz H. A., Gesamm. Math. Abh., Springer, Berlin (1890).
72. Scriven L. E., in Micellisation, Solubilization and Microemulsions, K.L. Mittal ed., 877-893, Plenum, N.-Y. (1977).
73. Singer S. J. A., Rev. Biochem, 43, 805 (1974).
74. Singer S. J. and Nicolson G., Science, 175, 720 (1972).
75. Stratton C.J. Tissue & Cell, 8, 693-712, 713-728 (1976).
76. Stratton C.J., Zasadzinski J.A.N. and Elkins D., The Anatomical record, 221, 503-519 (1988).

77. Suzuki Y., Fujita Y. and Kogishi K., Am. Rev. Resp. Dis., <u>140</u>, 75-81 (1989).
78. Tardieu A., Thesis, Univ. Paris 11, Orsay (1972).
79. Thomas E.L., Anderson D.M., Henkee C.S. and Hoffmann D., Nature, <u>334</u>, 598-601 (1988).
80. Thompson D. W. On Growth and Form, Cambridge Univ. Pr. (1917).
81. Webster F. de, J. Cell Biology, <u>48</u>, 348-367 (1971).
82. Weiss L., Cell and Tissue Biology, A Textbook of Histology, 6th edition , Urban & Schwarzenberg (1988).
83. Williams, M.C., J. Cell Biol., <u>72</u>, 260-277 (1977).
84. Williams R. The Geometrical Foundation of Natural Structure, Dover, N.-Y. (1972, 1979).
85. Winsor P.A., in 'Plastic Crystals and Liquid Crystals', G.W. Gray and P.W. Winsor eds., E. Horwood publ., Chichester, <u>1</u>, 199-287 (1974).
86. Zasadzinski J.A.N., Stratton C.J., Davis H. T. and Scriven L.E., J. Electron Micr. Techniques, <u>3</u>, 385-400 (1988).
87. Zasadzinski J.A.N., Kerins J. and Rudolphi R., The Anatomical record, <u>221</u>, 520-532 (1988).

Chapter V
NON-CRYSTALLINE ATOMIC STRUCTURES: FROM GLASSES TO QUASICRYSTALS

Remy MOSSERI and Jean François SADOC

NON-CRYSTALLINE ATOMIC STRUCTURES:
FROM GLASSES TO QUASI-CRISTALS

R. Mosseri †and J.-F. Sadoc ‡,
†Laboratoire de Physique des Solidesde Bellevue-CNRS
1 pl. A. Briand, 92195-Meudon Bellevue, France.
and
‡Laboratoire de Physique des Solides,
associé au CNRS (LA 02),
bat. 510, Université Paris-Sud,
91405 Orsay, France.

I-INTRODUCTION

Our present knowledge of glass structure is based on various experimental results from X-ray and neutron diffraction, EXAFS, NMR, Mossbauer spectroscopy, etc... These experiments give access to the local and medium range structure of atoms. Some other kinds of experiments give access to macroscopic properties, for instance density measurements, magnetic properties, mechanical properties, etc. A good model must describe all these experimental results. Current models do not fulfil this requirement but explain rather well some of the physical properties. There is a good historical example: twenty years ago X-rays and neutron diffraction by amorphous metallic alloys were sufficiently accurate to support models with an icosahedral local order leading to the well known shoulder on the 2nd ring of the interference function [1]. It was clear by observation of interatomic distances on a Radial Distribution Function (RDF), that icosahedral coordination was the main ingredient of the local order. (fig. 1). Positions and intensities of first peaks on a radial distribution correspond to an icosahedral order. So a model of a disordered structure could be obtained by adding new atoms to an icosahedron with respect to a simple rule leading to a local order as close as possible to an icosahedral one.

But at this time no such model was computed which had both the icosahedral local order and a macroscopic density as observed in experiment. So during several years two types of model have been investigated : the first one exhibiting good details of the interference and radial distribution function [1][2] ; the second one giving a better atomic density [3][4]. Only in the early 80's were both properties reproduced in a model of metal structure [5]. But, large models allowing simulation of macroscopic properties like mechanical plasticity of glasses are still difficult to

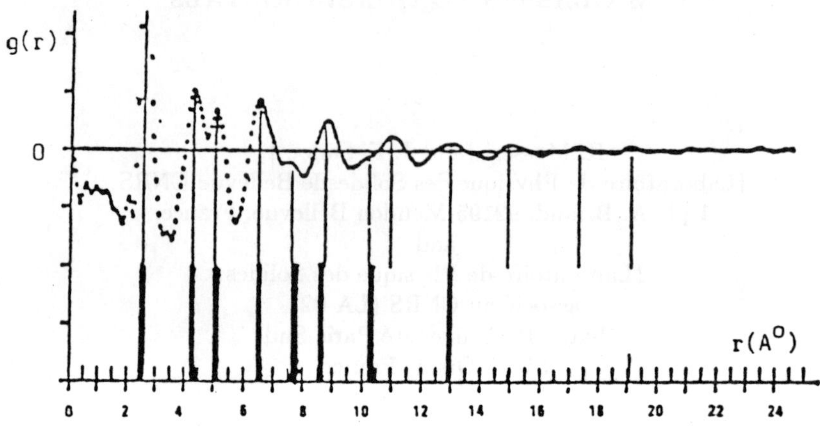

fig. 1

Radial distribution function $R(P(R) - 1)$ obtained from X-rays diffraction by amorphous iron. a-Fe prepared by glow discharge (courtesy of J.P. Lauriat). Position of maximum in first neighbours distance unit are very close to what it is shown on the figure 8.

obtain, and more work is needed in this field.

The complexity of the structure is related to the propagation of the local order, and consequently, it is not surprising that a good model have not only to reproduce the local order but also to explain how it is propagated through the whole structure leading to the longer range correlations which are manifested by the presence of a structured RDF at larger distances. It is well known that the icosahedral symmetry is incompatible with any crystallographic (periodic) symmetry group. So it was relatively reasonable to consider that the disorder in amorphous materials results from this lack of compatibility. But things are not so simple, and for example the observation of quasicrystals [6] shows that a sophisticated long range order could appear in materials with local icosahedral symmetry. Frank-Kasper structures [7] are also interesting examples of the natural solution to this problem. In these structures a large number of atoms have an icosahedral symmetry but they are arranged in a crystalline way, with very large unit cell. Note that, in the framework of the so-called approximant structures, quasicrystals could be seen as Frank and Kasper phases with an unit cell of infinite size.

The geometrical frustration is an important parameter in the organisation of all these atomic structures. The structure of condensed matter results from a competition between local interactions and topologic and geometric rules imposed by the space filling requirement. In main cases the structure results in a compromise. Sometimes the solution is unique, clearly there is deep minimum in the free energy function, but in other cases there are several solutions which are energetically possible even if these solutions may be very different one from the other (from a geometrical point of view).

Two simple 2D examples are helpful to get some feeling about the origin of the competition between local rules and geometry in the large. Consider a disc packing (which could model a 2D metal) on a plane ; the interaction, which is isotropic, locally tends to arrange disks in the densest way : it forms triangles with 3 disk centres. In order to understand the long range structure we have to study the properties of the plane to be filed by equilateral triangles. In this simple example, this can be done easily, leading to the well known triangular lattice : there is an entire compatibility between the two kinds of rules.

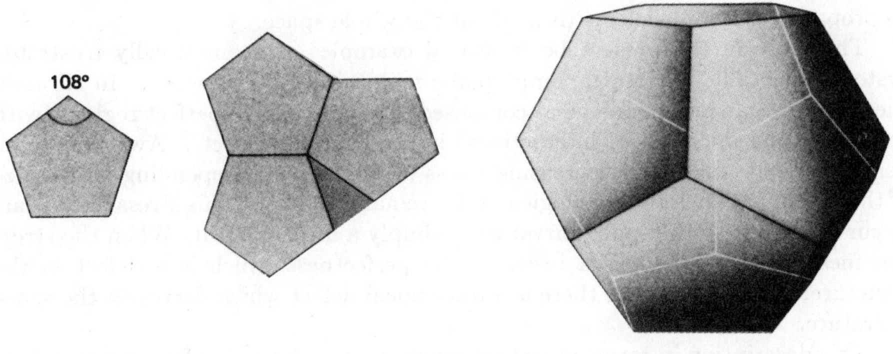

fig. 2

Pentagons do not tile the plane, but can tile a sphere, and form a dodecahedron.

In the second example in 2D, the interaction energy is supposed to be minimum when atoms sit on the vertices of a regular pentagon. When trying to propagate in the long range a packing of these pentagons sharing edges (atomic bonds) and vertices (atoms), one faces a difficulty due to the impossibility of tiling a plane with pentagons. It is this kind of discrepancy which is often called "geometric frustration" (Fig. 2).

There is one way to overcome this difficulty. Let the surface to be tiled be free of any presupposed topology and metrics, and build the tiling with a strict application of the local interaction rules. In this simple example, we observe that the surface inherits the topology of a sphere and so receives a curvature. The final structure, here a pentagonal dodecahedron, allows for a perfect propagation of the pentagonal order. It is called an "ideal" (defect-free) model for the considered structure. This is the scheme of the curved space approach of disordered structures. The second step consists of coming back to the plane by lowering to zero the curvature. This procedure requires the introduction of topological defects which locally perturbs the structure and breaks the local rules. But this is a needed compromise in order to flatten the 2D space.

In 3 dimensions, for example, amorphous metals can be well described by close packing tetrahedra [1] . A regular tetrahedron is the densest configuration for the packing of four equal spheres. The dense random packing of hard spheres problem can thus be mapped on the tetrahedral packing problem. The dihedral angle of a tetrahedron is not commensurable with 2π, consequently, a perfect tiling of the euclidean space R_3 is impossible with regular tetrahedra. Note that, at this local level, the "frustration" (deviation to perfectness) is of a metrical rather than of a topological nature. Then we define an unfrustrated structure by allowing, as in the 2D example, for curvature in the space, in order for the local configurations to propagate without defects throughout the whole space.

There are in the present book several examples of geometrically frustrated systems : metallic structures, amphiphilic molecules, blue phases ... In all cases the structure can be considered as composed of almost locally perfect regions (with respect to the microscopic interactions) interrupted by defects. Away from the perfect regions there is an increasing stress in the material depending on the size of the region in which the topological order remains perfect. This stress disappears in curved space, so the space curvature is simply a measure of it. When the stress becomes too large there is a break in the perfectness which is a defect in the structure, or in other terms there is a topological defect which decreases the space curvature.

This description in terms of curved space and topological defect has been first developed for amorphous metallic [8][9] structures, but it can be applied also the quasicrystals, which are usually described using the cut and projection method[10]. In both descriptions there is an intermediate step in the structural description which consists to use a structure in a space which is flat only on average. In the curved space approach this space is a corrugated space which is positively curved where the structure is "perfect" (icosahedral) and is negatively curved on topological defect, the negative curvature compensating the positive curvature so that the space in flat is average. In the cut and projection method, there is an intermediate space obtained by the cut using a "strip" into a 6-Dimensional space. The resulting space is a 3D space, but this space is not flat. In the 1D quasicrystal case obtained from a 2D space, the intermediate space is an irrational staircase enclosed in the strip.

II-CURVED SPACE MODEL

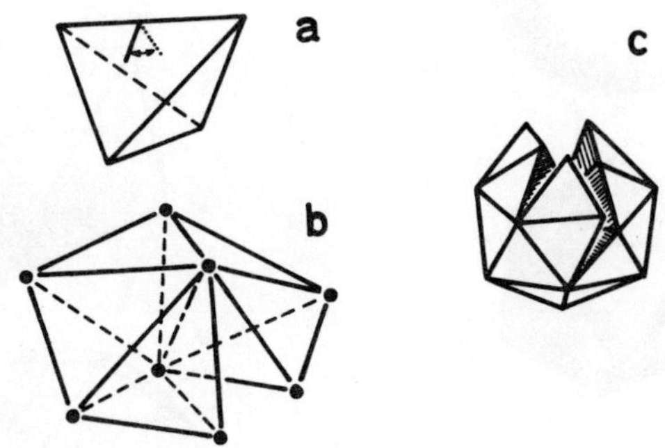

fig. 3
Regular tetrahedra do not fill perfectly the euclidean space R_3. When
20 tetrahedra are packed around a common vertex, the 12 outer vertices
form an irregular icosahedron

In the case of polytetrahedral structures the frustration is due to the
impossibility to tile the space with regular tetrahedra.

20 tetrahedra pack with a common vertex in such a way that the 12 outer
vertices form an irregular icosahedron (figure 3). Indeed the icosahedron edge
length l is slightly longer than the circumsphere radius r ($l \simeq 1.05r$). It is
possible to make the icosahedron regular by first shortening accordingly halves of
the tetrahedra edges. The tetrahedra regularity can then be recovered by giving
to the central site a small 4th coordinate. One then tries to add new shells of sites
while keeping the same icosahedral environment to the sites. It is possible if these
new sites are also given at 4th coordinate with opposite sign. This procedure can
be continued for several shells until one sees that no new sites are needed. Indeed
a finite set of points has been generated such that each vertex has 12 neighbours
in perfect icosahedral configuration. Let us describe this set :

- There are 120 vertices which all belong to the hypersphere S_3 with radius,
the golden number ($\tau = (1 + \sqrt{5})/2$) if the edge are of unit length.

- The cells are regular tetrahedra grouped by 5 around a common edge and by
20 around a common vertex.

- This structure is called a polytope [11] which is the general name in higher
dimension in the series polygon, polyhedron, ... The standard notation for a
polytope on S_3 is $\{p, q, r\}$ where $\{p, q\}$ denotes the polyhedral cell (such that
there are q p-gons around each vertex) and r is the number of such polyhedra

240

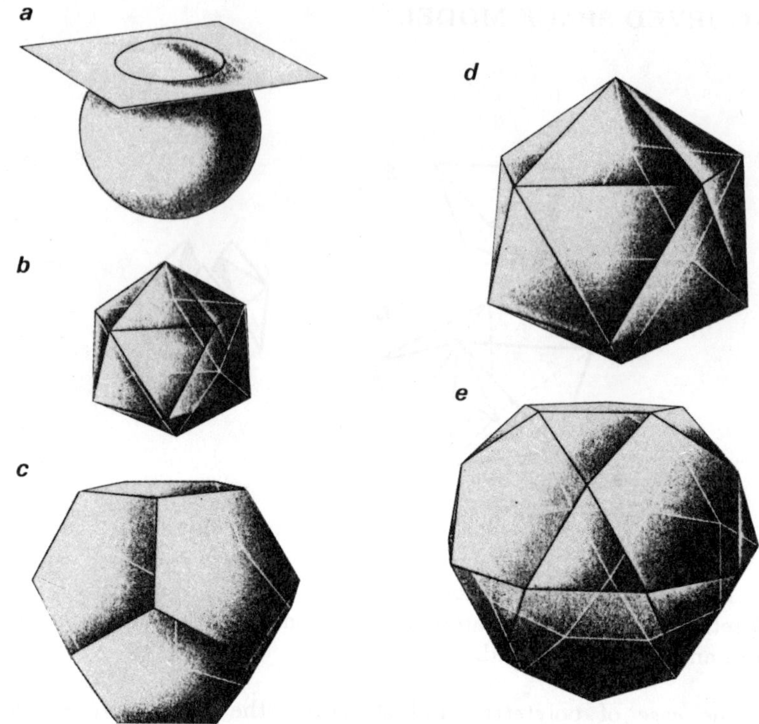

fig. 4

"Horizontal" sections of the polytope $\{3,3,5\}$. The section of S_3 by a hyperplan is a sphere. With one $\{3,3,5\}$ vertex at the "north pole, the successive sections are: b) an icosahedron, c) a dodecahedron, d) an icosahedron, e) an icosidodecahedron. The figures are reproduced from 'Pour la Science', january 85

sharing a common edge. Hence this polytope is called $\{3,3,5\}$ ($\{3,3\}$ denoting a regular tetrahedron).

- We can do a shell by shell description of the $\{3,3,5\}$ if one recalls the hyperpolar coordinate system R, θ, ϕ, ω

$$(1) \quad \begin{aligned} x_1 &= R\cos\omega, \\ x_2 &= R\cos\mu\sin\omega, \\ x_3 &= R\sin\mu\cos\phi\sin\omega, \\ x_4 &= R\sin\mu\sin\phi\sin\omega \end{aligned} \quad \text{with} \omega, \mu \in [0,\pi] \quad \text{and} \quad \phi \in [0, 2\pi[$$

Indeed fixing ω gives a sphere S_2 embedded in S_3. It is the section of S_3 by a "horizontal" hyperplane, similar in lower dimension to the "parallels" circles on S_2 which are sections of S_2 by horizontal planes. Let us fix one polytope vertex at the "north" pole $x_1 = R, x_2 = x_3 = x_4 = 0$. We get for the following value of ω (fig. 4).

$\omega = \pi/5$: 12 vertices which form a regular icosahedron. The vertices are first neighbours in the polytope.

$\omega = \pi/3$: 20 vertices, a regular dodecahedron. Each vertex is "above" one triangle of the icosahedron at $\omega = \pi/5$. It is exactly above if one orthogonally map these two shells on the same horizontal hyperplane.

$\omega = \pi/5$: a regular icosahedron whose 12 vertices are second neighbours in the polytope.

$\omega = \pi/2$: the "equatorial" sphere is tiled by 30 vertices which form a regular icosidodecahedron. The situation is then symmetrical with respect to the equatorial sphere.

$\omega = 3\pi/5$: an icosahedron

$\omega = 2\pi/3$: a dodecahedron

$\omega = 4\pi/5$: an icosahedron

$\omega = \pi$: one vertex at the south pole $x_1 = -R, x_2 = x_3 = x_4 = 0$

II-1. Symmetry properties

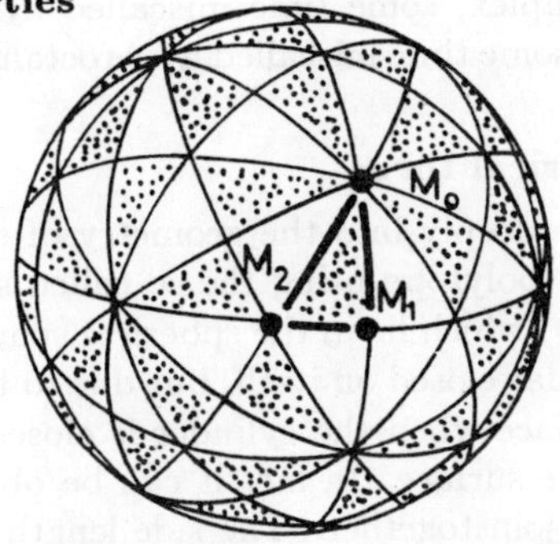

fig. 5

Tiling of a sphere with an orthoscheme triangle having angles $\pi/2$, $\pi/3$, $\pi/5$, which correspond to the ordinary icosahedral group of order 120 (containing both direct and indirect symmetries). The icosahedral and dodecagonal polyhedra vertices can be easily recognized as the orbit under reflections of special points in the orthoscheme.

Let us first mention an important property of polytope $\{3,3,5\}$. Recall that S_3 is a topological group, isomorphic to Q, the group of unit quaternions (some properties of quaternions are recalled further in this book, in the contribution by E. Dubois-Violette and B. Pansu). It happens that polytope $\{3,3,5\}$ vertices, considered as a set of 120 unit quaternions forms a group isomorphic to a discrete subgroup of Q, the binary icosahedral group Y'. One necessary condition is that it contains the identity (1,0,0,0) which means that polytope $\{3,3,5\}$ is placed on S_3 such that one vertex is on the "north" pole (1,0,0,0). Y' is the lift in Q of the usual icosahedral group Y [12][13]:

242

$$Q = SU(2) \longrightarrow SO(3) = Q/Z_2$$
$$Y' \subset Q \longrightarrow Y = Y'/Z_2 \subset SO(3) \qquad ,$$

(2)

where Z_2 is the two-element group. Now the group G' of orientation preserving symmetry operations of polytope $\{3,3,5\}$ is a sub-group of $SO(4) = Q * Q/Z_2$. It is clearly the group $G' = Y' \times Y'/Z_2$ of order 7200. The full symmetry group G contains in addition indirect orthogonal transformations and has order 14400 . This number can be calculated easily using geometric arguments. Indeed the order of the symmetry group of a regular polytope is equal to the number of fundamental regions or orthoschemes defined as follow. Given a cell of the polytope, the orthoscheme is the trirectangular tetrahedron whose vertices are one cell vertex, the cell centre, a face centre and the midpoint of an edge adjacent to this face[11] (on the fig.5 a 2D example of an orthoscheme triangle is given). Now the symmetry group of the polytope is generated by reflections in the orthoscheme faces. Polytope $\{3,3,5\}$ has 600 tetrahedral cells. Each cell is trivially decomposed into 24 orthoschemes, which amounts to be required total number of 14400. Using this argument one gets for the two other regular polytopes with tetrahedral cells, the $\{3,3,3\}$ (the simplex, some time miscalled hyper-tetrahedron) and $\{3,3,4\}$ (the cross polytope, some time miscalled hyperoctahedron), respectively the orders 120 and 384.

II-2. The spherical torus

Let us look more deeply into the geometry of the polytope. It is possible to describe the $\{3,3,5\}$ polytope using the so-called spherical torus. The spherical torus is a surface (2D) which lie in the spherical space. Like a classical torus it can be built from a cylindar closed on itself, but due to the space curvature there is no distortion of the surface when the cylinder is closed onto itself. So the spherical torus is a developable surface. In fact it can be obtain from a square sheet, two opposite sides being join together. The side length of the square is $\sqrt{2}\pi R$ where R is the radius of curvature of S_3.

Any line parallel to a diagonal of the square corresponds to a great circle of the 3-sphere (a geodesic line). There are two other important lines, the "axes" of the torus. For a classical torus one is a straight line, the other being a circle. For the spherical torus the two lines are great circles of the 3-sphere. There exists a decomposition into toric layers of the S_3 spherical space which is simply given in term of "toroidal" coordinates :

(3)
$$\begin{aligned} x_1 &= R\cos\theta\sin\phi, \\ x_2 &= R\sin\theta\sin\phi, \\ x_3 &= R\cos\omega\cos\phi, \\ x_4 &= R\sin\omega\cos\phi \end{aligned} \quad \text{with} \theta, \omega \in [0, 2\pi[\text{and} \phi \in [0, \pi/2].$$

which are coordinates more symmetrical than the spherical coordinates presented before. Fixing ϕ gives a torus in the 3-sphere ; the spherical torus corresponds to $\phi = \pi/4$.

In order to understand how the spherical torus appears in the 3-sphere it is helpfull to study a very simple polytope : the hyper-cube. There is a usefull representation of a cube in a 2D plane : a large square and a smallest one inside. In this representation the outside of the large square is taken as a face of the cube. (as if we look at a cube with an eye in the middle of a face) . There is a similar representation of a hypercube : a large cube and a smaller one inside (fig. 6).

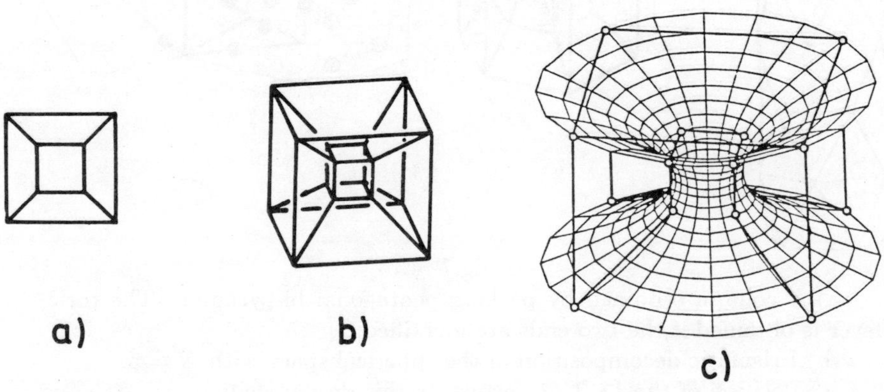

fig. 6
Representation of a hyper-cube. a) a classical cube; b) an hypercube; c) a torus from a cube.

Using this representation we consider the small central cube as a hole forming a tunnel connecting the upperpart of the "outside" cube to the lower part. Then we can observe how four cubic cells of the hypercube are assembled in order to enclose a volume which is topologically identical to the one bounded by a torus. This surface is tesselated by 16 squares which are 2D-faces of the hypercube $\{4,3,3\}$. Each vertex of this surface is common to 4 squares on the surface ; consequently this surface can be tiled by a piece of a $\{4,4\}$ tesselation, as the euclidean plane does : the spherical torus is a developpable surface. In the $\{3,3,5\}$ polytope the spherical torus separates two identical torus having the same "axis" than the spherical torus. On each of these two torus parallel to the spherical torus there are 50 vertices drawing a triangular tiling of this surface. There are 10 vertices regularly disposed on each "axis" of the spherical torus to complete the set of vertices to 120. We can represente one torus containing vertices by a cylindar, the two circular basis having to be identified. The vertices form a column which can be obtained by piling up pentagonal antiprisms with respect to an icosahedral environment for the vertices on the axis (fig.7-a).

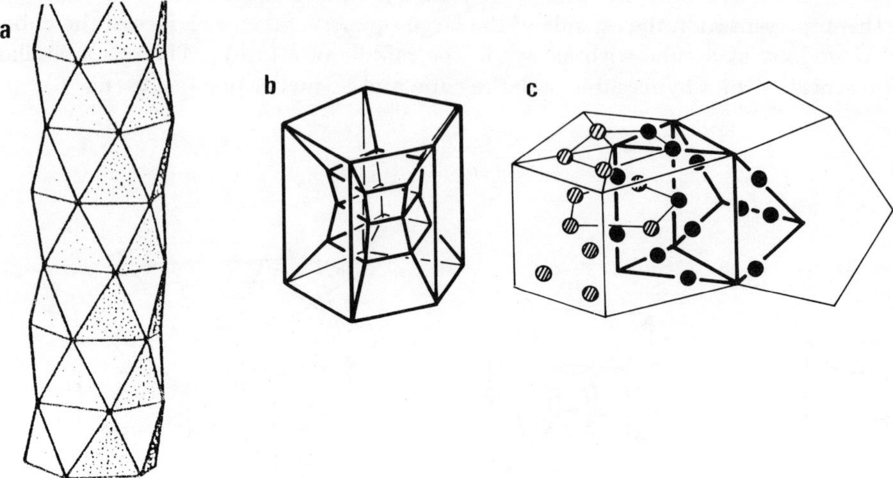

fig.7

a)-A column obtained by packing pentagonal bi-pyramid. The toric layer is obtained if the two ends are identified.

-b) Prismatic decomposition of the spherical space with $N = 5$.

-c) Position of the $\{3, 3, 5\}$ vertices in this decomposition.

It is possible to divide the spherical space into $2N$ prisms having N-sided polygonal basis. The figure 6 shows such a representation of this partition with $N = 4$. The case $N = 5$ is shown in figure 7. This representation is similar to the one shown on fig. 6 : the outside of the figure is a pentagonal regular prism exactly like other prisms appearing on this figure. To achieve the partition each prism is divided into $2N$ triangular prisms as shown on figure 6-c. Each prismatic unit contains some points of the $\{3, 3, 5\}$ polytope as shown on the figure.

II-3. Polytope $\{3, 3, 5\}$ and amorphous metals

The $\{3, 3, 5\}$ has become a template for amorphous metals under successive idealizations where parts of the reality have been lost :

- Most of amorphous metals are alloys. We have chosen to describe an ideal monoatomic structure.

- Metallic atoms have complex electronic structures with different types of orbitals whose hybridization may play important role. Here we have focused on isotropic "s" type of bonding. The atomic arrangement is coded into a sphere packing.

- Finally the template itself lives in an unphysical space (S_3).

The richness of this approach is obvious if there exists deep intrinsic properties of the materials which survive these simplifications. We shall see that this approach allows to generate a general picture of amorphous solids, with qualitative and quantitative results, at geometrical and topological levels. The main results is probably to split the structure into ordered regions and defects, as it will be explained later. Let us first look to two simple quantities.

- The *intersite distance distribution* in the polytope. It is easily calculated since, as a regular polytope, all the sites are equivalent and one can look to the environment of the "north" pole ($\theta = 0$ in the polar coordinate system of eq.1). The geodesic distance to its neighbours is then simply given by $R\omega$. We have seen above that the 119 other vertices are gathered in 8 shells, which lead to the distance distribution function. At this stage, a comparison with experiment already shows the interest of the polytope. Figure 8 gives the Radial Distribution Function of the polytope with the geodesic distance r scaled to the experimental values. Except for one far distance at about 11Å there is an obvious one to one relation between the experimental data (fig.1) and theoretical distances, with slight metrical differences. This relation indicates (if not proves) that the medium range atomic arrangements in the amorphous material bears strong similarities with the polytope order.

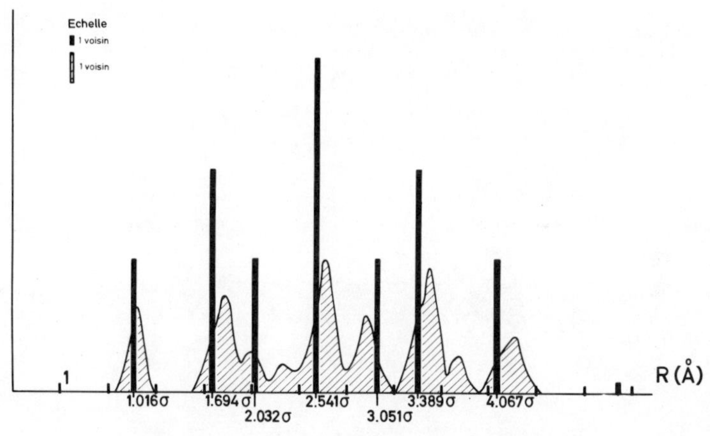

fig. 8

Radial Distribution Function for the $\{3,3,5\}$ polytope (delta functions in dark and a Radial Distribution Function obtained after mapping of the polytope.

- The *packing fraction* of the polytope can be easily calculated. One consider that hard spheres are placed at the polytope vertices with radius such they are in contact with neighbouring spheres. All the calculation is done in the spherical space S_3 (the hard spheres are "curved" such that their interior belongs to S_3). The first neighbour angular distance is $2\theta_1 = \pi/5$, which leads to a radius $r = R\theta_1$. The packing fraction f is : 0.774. This is a very high value which exceeds the value 0.74 for the densest sphere packing in R_3 (realized for example by the fcc or the hcp packing). The reason is rather simple. The fcc structure is a 3D regular packing of tetrahedra and octahedra while polytope $\{3,3,5\}$ contains only tetrahedra. The packing deficit of the FCC is due to the octahedra which are less efficient packing configurations than tetrahedra. Now amorphous metals are less dense than their crystalline counterpart. This apparent discrepency with the model disappears when one realizes that a decurving procedure from S_3 to R_3 implies a lowering of the packing efficiency of the mapped structure. A simple estimation of the mapped structure density gives very reasonable values compared to both numerical simulation of sphere packings and experimental trends in density variations.

The $\{3,3,5\}$ polytope is an important template for metallic structures, but exactly as there are, in crystalline phases, a large number of structures derived from the fcc dense structure, the $\{3,3,5\}$ polytope gives rise to several interesting derived structures, some among them being described below.

III- CURVED SPACE APPROACH: OTHER EXAMPLES OF STRUCTURES

We present now several structures in spherical space which are the frustration-free solutions to the propagation of given local orders.

III-1. Tetracoordinated structures

Tetrahedrally coordinated covalent structures (diamond, silicon, germanium, silicates, etc.) form ordered structures based on hexagonal rings of bonds. In most cases, boat- or chair-like rings are present. Neither are planar and they both have equal bond lengths and tetrahedral angles at their vertices. Configurations can also be characterized by the orientation of tetrahedra along each bond. In these ordered structures one usually find eclipsed or staggered configuration. The chair has all its bond staggered, whereas the boat has two eclipsed and four staggered bonds. These basic rings are the fundamental units in certain periodic structures. Diamond, for example, consists only of chairs rings while the wurtzite structure contains both chairs and boats. The more complex structures of zeolites, like faujasite (see, for example, Wells [14]), which contain larger rings surrounding empty channels, provide other examples of tetracoordinated structures.

In disordered covalent structures, like amorphous a-Si or a-Ge, short range order is found to be an almost perfect tetrahedral connectivity. But this local order can lead to numerous different structures in Euclidean or in curved spaces[15]. It is necessary to have informations about the order at a scale larger than the first neighbour distance. The important topological ingredient in such local order description is the local ring configuration. A ring configuration is mainly characterized by the number of edges in the ring (and its parity), the twist of rings and also by the number of rings sharing a vertex. In dense and tetracoordinated structures, the similarity of the local symmetry (tetrahedral interstices and tetrahedral binding) leads to geometrical relations between them. In curved space also, symmetry groups of both models have common sub-groups. This explain why we can obtain models for covalent structure starting from the $\{3,3,5\}$ polytope.

III-1.1 The polytope $\{5,3,3\}$

This is the polytope dual to polytope $\{3,3,5\}$ (observe in the formula $\{p,q,r\}$ the inversion between p and r). Geometrically it is obtained by centering the 600 tetrahedral cells of polytope $\{3,3,5\}$ and joining two such centres by an edge whenever their tetrahedral cells share a face. This polytope has 600 tetravalent vertices, 1200 edges, 720 faces and 120 cells. The duality is obvious in that the number of i-dimensional cells of the $\{3,3,5\}$ equals the number of (3-i)-dimensional cells of the $\{5,3,3\}$. Since each edge of the $\{3,3,5\}$ is shared by 5 tetrahedra, then each face of the $\{5,3,3\}$ is a pentagon and the cells are regular dodecahedra. On figure 9 several cells of the $\{5,3,3\}$ has been mapped and the outer dodecahedra already appear distorted.

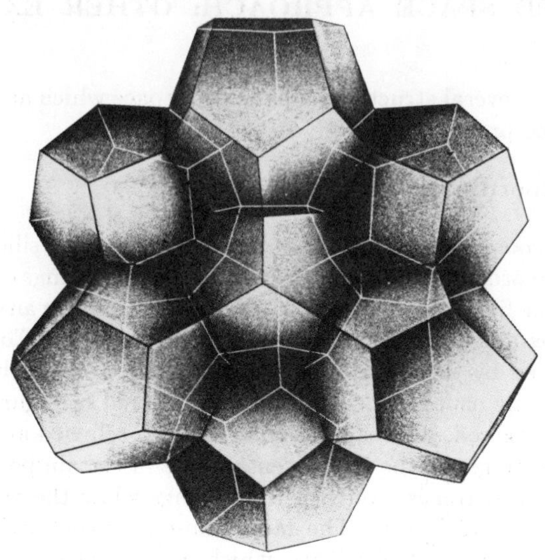

fig. 9

Local arrangement of several dodecahedral cells of polytope $\{5,3,3\}$ after mapping on a tangent R_3 space

The $\{5,3,3\}$ and $\{3,3,5\}$ share the same symmetry group G. The $\{5,3,3\}$ vertices are the image under G of the orthoscheme vertex located at a $\{3,3,5\}$ tetrahedral cells vertex. Now scaled to the first neighbour distance, the $\{5,3,3\}$ is less curved than the $\{3,3,5\}$. This is related to the smaller angular mismatch in R_3 when gluing 3 dodecahedra around a common edge compared to 5 tetrahedra. As a consequence, with the same metrical distortion, it is possible to map more $\{5,3,3\}$ vertices than $\{3,3,5\}$ vertices.

The $\{5,3,3\}$ is a clathrate-like "caged" structure. Indeed such dodecahedral cells are present in the clathrate crystal structure either in silicate compound [16] and in some ice structures [17]. The polytope $\{5,3,3\}$ may be usefull as a template for the non crystalline versions of these materials. In fact we shall show that even the crystalline clathrate structures (and their dual Frank-Kasper metallic phases) have a relationship with template polytopes and can be described as a polytope-like structure threaded by a periodic array of disclination lines.

As far as amorphous tetracoordinated semiconductors are concerned (like a-Si, a-Ge, a-GaAs,...) the polytope $\{5,3,3\}$ does not seem to describe well their local order. Indeed the $\{5,3,3\}$ contains a very high number a pentagons, whose presence leads to definite intersite distances which are not experimentally observed (at least significantly). Furthermore the expected density, with $\{5,3,3\}$ as template, is much lower than the measured one. These materials are much better described with polytope "240" which is presented below.

III-1.2 The polytope "240"

We now describe a polytope which does not present all the symmetries of the $\{3,3,5\}$. We shall use a building rule similar to that which leads to the diamond structure starting from the fcc structure : a new vertex is placed at the center of some tetrahedral cells of the compact structure. One tetrahedron over five will be centred, which has the consequence of braking the 5-fold symmetry of the $\{3,3,5\}$, only a ten-fold screw axis being preserved. The direct symmetry group is :

$$G_{240} = Y' * O'/Z_2$$

where O' is the lift of the octahedral group. Note the two following points : O' is not a subgroup of Y', polytope "240" has new symmetries absent in the $\{3,3,5\}$ (in particular a 40-fold screw axis).

The polytope "240" is chiral : it cannot be superimposed to its mirror image. This is manifest in the disymmetry between the groups acting from the left and from the right. The polytope "240" with opposite chirality has $G = O' * Y'/Z_2$ as symmetry group.

Another way to describe the "240" is to say that as for the relation fcc-diamond, one has added to the $\{3,3,5\}$ a second replica of itself, displaced along a screw axis of S_3 as it was displaced in the (1,1,1) direction in the crystalline case. It is not a regular polytope in the Coxeter sense, and so cannot be named by a triplet $\{p,q,r\}$.

Let us look to its structure more in detail. Each vertex of the first $\{3,3,5\}$ replica is surrounded by 4 vertices from the second replica. The polytope is then an alternated, bicolor structure and contains only even-membered cycles; the smallest are hexagons in a twisted boat configuration which brings the dihedral angle into a value intermediate between the eclipsed and the staggered case without modifying the "bond" angles and the first-neighbour distances. The local configuration is perfectly tetrahedral as can be seen on figure 10 which shows orthogonal mapping of subsets of increasing size of the polytope. 3 perfect boat rings can be arranged such as to form an 8-vertex cluster called the "small barrelan" (figure 11). The twisted hexagons of the "240" are also arranged by three such as to give twisted small barrelans, the twist being along the barrelan 3-fold axis respecting the tetracoordination of the sites) and are threaded by "channels" somewhat similar to that present in the diamond or wurtzite crystals. These channels follow great circles of S_3 and are bounded by a triple arrangement of chain of sites (a triple helix) close to the (1,1,0) chains of the diamond structure.

We have not described yet the frustration problem to which the polytope "240" provides an anwer. Suppose that one tries to construct a tetravalent cluster in R_3 in such a way as to maximize the interconnectedness of the structure, e.g. the number of rings sharing a common vertex. Looking to structures with hexagons, the diamond and wurtzite crystalline structures have equivalent interconnectedness: 12 hexagons through each point. In fact these two structures are equivalent to a much higher degree : the numbers of returning walks of n step in the network

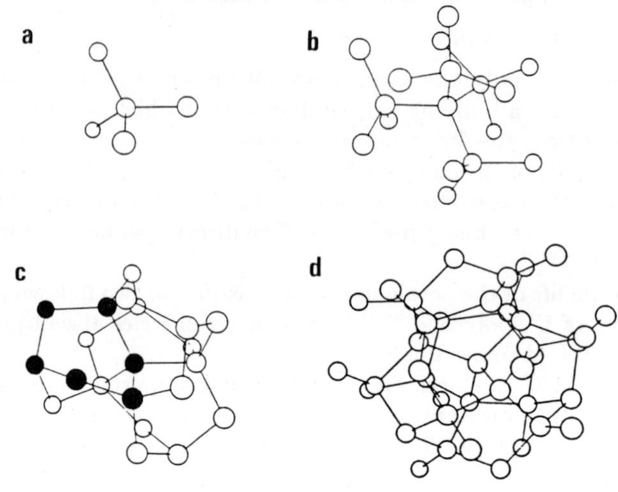

fig. 10

Local view of the polytope "240".
a) 5 vertices, the tetrahedral symmetry is clearly visible;
b) 17 vertices;
c) 21 vertices, a sixfold ring is distinguished;
d) 39 vertices.

for any value of n, are the same. In a simple tight binding model for elementary excitations they are "isospectral". Now it is locally possible to put more hexagons around a common vertex. By twisting boat-like hexagons, it is possible to have as much as 18 rings passing through a vertex. This configurtation cannot however propagate freely in R_3 ; it is geometrically frustrated. But it can be realized on S_3 in the form of the polytope "240". Note that this local order minimizes a tight binding form of the energy for small clusters. Indeed the number of dangling bonds (which cost energy) per atom is lower than for small clusters of diamond or wurtzite structure. Finally let us say that polytope "240" is locally more dense than diamond.

Polytope "240" has a very close relationship with a well known hand built model, the Connel-Temkin model [18]. The C-T model plays a very special role among CRN because it contains only even-membered rings. Dixmier et al [19] have given a decisive argument to experimentally distinguish the C-T model from more generic continuous random networks (CRN) like the Polk model [20], involving relative positions of the Interference Function peak. The C-T model is an extreme case in the class of CRN, and is strongly related to polytope "240". Indeed as in

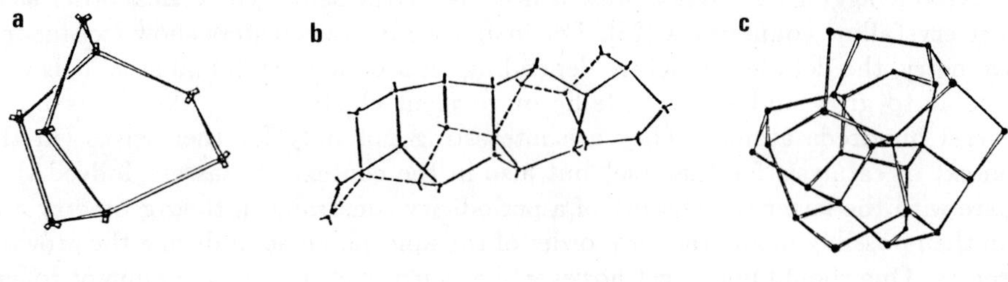

fig. 11

a) A small barrelan

b) The local bonding arrangements of sites around the 30/11 channeling axis.

c) An assembly of four small barellans around a central atom forming a 27-vertex cluster. This 4-valent configuration presents a high degree of interconnectedness.

the polytope case, the C-T contains many twisted boat-hexagons forming small barrelans. Even more an inspection by eye of the C-T model [21] has shown a region characterized by a given sign of chirality separated from a region of opposite sign by a thin interface defect region. If polytope "240" should be a good model for a-Si, an important question is whether the local chirality can be observed. Note that a positive experimental answer would provide a spectacular confirmation for the Curved Space Model. However there are several difficulties. It is for instance possible that the actual material contains both sign of chirality, with the same amount, leading to an overall racemic material, difficult to detect. Nevertheless it has been suggested that third harmonic generation in optical experiments may be sensitive to the local chirality [22].

III-2. Description of finite clusters

Atomic aggregates do not present necessarily the same geometrical order as in their crystalline counterpart [23]. For instance rare gas clusters show icosahedral symmetry, the detailed structure depending on atomic species and size. It is now possible to grow and study metallic or covalent clusters from few atoms up to several hundreds atoms. They are interesting not only for themselves (in the context of catalysis for example) but also in the context of glasses. Indeed they share with the latter the absence of a periodicity constraint in their geometry and can then possibly model the local order of the amorphous solid during the growing process. One should not forget however that surface effect play a dominant role in cluster stability while its occurence in glasses could only be invoked for a dynamical process. Several types of clusters have been proposed in the past. Some of them have a close relationships with mapped polytopes. The main results have been summarized in ref.[23].

III-2-1. Cluster indexation

We want to index finite clusters which can be derived from a given polytope P. A finite portion of P is mapped onto a tangent hyperplane R_3. We need to specify the tangent point T and the polar angle α which limits the region to be mapped. This is shown schematically in Figure 12. The metrical properties depend on the precise location of the projection point Q which is defined by its distance d from T. Note that this position may not necessarily be unique for all the values of β ($\alpha = max(\beta)$). The general notation for the cluster is

$$(4) \qquad\qquad \mathcal{C} = [P, T, \alpha, d].$$

The first three quantities completly define the cluster from a topological point of view, while the last one is of metrical nature.

III-2-2. Compact clusters

<u>The polytope $\{3, 3, 5\}$ and icosahedral local order in R_3</u>

Clusters mapped from the $\{3, 3, 5\}$ should be the closest to perfect icosahedral order. The price to pay is that the clusters are fairly small (recall that the polytope has 120 vertices). Table 1 give a description of these clusters as well as their energy per site calculated within the simplest Lennard-Jones potential ($V(r) = 1/r^{12} - 2/r^6$).

E_1 correspond to a unique d optimized for each cluster as a whole, while for E_2 d is shell-dependant. The cases $\alpha \geq 2\pi/3$ are not considered because unrealistic. The first 4 clusters, shown on Figure 13 are very familiar in this field and have been used to model the growth of quasicrystals [24]. Note that by rotating S_3 we could obtain clusters with the same local order but with tetrahedral global symmetry. This is true for any of the clusters described below.

<u>Geodesic hyperdomes</u>

Regular triangulation of S_2 based on the icosahedron have been popularized by B. Fuller and are called geodesic domes. It is possible to generalize this idea

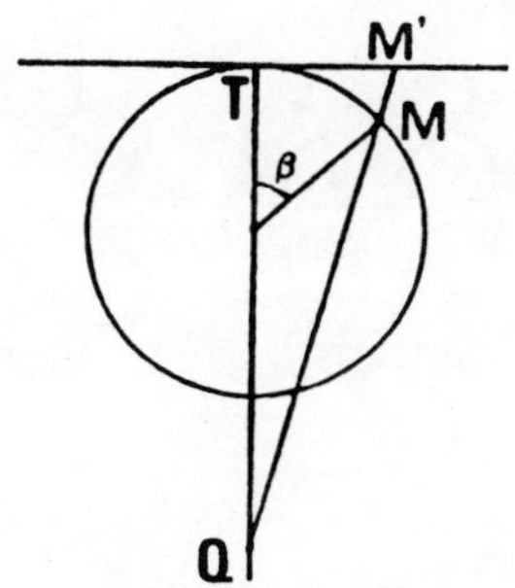

fig. 12

Partial mapping. The polytope is embedded in S_3. M (with polar angle β) is mapped onto M'. T is the tangency point; Q is the projection point and d the distance TQ.

$\alpha\,(rd)$	N	$E1$	$E2$
$\pi/5$	13	-3.41	-3.41
$\pi/3$	33	-4.28	-4.32
$2\pi/5$	45	-4.15	-4.5
$\pi/2$	75	-4.16	-4.54
$2\pi/3$	87	-3.06	-4.22
$3\pi/5$	107	-3	-4.3

Table 1

Clusters mapped from the $\{3,3,5\}$ polytope and their energy per site.

for higher dimensinal spheres and here to S_3 [25]. The $\{3,3,5\}$ tetrahedral cells are iteratively filled by larger and larger portions of fcc lattice. We call G_i these successive polytopes (G_0 being the $\{3,3,5\}$). Table 2 describes clusters mapped from G_1 and their corresponding energies. The 55 atom cluster is often called the 'Mackay icosahedron' in the field of quasicrystals.

III-2-3. Tetravalent clusters
The polytope $\{5,3,3\}$ and dodecahedral local order in R_3

Tilton [26] described dodecahedral packings of increasing size. He noted that, with one central dodecahedron, the more symmetrical clusters contain 13 and 45 units. If we call C the centre of a $\{5,3,3\}$ cell, they correspond to:

$$[\{5,3,3\}, C, 0.388rd, d] : 1 \quad \text{dodecahedron,}$$
$$(5) \qquad [\{5,3,3\}, C, 0.962rd, d] : 13 \quad \text{dodecahedra,}$$
$$[\{5,3,3\}, C, 1.435rd, d] : 45 \quad \text{dodecahedra.}$$

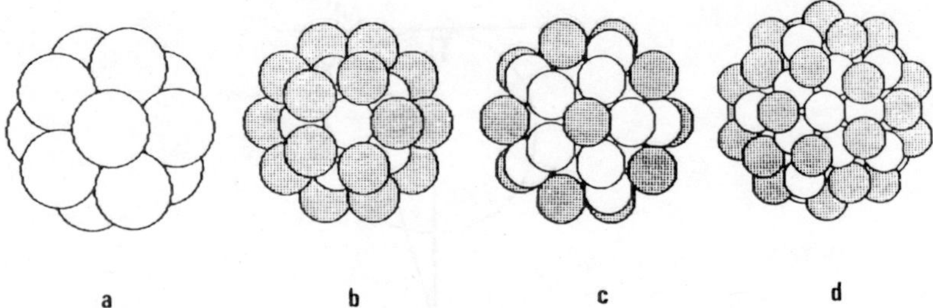

a   b   c   d

fig. 13a-d

The first 4 clusters ($N = 13, 33, 45, 75$) obtained by mapping the $\{3,3,5\}$.

$\alpha\,(rd)$	N	$E1$
$\pi/10$	13	-3.37
0.56	43	-4.45
$\pi/5$	55	-4.83
0.82	115	-5.36
$3\pi/10$	127	-5.19
1.02	157	-5.26
$\pi/3$	177	-5.32
1.14	237	-5.1

Table 2

Clusters mapped from the G_1 polytope and their energy per site.

d is not known, which means that our mapped clusters are only topologically equivalent to those built by Tilton. He also described a cluster centred on a vertex with 54 dodecahedra. It is equivalent to $[\{5,3,3\}, V, 1.475rd., d]$ where V is a vertex of the $\{5,3,3\}$. Note that none of the clusters have $\alpha > \pi/2$. Tilton noted that larger clusters would have too much strain which is evident from the the polytope point of view since the next shells preimage would belong to theS_3 south hemi(hyper)-sphere. A more metrical comparison can be done with the work of Coleman and Thomas [27] who gives not only the number of sites but also the distances from a central vertex. Their dodecahedral cluster corresponds to $[\{5,3,3\}, V, \pi/3, d]$. Table 3 compares their data with those calculated for the polytope (with geodesic distances scaled to a first-neighbour of 2.35, appropriate to silicon atoms). The chosen value of d is $2.75 \times R$. The small differences between the hand-built and polytope-mapped models simply means that one should not use a constant d value to fit the data.

Decorated geodesic hyperdomes

It is possible to decorate the above fcc regions in the G_i polytopes such as to give diamond lattice regions. We call T_i the resultant tetracoordinated polytopes. The matching around the $\{3,3,5\}$ original edges involves pentagonal rings. The first two stages of the decoration are shown on Fig.14. Now it is possible to

Number of neighbour	4	12	24	12	4	24	24	28
Coleman-Thomas	2.35	3.83	5.45	6.3	6.67	7.53	8.6	9.12
Polytope (on S3)	2.35	3.82	5.45	6.27	6.62	7.43	8.48	9.08
Mapped polytope	2.35	3.83	5.47	6.3	6.66	7.49	8.56	9.17

Table 3

Comparison between the Coleman-Thomas cluster and the $\{5,3,3\}$ based cluster.

identify the large cluster built by Dandoloff et al. [28] as mapped from T_1 and those studied by Gaskell [29] as mapped from T_2. More precisely in the latter case, the 280 vertex model with icosahedral symmetry is $[T_2, C, \pi/5, d]$ while the 798 vertex model with tetrahedral symmetry is very close to $[T_2, V, 0.776rd, d]$ (Figure 15).

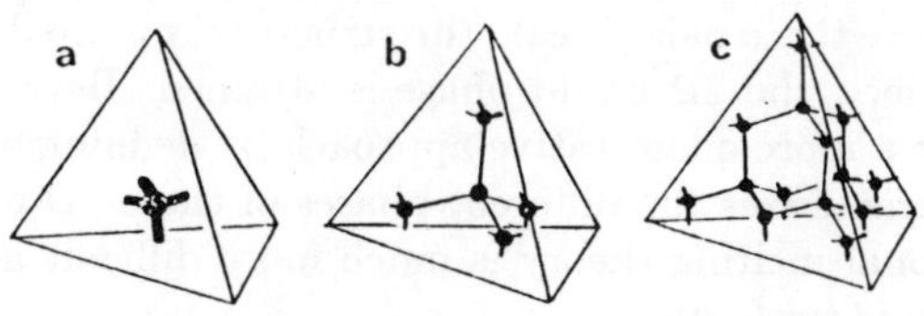

Figure 14a-c

First stages of decoration of the $\{3,3,5\}$ tetrahedral cells. a) 1 vertex, which generates the polytope $\{5,3,3\}$, b) 5 vertices, c) 14 vertices.

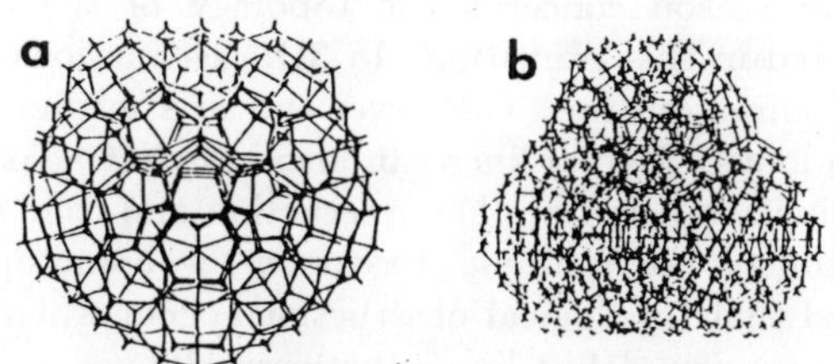

Figure 15a-b

Tetravalent clusters described by Gaskell (Ref.29). a) the 280-vertex model with icosahedral symmetry, b)the 798-vertex model with tetrahedral symmetry.

IV-DISCLINATIONS AND CURVATURE

IV-1 Disorder and defects

It has been proposed in the past to describe amorphous structures as regular crystals containing dislocation defects. However the number (or density) of such defects should be very high, of the order of one defect per atom, in such a way as to describe reasonable non crystalline structures. Therefore the dislocated crystal approach becomes meaningless.

Nevertheless these defects play a crucial role in the theoretical description of 2-Dimensional melting as proposed by Halperin and Nelson [30] in continuation of the work done by Kosterliz and Thouless [31]. Starting with the low temperature crystal phase, and raising the temperature, this model leads first to a transition to an "hexatic" phase. Dislocations, which appear as pairs and then become unlinked, destroy the translational order of the crystalline phase while keeping the orientational order. 2D dislocation points are disclination pairs (of opposite sign). At higher temperature these pairs break, the structure is more disordered and loses its orientational order : the 2D liquid phase is attained. Beyond this description it is possible to give a more quantitative approach by defining an order parameter whose behavior characterizes the different phases of these 2D materials.

The 3 Dimensional melting theory is much more difficult and is not yet fully derived. The reason is twofold :

1) In 3D the locally more compact structures (tetrahedral packings) cannot propagate into compact crystals, as we saw in the preceding chapter. Recall that in 2D the local and global best packing of disks is achieved in the triangular lattice. It is this periodic structure which serves as initial crystal in the 2D melting theory and its numerical study. In 3D the more compact crystals, fcc or hcp, do not fulfill a local stability criterion as it is well known from the physics of clusters.

2) The second reason concerns the topology of the defects which would naturally be invoked in this transition. In 3D, disclinations and dislocations are linear defects. Disclinations lines can never end in a medium ; they form infinite lines or close into loops. Several lines can also intersect at points which form the vertices of a disclination network. The interaction and combination of these lines are difficult to model. For example, based on the homotopy theory of defects, it has been argued that topological obstructions occur when two lines cross each other, giving rise to a new defect line in between [32] .

So the type of problems that makes the 3D melting theory difficult are similar to that encountered in amorphous systems. The curved space model provides a theoretical solution to the first above mentionned difficulty by invoking an ideal local and global close packing on the hypersphere S_3. We shall see now that disclination lines enter very naturally in the process of decurving this ideal structure back to the euclidean space R_3. The complexity of the final structure will be encoded in the geometry of the disclination lines, either periodic, hierarchical or more disordered.

IV.2 What is a disclination ?

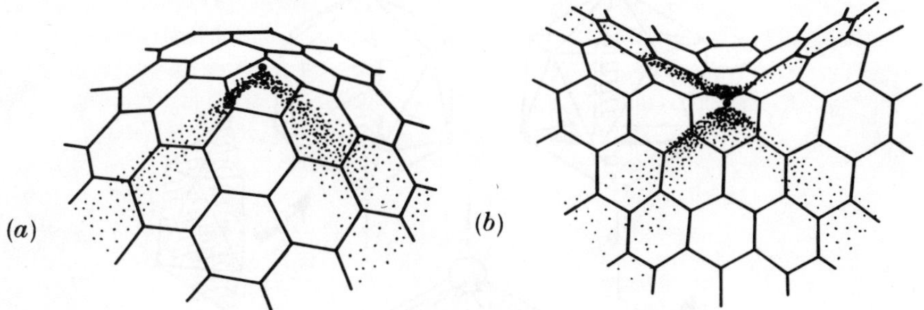

fig. 16

Disclinations in a hexagonal structure:
a) Positive disclination.
b) Negative disclination.

A disclination is a defect involving rotation operations, as compared to usual dislocations associated to translations. It can be generated via a Volterra process by cutting the structure and adding (or removing) a wedge of material between the two lips of the cut. The defect is point-like in 2D and linear in 3D. The two faces of the wedge should be equivalent under a rotation belonging to the structure symmetry group in order to get a pure topological defect which is then confined near the axis of the cut. It is then possible to describe the defect, and the induced deformation, as a concentration of curvature (figure 16). Starting with a regular structure defined in curved space, it is then possible to lower its curvature by introducing disclinations of the appropriate sign.

Clearly visible in figure 18 is the topological consequence of this defect. If the disclination goes through a vertex or an edge, it changes its coordination ; otherwise it modifies the tiling element (here the ring size). So introduction of disclinations not only allows for decurving but also generates slight modifications of the local configurations, as expected in amorphous solids. This makes the disclination decurving mode very attractive in the curved space approach. Furthermore we shall see later that intrinsic disclination networks can be found in some complex crystalline structures. Note the following important point: the term "defect" is used with reference to the ideal (defect-free) curved space structure. Its occurence in ordered structures as intrinsic ingredients does not lead therefore to a paradox, but is just a matter of definition.

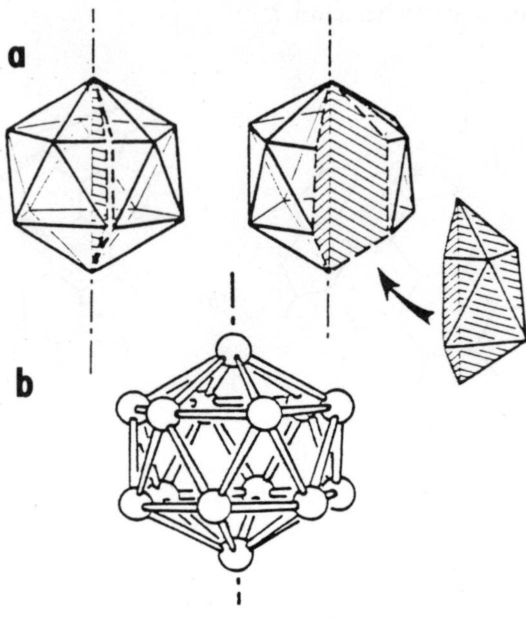

fig. 17

a) Procedure to insert a disclination.

b) Effect of a disclination on an icosahedral order: a Z_{14} coordination polyhedron.

IV-2.1 Disclinations in a polyhedron

We now show how to generate disclinations in the simple case of a polyhedron. We do not consider the solid object but the 2D tiling on a topological sphere. The cut axis is aligned with one of the polyhedron symmetry axis and the Volterra construction is done. Figure 18 illustrates the case of a tetrahedron with disclinations along the 3-fold and 2-fold axes. The axes cuts the polyhedral surface in two points, the defect being thus a couple of disclination points ; the 3-fold axis case changes the tetrahedron into a square pyramid. The two point defects are a vertex (which becomes 4-fold coordinated) and the center of a triangle (which becomes a square). The 2-fold axis case is remarkable in that its transforms the tetrahedron into another regular (Platonic) polyhedron : the octahedron. In that case two opposite edges are transformed into triangles and the vertices become 4-fold coordinated.

IV.2.2 A pair of disclinations in polytope $\{3,3,5\}$

We are going to generate a pair of $2\pi/5$ disclination lines in the $\{3,3,5\}$. The local order around such a line is shown on figure 19. In the $\{3,3,5\}$ the coordination polyhedron of a point is a regular icosahedron. The cut axis is aligned with a 5-fold icosahedral axis and a $2\pi/5$ wedge is inserted. The new coordination polyhedron is a 14-vertex triangulated structure and the central vertex is called a

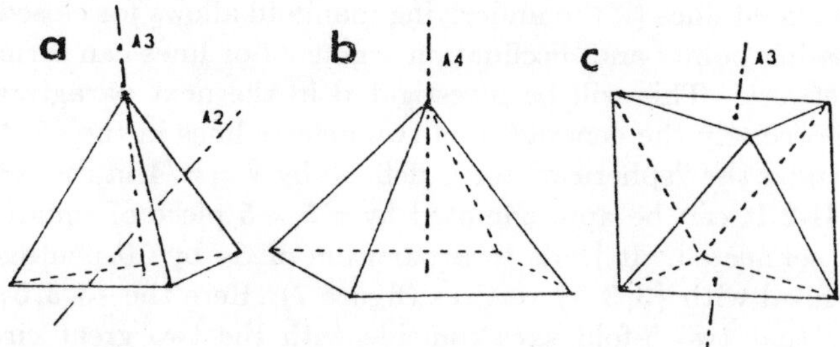

fig. 18

Disclination in a tetrahedron.
a) The initial polyhedron
b) The defect line along a 3-fold axis gives rise to a square pyramid.
c) The defect line along a 2-fold axis gives rise to an octahedron.

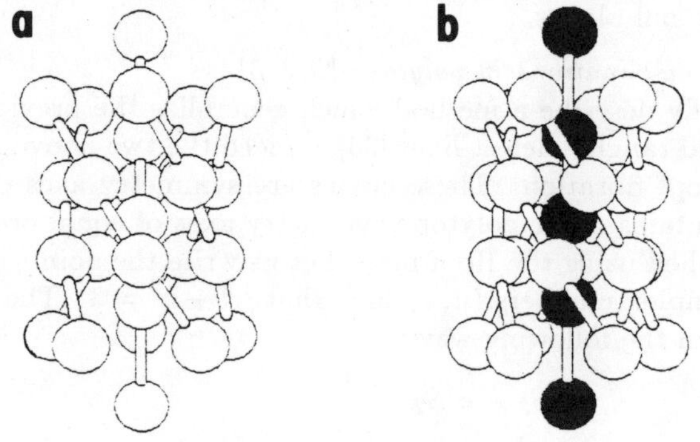

fig. 19

Local configuration of sites in S_3. Only intra-fibre edges are drawn.
a) In the $\{3,3,5\}$ each arrangement of neigbouring sites is surrounded by
5 similar sets. Note that the mapping has transformed the central great
circle subset into a straight line and other circles into helices. b) When
a disclination is introduced, the line defect (with sites in dark) is now
surrounded by 6 similar sets.

Z_{14} site in standard notation. The central site and the two opposite vertices on
the cut axis belong to a $2\pi/5$ disclination line. It is possible to show that such

line cannot stop in the material and either ends at the surface, splits at crossing points or form closed lines (if the underlying manifold allows for closed geodesics). The set of crossing points and disclination segments or lines can form a so-called disclination network. This will be investigated in the next paragraph. We first describe more precisely the generation of two defects lines in the $\{3,3,5\}$.

Let us consider the "spherical" torus defined by $\theta = \pi/4$ in the torus foliation of paragraph II-2 It can be approximated by a 5×5 piece of square tiling with opposite sites connected. It leads to a partition of S_3 by 10 pentagonal prisms which can be filled with $\{3,3,5\}$ vertices (figure 7). Here the $\{3,3,5\}$ is oriented in such a way that two 5-fold axes coincide with the two great circles defined by $\theta = 0$ and $\theta = \pi/2$. Each pentagonal prism can be still cut into five trigonal prisms. A $2\pi/5$ disclination line can be generated by inserting a sixth such trigonal prism. This can be done simultaneously for the two sets of five pentagonal prisms, leading to two sets of hexagonal prisms. Only the sites located on the great circles $\theta = 0$ and $\pi/2$ have their local environment modified ; they become Z_{14} sites. The new polytope contains 168 sites, 144 12-fold coordinated vertices (Z_{12} sites) and $24 Z_{14}$ sites. The combination of two disclinations along opposite great circles can be called a screw disclination since it involves simultaneous rotations around two completly orthogonal planes.

IV-2.3 More disclinations in polytope $\{3,3,5\}$

We now briefly describe a method which generalize the previous approach to the case of several tangled defect lines [33]. Indeed the two above great circles are members of a Hopf fibration. These circles are symmetry axes of the polytope. The fibration contains other polytope symmetry axes of equal order. These axes can be distinguished using the Hopf map. Let us write the points on a unit radius S_3 as pair of complex numbers (u, v) such that $u^2 + v^2 = 1$. The Hopf map may then be defined in the following way :

$$
\begin{aligned}
\psi : S_3 &\longrightarrow S_2 \\
(u, v) &\longrightarrow (\omega, \phi) \\
\omega &= 2cot^{-1}(|\frac{u}{v}|) \qquad \omega \in [0, \pi] \\
\phi &= arg(u/v) \qquad \phi \in [0, 2\pi[
\end{aligned}
$$

(6)

where ω and ϕ are polar coordinates on S_2.

This map is such that the inverse image of a point on S_2 is a great circle on S_3. The full inverse image of S_3 gives a fibration of S_3 with great circles, called the Hopf fibration, similar to the one considered in the blue phase problem (see the paper by E. Dubois-Violette and B. Pansu in this volume). Each fibre is tangled with any other fibre. Now if a polytope symmetry axis is a member of the fibration, other axes, conjugates to the previous one also belong the fibration. In the case of the 5-fold axis the great circle contains 10 vertices (it is also a 10-fold screw axis). There are 11 other axes belonging to the fibration, each polytope vertex being on exactly one such axis. The Hopf map of these 12 axes leads to 12 points on S_2 and its five neighbours. The pre-image on S_3 is a great circle (containing 10 polytope vertices) surrounded by five silimar great circles as shown in figure 19. Disclinations on S_3 can be introduced by first disclinating on S_2 and then inverting the Hopf map. In the case considered on figure 19, two $2\pi/5$ disclination points are created on $S2$ which leads to two $2\pi/5$ disclination lines. One recovers the above discussion of polytope "168". The local order along one such disclination in S_3 is shown on figure 19. The line defect (with dark sites) is now surrounded by 6 lines of the fibration. It is further possible to generate 3 or 4 defect points on S_2, leading to 3 or 4 defect lines and the polytopes "180" and "192" according to their number of vertices. These two polytopes have also been called D_{15} and D_{16}. This is because they give, under a Hopf map, a Z_{15} and a Z_{16} polyhedron on S_2. For the same reason polytope "168" is also called D_{14}. The RDF of these polytopes is shown in figure 20 and compared to the original $\{3,3,5\}$ RDF. New distances have appeared which can be interpreted as a broadening of the $\{3,3,5\}$ distribution. Scaled to the first neighbour distance, the radius of S_3 has increased from the $\{3,3,5\}$ to the disclinated polytopes values which shows that it is a decurving process.

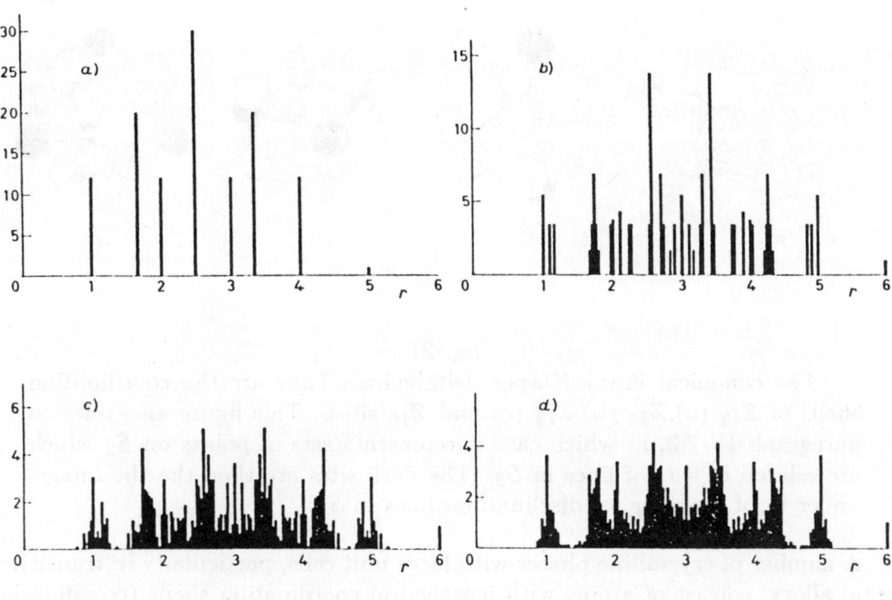

fig. 20

Radial Distribution function of disclinated polytopes : a)$\{3,3,5\}$, b) D_{14}, c) D_{15}, d) D_{16},

IV-3 Disclinations in other polytopes

- polytope $\{5,3,3\}$

To each disclinated $\{3,3,5\}$, it corresponds a disclinated $\{5,3,3\}$ obtained by taking the dual. For example the polytope "864" with a pair of $2\pi/5$ disclination has been previously described and its excitation spectrum calculated [34].

- polytope "240"

It is possible to introduce a pair of disclinations along two opposite 30/11 axes. Since the defect lines passes through the middle of the channels, they avoid the sites which remains 4-fold coordinated. The defects are twisted disclinations and according to their strength, one obtains different polytopes. We have presently generated 2 polytopes with respectively 384 and 360 vertices. The latter is very interesting in that the two defect lines thread 7-fold rings, while all the remaining rings are even-membered. These lines are therefore instances of Rivier lines [35] whose effect on the excitation spectrum has been investigated [36].

IV-4 Disclination networks in periodic complex structures: The Frank-Kasper phases

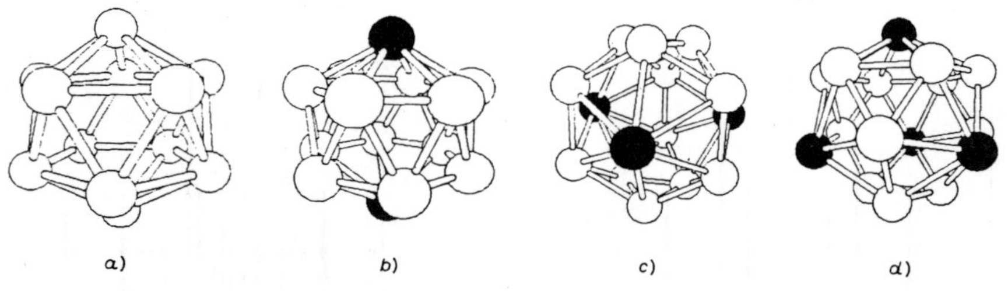

fig. 21

The canonical Frank-Kasper deltahedra. They are the coordination shells of Z_{12} (a),Z_{14} (b),Z_{15} (c) and Z_{16} sites. This figure also refer to paragraph IV-2.3, in which case it represents sets of points on S_2 which are related to sets of lines in S_3. The dark sites are then the the image, under Hopf mapping, of disclinations lines in S_3

A number of crystalline phases with large unit cells, particularly in transition-metal alloys, consist of atoms with icosahedral coordination shells (coordination number $Z = 12$) interrupted by atoms with $Z = 14, Z = 16$ or even more. Since these structures are tetrahedral packings, the first neighbour shell of any vertex is a fully triangulated polyhedron, called a deltahedron. Frank and Kasper described these structures in term of canonical deltahedra (figure 21) and showed that their complexity can be encoded in a skeleton whose edges connect neighbouring sites which are not surrounded by an icosahedron[7]. This skeleton can be considered as a disclination network [37][38]. Indeed consider a Z_p site (a site with p neighbours). The p neighbours belong to a deltahedral shell. We can consider successively all the edges joining the given site to its neighbours. Each such edge is the axis of a t-fold bipyramid, where t is coordination of the neighbour on the deltahedral shell. This edge is called a disclination edge when $t \neq 5$ (figure 22). This defect is referred to the polytope $\{3,3,5\}$ where any edge is the axis of a pentagonal bipyramid. The above skeleton consists of the t = 6 edges and thus avoids the sites with an icosahedral environment. The disclination network has two basic properties :

- A defect edge cannot simply terminate at a site where all the remaining bonds are fivefold. Indeed such a site would have coordination number $Z = 13$ with one 6-fold and twelve 5-fold coordinated vertices on its coordination shell. It is known that such a deltahedron does not exist . When disclinations are interpreted as lines carrying curvature, the fact that a line cannot terminate is a discrete consequence of Bianchi identities [39]. So the defect network consists in linear parts (passing trough Z_{14} sites) which may ramify at 3-fold nodes (Z_{15} sites), 4-fold nodes (Z_{16} sites), etc...

- Since the complex structure is crystalline, the disclination network is itself an ordered periodic array.

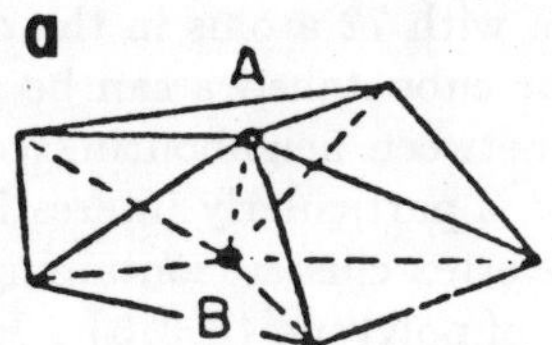
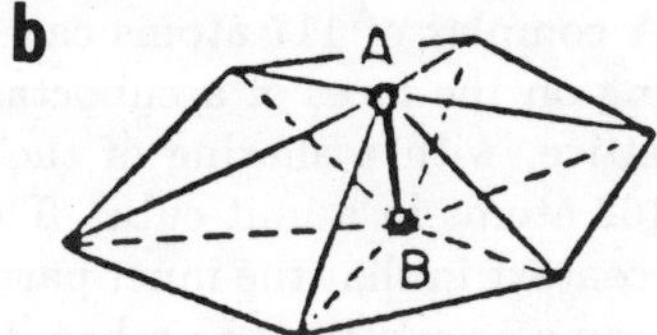

fig. 22
Description of a disclination in a tetrahedral medium:a) a pentagonal bipyramid, no disclination. b) a 6-fold bipyramid. The edge AB is part of a $2\pi/5$ disclination line.

Let us describe three cases :

- The tungsten (A15) structure.

There are 8 atoms per unit cell, two of them with $Z = 12$ (and an icosahedral shell) and the remaining with $Z = 14$. The latter form a triple periodic array of infinite non interesting lines (figure 23).

- The Laves phases structure

There are twelve atoms per unit cell, 8 Z_{12} and 4 Z_{16} sites. Four disclination edges connect at a Z_{16} site at tetrahedral angle and form a diamond-like structure.

264

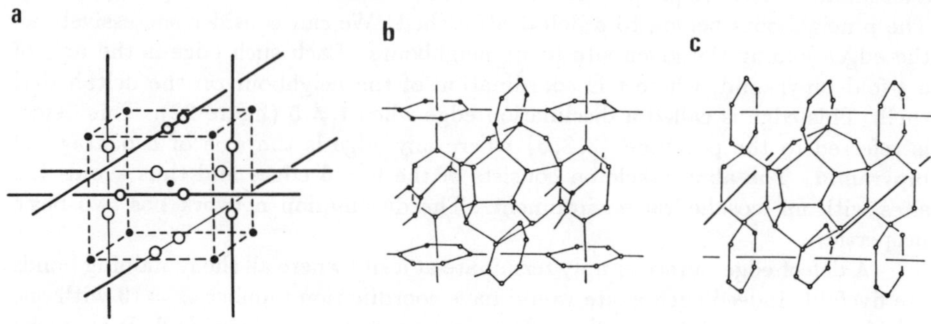

fig. 23

The disclination networks (major skeletons) of A15 (a), Bergman (b)
and SM (c) structures.

- The structure of $Mg_{32}(Al, Zn)_{49}$ [7]

This is a very interesting though rather complex structure. It represents the
result of maintaining icosahedral packing. starting with an icosahedron, spheres
are added in successive shells in a manner always to center the triangles of previous
shells. A complex of 117 atoms can be built up, with 72 atoms in the outermost
shell lying on the faces of a cuboctahedron. The cuboctahedra can be packed in
a bcc lattice, with a sharing of the 72 atoms between neighbouring complexes,
giving 162 atoms in a unit cube. This structure is particularly interesting in the
present context in that the inner part of the 117-vertex cluster, containing 45 sites,
has the same geometry as in a hemihypersphere of polytope $\{3,3,5\}$. Indeed the
successive shells surrounding the central vertex are an icosahedron, a dodecahedron
and a larger icosahedron. So this crystalline structure realizes a mapping of the
$\{3,3,5\}$ onto R_3, with large mapped polytope regions interrupted by defects. It is
in fact possible to derive another structure sharing the same atomic shells, which
was called SM structure elsewhere [44]. It is a simple cubic structure with the
same cuboctahedral clusters of atoms centred at the cubic cell vertices and at
its centre, but with a different orientation for the latter. Both structures can be
described as $\{3,3,5\}$ decurved by an infinite network of disclinations (figure 24).
The SM structure can be easily obtained from an A15 structure, applying on it
the iterative construction described in the next paragraph.

IV-5. Hierarchical disclination networks

We now show how it is possible to complertly flatten a polytope and get an infinnite structure in R_3. This is done by an iterative process whose main effect can be put in the follwing matrix form:

$$(7) \qquad \mathcal{N}^{(i)} = \mathcal{T}\mathcal{N}^{(i+1)}$$

$\mathcal{N}^{(i)}$ is a 3D vector whose components ($\{N_p\}$) are the total number of Z_p sites after the ith iteration. The matrix $\mathcal{T}$ can be interpreted as a transfer matrix for the iteration. There are many distinct types of iterations [25][40][41] and we just focus here on the simplest.

The Voronoi cell of the $\{3,3,5\}$ is a dodecahedron $5,3$, the basic cell of the dual polytope $\{5,3,3\}$. A new polytope is generated by filling in these Voronoi cells by with suitably oriented centred icosahedra (figure 24).

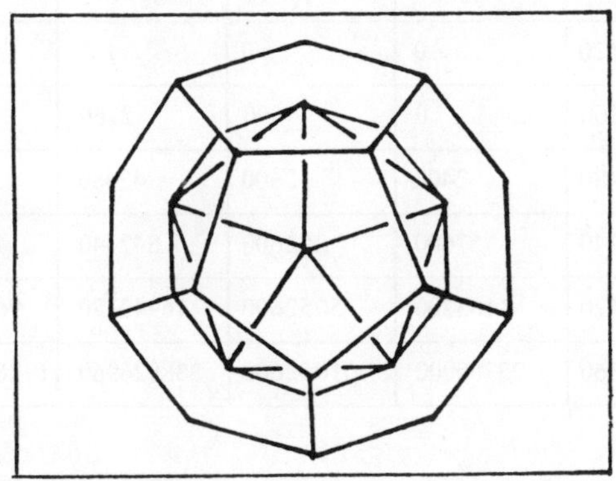

fig. 24

Decoration of the dodecahedral Voronoi cell by a centred icosahedron

The new structure has 2160 vertices , distributed as follows: there are 600 vertices as the Voronoi cell vertices (the number of sites in the $\{5,3,3\}$). Each of the 120 dodecahedral cell is filled with 13 vertices (the centred icosahedron) which add up to the remaining 1560 sites. A closer look to the structure shows that the 600 sites are 16-fold coordinated while the others are 12-fold coordinated. The transformation can repeated by considering the Voronoi decomposition of the previous structure and filling in again with appropriate centred deltahedra. At the second step, 14-fold (Z_{14}) sites are generated. When the process is repeated again and indefinitely, only Z_{12}, Z_{14} and Z_{16} are generated. The transformation can be written formally :

$$(8) \qquad \begin{array}{lllllll}
Z_{12} & \longrightarrow & 13\ Z_{12} & + & & + & 5\ Z_{16} \\
Z_{14} & \longrightarrow & 12\ Z_{12} & + & 3\ Z_{14} & + & 6\ Z_{16} \\
Z_{16} & \longrightarrow & 12\ Z_{12} & + & 4\ Z_{14} & + & 8\ Z_{16}
\end{array}$$

Thus the above $\mathcal{T}$ matrix reads

$$(9) \qquad \mathcal{T} = \begin{pmatrix} 13 & 12 & 12 \\ 0 & 3 & 4 \\ 5 & 6 & 8 \end{pmatrix}.$$

The iteration begins with the $\{3,3,5\}$ and so $\mathcal{N}^{(0)} = (120,0,0)$. Multiplication by $\mathcal{T}$ gives informations about the successive polytopes as summerized below on table 4

	N_{12}	N_{14}	N_{16}	N	T	$\overline{Z}$
P_0	120	0	0	120	600	12.
P_1	1560	0	600	2160	12000	13.111
P_2	27480	2400	12600	42480	240000	13.2999
P_3	537240	57600	252600	847440	4800000	13.328
P_4	10706520	1183200	5052600	16942320	96000000	13.3325
P_5	214014360	23760000	101052600	338826960	1920000000	13.3332

Table 4

Informations about the iterative polytopes denoted by P_i: N_p is the number of p-fold coordinated vertices, N the total number of vertices , T the number of tetrahedra and Z the mean coordination number .

Note that N, the total number of sites, increases very rapidly. Scaled to the first neighbour distance, the radius of S_3 increases accordingly and tends to infinity: the curvature vanishes for the asymptotic polytope P_∞, whose mains characteristics can be derived from the properties of $\mathcal{T}$. To the largest eigenvalue of $\mathcal{T}$ (the so-called Perron root) corresponds an eigenvector whose components described the average properties of P_∞. Indeed it refers to an asymptotic situation where the relative fraction of different types of sites remains constant under iteration. In the present case the Perron root is $\lambda = 20$ and the P_∞ mean coordination number , $40/3$, is closely approached after only very few iterations. This number is also not far from that of the Coxeter "statistical honycomb" discussed below.

Up to now we have said nothing about the geometry of the disclination network. Here again it contains all except the Z_{12} sites. At the first iteration the Z_{16} sites are connected along the edges of the dual polytope $5, 3, 3$. Thus the disclination network is a packing of dodecahedra and has the same symmetry as the polytope itself. At the second iteration, the preceding network is still present (with larger edge length) and is interlaced with a new one having shorter edges. The Z_{16} sites are nodes in both networks and the Z_{14} sites sit on the edges of the former. A local view of the disclination network is shown on figure 25. Upon iteration one gets a hierarchy of interlaced disclination networks.

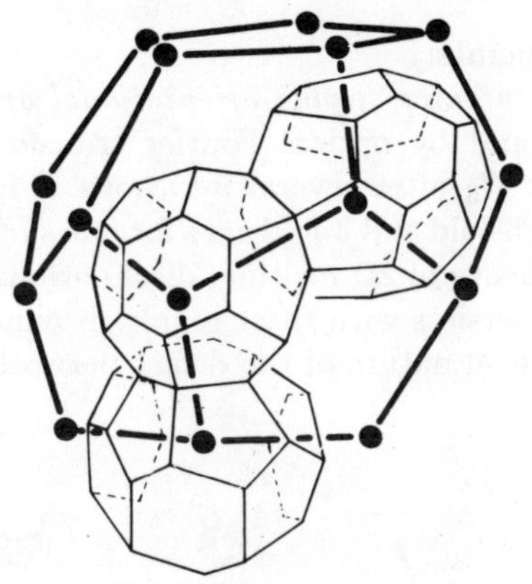

fig. 25

Local view of the interlaced disclination network after two iterations

At each iteration a network of disclinations is introduced. This network is formed by edges of the dual of the previous structure, but the networks of disclinations obtained by all previous iterations remain unaffected. Upon iterating an infinite number of times, an infinite number of interlaced disclination networks lies inside the structure. Each one has its own length scale : the distance between two nodes in the network (Z_{16} sites). In the structure obtained at iteration i the network introduced by a previous iteration j has an internode distance equal to 3^{i-j} , in interatomic distance units. The networks are all tetravalent, but not exactly self-similar, the latter property being obtained only at the infinite limit. The above decoration could be as well described by saying that the tetrahedral cells of the $\{3,3,5\}$ polytope have each been separated into four smaller tetrahedra and one Friauf-Laves (FL) polyhedron by dividing edges into three segments ($\lambda = 3$). This procedure can be extended to any division by odd-number splitting tetrahedra into smaller tetrahedra and FL polyhedra[41]. We have seen above that, for $\lambda = 3$, the decorated structure is a polytope (P1) which can be characterized by a disclination network formed by edges of the $\{5,3,3\}$ polytope. When λ is greater than 3, one obtains also a disclination network, but with a different geometry. It

268

can be shown that, upon a suitable decoration of the FL polyhedra, there remains only Z_{12}, Z_{14} and Z_{16} sites. A transfer matrix can also be introduced depending on the parameter λ

$$(10) \qquad \mathcal{T}_\lambda = \begin{pmatrix} 10A + \lambda & 12A & 14A - (\lambda - 1) \\ 0 & \lambda & 2(\lambda - 1) \\ 5A & 6A & 7A + 1 \end{pmatrix}.$$

with $A = \lambda(\lambda^2 - 1)/24$.

All such decorations of the $\{3,3,5\}$ polytope yield structures that are packings of tetrahedra.

Note the following points:

-*The hierarchical structures display orientational order.* This has been tested both numerically [42] and by optical Fourier transform [43] of large clusters, mapped onto a tangent R_3 after several iterations. Figure 26 shows the optical Fourier transform along 5-fold and 3-fold axes for one such cluster. Note that in the disclination mediated theory of 2D melting, disclinations destroy the orientational order. The fact that it persists with these highly disclinated networks is probably linked with the hierarchical nature of the defect networks.

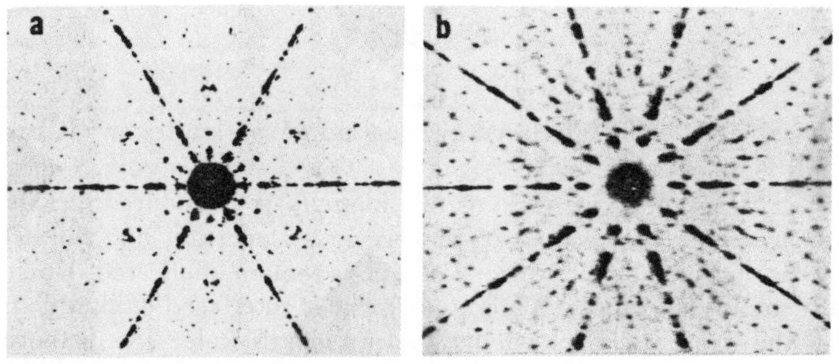

fig. 26

2D optical Fourier transform of a finite piece of the hierarchical polytope P_2 orthogonally mapped onto a tangent R_3: a) 3-fold symmetry axis. b) 5-fold symmetry axis.

-*Hierarchical clathrate-like models.* It is possible to decorate the above close-packed structures ans get several infinite 4-fold networks with very intricate disclination networks, the simplest ones consisting in taking the dual of the polytopes. Some informations about these structures can be found in Ref.47. They also display long range orientational order and would provide very reasonable models for an hypothetical tetravalent quasicrystals. Note that we have already described in the past a 3-Dimensional tetravalent Penrose like structure[48]. More recently an interesting such structure was obtained[49] through a suitable decoration of the 3D Penrose rhombohedra.

- The tranformation can be implemented directly in euclidean space by considering any tetrahedral close packed system and its associated Voronoi partition. If one apply the above transformation to the Frank-Kasper A15 structure, the first iteration gives a new crystalline structure with 162 atoms in the unit cell. This is exactly the same number as in the case of $Mg_{32}(Al, Zn)_{49}$. In fact the two structures are almost identical (even though different) from the point of view of a shell by shell analysis of the local environment. In fact a recent closer analysis of the experimental data and of the defect symmetries[46] seems to contradict the old structural assignement and favours a description in term of the A15 structure with one iteration.

- It is possible to choose among different transformations at each step [25][41]. The amount of disorder can thus be monitered at a higher level according to the periodic, quasiperiodic or random character of the sequence of iterations. Even the global character of an iteration step can be broken ,leading to more disorder[41] as described below.

-Non-uniform decoration and spatial disorder[41]. Up to now, we have supposed that, at a given stage of the decurving, all tetrahedral cells of the structure are decorated in the same fashion. But different decoration procedures at different places in the structure, if they were compatible, would constitute a very efficient way to generate disorder. It is clearly impossible to divide a tetrahedron by using a decoration defined by λ_1 and another nearby tetrahedron with a decoration defined by λ_2 if both tetrahedra share a face. Distances between neighbours in the two tetrahedra are different, and an ambiguity appears on common edges : whether to divide the edge into λ_1 or λ_2 equal segments. But it is possible to commute the order of two decurving operations at different points in space. The structure can be patched up geometrically, at the expense of defects that are still disclinations, but of a different kind from those resulting from the original decurving algorithm. These defects are more or less extended in space, and the energy barriers between configurations are more or less important, depending upon whether the commutation occurs earlier or later in the succession of decurvings. Earliest decurvings are well frozen-in, whereas the last ones can be explored by tunnelling, which only involves flipping (breaking and reconstructing) bonds. Disorder, and ergodicity breaking, occur at all length scales.

IV-6 Quasicrystals, Polytopes and the Cut and Projection Method

We have already mentionned that the hierarchical polytopes, initially generated to flatten the $\{3,3,5\}$ and get suitable amorphous models, present too much order probably better apply to quasicrystals. The general question regarding which role could the polytopes play in the description of quasicrystals is still presently under study. At least two partial answers have already been proposed, which we propose to summarize now. In both case we shall make use of the Cut and Project (C.P.) approach [10].

IV-6.1 Propagation of the $\{3,3,5\}$ polytopes in 4 Dimensions[45]

The C.P. algorithm provides quasiperiodic arrangements with given types of volumes which are low dimensional (usually 2D or 3D) faces of some periodic lattice in higher dimensional space. The choice of this space and the relative position of the lower dimensional "physical" space is usually imposed by point group symmetry considerations. This has proved extremelly usefull in providing simple indexing schemes for diffraction patterns[50] and in generating polyhedral (mainly rhombohedral) space packings. More recently the question of atomic location has been addressed[51][52][53] in the framework of the so-called atomic surfaces. Nevertheless the standard method is more a "global" geometrical method rather than a "local one", even though the "acceptance region" concept[10] allows precise insights in this problem. The polytope model , on the other hand, can mainly be regarded as local approach. The local atomic order is known or guessed and barriers to its propagation is passed around upon curving the space. Moreover the polytopes, considered as 3-D manifolds, are more easily handled(they can be locally mapped onto 3-D tangent euclidean space) than for example the infinite atomic surfaces (one associated with each atomic position) of the kind discussed for example in Ref.54. It is therefore of interest to ask whether these two approaches can be combined. The answer is yes and the answer is the following: instead of using tiling of the hyperspace with hypercubes, we try to pack hyperunits which have from the start a local order similar to the guessed atomic arrangement. In the case of metallic structures, the simplest thing to do would be to pack the $\{3,3,5\}$ in 4-D. No such regular honeycomb can be found, but it is however possible to realize such an infinite tiling if one allows for deformations, as discussed in Ref.45. Let us briefly recall this approach.

The basic idea is to use the particular relationship between the $\{3,3,5\}$ and a simpler (self-dual) polytope, the $\{3,4,3\}$, which contains 24 vertices and (octahedral) cells. These two polytopes share a common symmetry subgroup. It is possible to suitably decorate each $\{3,4,3\}$ edge by one site (dividing the edge in the golden mean ratio τ). One gets 96 vertices which, together with the 24 vertices of the dual $\{3,4,3\}$ adds up to 120 and realize a $\{3,3,5\}$ polytope. This is called the Gosset construction for the $\{3,3,5\}$. Vertex coordinates are given by Coxeter[11]:

- even permutations of $\left(\pm\tau, \pm1, \pm\tau^{-1}, 0\right)$ for the 96 vertices,

- permutations of $(\pm 2, 0, 0, 0)$ and $(\pm 1, \pm 1, \pm 1, \pm 1)$ for the remaining 24 vertices.

Note that the above $\{3, 3, 5\}$ is inscribed in a hypersphere S_3 of radius 2. Now $\{3, 4, 3\}$ polytopes can be regularly packed in a 4-Dimensional space and form the so-called $\{3, 4, 3, 3\}$ honeycomb. A theorem states that, due to the evenness of the second number, it is possible to devise a decoration of the honeycomb edges such as to divide them in the ratio 1 and τ and to obtain the same 96-vertex polytope (denoted by $s - \{3, 4, 3\}$. This gives a honeycomb called the $s - \{3, 4, 3, 3\}$. In each cell the 24 vertices of the dual $\{3, 4, 3\}$ can be added and a packing of $\{3, 3, 5\}$ is obtained (with additional $\{3, 3, 3\}$ and $\{3, 3, 4\}$ at the vertices or along the edges). The distortion comes, for each cell, from these 24 additional vertices. Indeed, these points are not added on the circumsphere of the $\{3, 4, 3\}$, but on the "euclidean faces" of the honeycomb units. Work is in progress to see whether such a decorated honeycomb can be of some use in the context of quasicrystal. In particular it is hoped that some among the Frank-Kasper phases could be derived from this 4-Dimensional structure instead of going to 6 Dimensions.

IV-6.2 Hierarchical polytopes and the E8 lattice[55]

A relation between the C.P. method and the polytope model can be found from the interesting work of Elser and Conway[56], who build a 4-D quasicrystal with the $\{3, 3, 5\}$ point group symmetry. They use as starting structure the $E8$ lattice ,which leads to the densest packing in 8 Dimensional space [57]. They generate then 3-D quasicrystals with icosahedral symmetry by suitable 3-D cuts in this 4-D quasicrystal. For the hierarchical polytope purpose we shall stay in 4-D space and modify the algorithm for the selection of points. Let us first briefly describe the geometry of $E8$.

$E8$ is a non-primitive cubic lattice whose points are

$$(11) \qquad \mathbf{R} = \sum_{i=1}^{8} v_i \mathbf{e}_i$$

where $\sum v_i$ is even and either all $v_i \in \mathbf{Z}$ or all $v_i \in \mathbf{Z} + 1/2$. $E8$ belong to the so-called "laminated" lattices and present some similarities with the 3-D fcc lattice. The latter has two types of interstices, tetrahedra and octahedra. Similarily, $E8$ has two types of 8-D analogous interstices, the 8-D simplices and cross polytopes; all 2D-faces are either triangles or squares. The first coordination shell is a 8-D semi-regular polytope with 240 vertices, the Gosset polytope denoted by 4_{21} [11]. The radius and edge-length of this polytope are equal.

-<u>Projection to 4-D</u>. The $\{3, 3, 5\}$ symmetry group is a sub-group of the $E8$ full group. Elser and Conway gives a 4-D subspace invariant under the action of the former group. By projection of the 4_{21} on this 4-D space, two concentric $\{3, 3, 5\}$ polytopes are obtained. The full $E8$ lattice gives by projection a dense set of points. Any such point is given by an integral combination of vectors taken among the 240 vectors which are the projection of the 8-D vectors joining the origin to the 4_{21} vertices. The 4-D quasicrystal is then generated with the standard C.P. algorithm by keeping only those points (within the dense set) which fall , by projection onto an orthogonal 4-D space, inside the projection of the $E8$ lattice Voronoi cell.

-<u>Concentric hyperspheres in $E8$ and the hierarchical polytopes.</u> All concentric coordination hyperspheres in $E8$ have the Gosset polytope symmetry. The second and third such shells contain respectively 2160 and 6720 sites. By projection onto the 4-D "physical" space, these shells give several polytopes with the $\{3,3,5\}$ symmetry (either $\{3,3,5\}$, $\{5,3,3\}$ or decorated versions) but with different radius. Now the selection of points is done as follow: we only map a finite number of shells from $E8$ and, moreover we only keep, after projection, points which fall inside a 4-D shell of given width.

For example if we define a spherical shell (in 4-D) of radius between 1.2 and 1.4 , one selects a $5,3,3$ of radius 1.202, a polytope with two points per $\{3,3,5\}$ edge of radius 1.312 and a $\{3,3,5\}$ of radius 1.376. If all these points are radially mapped onto the same hypersphere, one gets a polytope which is topologically equivallent to the above first iterated polytope P_1 with 2160 vertices (some slight metrical differences). It is guessed that any member of the above hierarchical polytope series (at least for $\lambda = 3$) could be generated that way.

IV-7 Coding the complexity in the disclination geometry

We have seen that the topology of tetrahedrally close-packed structures can be encoded in their intrinsic disclination network. In doing that the complexity of these structures is lowered. Indeed the exact location of the individual atoms is no more needed and can be derived (up to small displacements) from the knowledge of disclination lines loci. A general classification can then be proposed according to the defect geometry :

- defect-free structure. The structure is the ideal template defined in curved space.

- Periodic array of disclination lines. This is the case of Frank-kasper phases (note that it is a general name to which one should associates others like Laves, Friehauf ...).

- Hierarchical disclination network. This domain covers the field of quasicrystals as well as some other orientationally ordered structures (like the hypothetical "icosahedratic" phase [58]).

- Disordered disclination network. This is the case of amorphous materials.

It is possible that the gap between these three types of structures be filled. Indeed we have mentioned the possibility for complex crystalline structures with large unit cells to present already a limited hierarchical disclination geometry, and maybe described as "rational" approximants of an ideal quasicrystal. The domain between quasicrystals and amorphous structures may also present a rich variety of more and more disordered structures .

V MISCELLANEOUS PROPERTIES

V-1 Coordination number, disclination density and Regge calculus.

In 2-Dimensional space, thank to Euler-Poincare and Gauss- Bonnet relations, any tiling is subject to topological and metrical constraints [59][60]. No such exact result can be derived in 3D, but it is nevertheless possible to do some good approximations.

- Coordination number in sphere packings

The coordination number of a site is equal to the number of faces of its associated Voronoi polyhedron. Since the packing units are supposed to be tetrahedra, the Voronoi froth vertex is tetravalent. Each such vertex belong to 4 Voronoi polyhedra and has 3 neighbours on each. Any ring is common to two Voronoi polyhedra associated with two neighbouring sites. The ring size is equal to the number of tetrahedra sharing the two sites. If the tetrahedra are regular, there is room for about 5.1 tetrahedra around a common edge. If we use this value for the mean ring size $\bar{p}$, it is possible to get the approximate number of faces of the Voronoi polyhedron. Indeed the following relation holds [60]:

$$(12) \qquad \sum_p (6 - p)F_p = 12.$$

Introducing $\bar{p} = (\sum pF_p)/F$ and assuming that the Voronoi cells are all equivalent one gets :

$$(13) \qquad \bar{F} = 12/(6 - \bar{p}).$$

With the above $\bar{p} \simeq 5.1$, this gives $\bar{F} \simeq 13.4$. This approximation corresponds to Coxeter "statistical honeycomb" $\{3, 3, \bar{p}\}$ [61].

It is interesting to look to the values that F takes in numerous examples. We reproduce in figure 27 the compilation done by Nelson [32]. It is striking that the ideal 13.4 value fall in the same small range than the Frank-Kasper phases. In fact one expects any structure related to the $\{3, 3, 5\}$ by a disclination decurving procedure to have an average coordination number very close to this value. As an example we have seen that in the above considered hierarchical polytopes the coordination tends to 13.33 in the flat limit. It is tempting to relate deviation from the ideal value to geometrical and physical characters. Such arguments have been tentatively given in term of the Voronoi cell asymmetry [62] or the geometry of the disclination network [63].

- Disclination density and Regge calculus [59][63].

We are now trying to approximate the disclination density in the 3 Dimensional case. Recall first that in 2D, the problem is exactly solved[60]. For a given coordination number, some polygons do not "carry" curvature and are called "neutral charges" (eg, hexagons in 3-fold coordinated 2-D tilings). As a result, on a

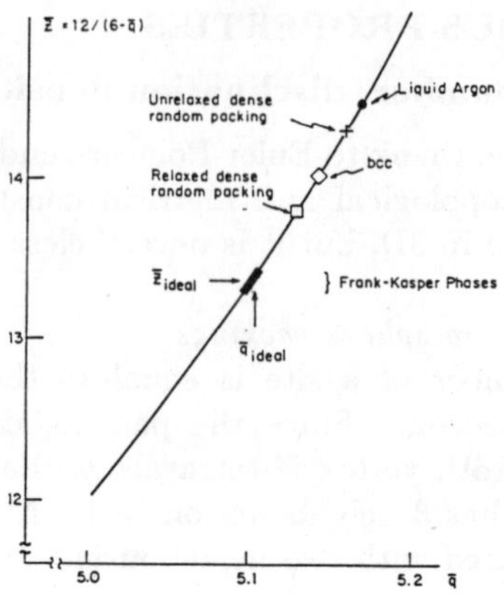

fig. 27

Coordination number $\bar{Z}$ with $\bar{p}$ for different kinds of dense atomic configurations.

unit radius spherical surface, summing the contribution of disclinations (identified to polygons different from the "neutral charges") gives the area of the sphere. This can also be stated in term of deficit angle as in the approach proposed by Regge [64][39], in order to discretize the curved manifolds of gravitation theory.

Let us now consider the hypersphere S_3. Its 3D volume (and not the hypervolume of its interior) is $2\pi^2 R^3$. As the usual sphere S_2 can be approximated by polyhedra, S_3 can be approximated by polytopes. Following Regge , the curvature is then concentrated on the edges (while it was concentrated at the vertices in 2D).

To each edge it is possible to associate a cell, called a E-cell, which contains all points closer to this edge rather than to any other edge (figure 28). This is a natural extension to the Voronoi cells associated to sites. In the case of a regular polytope $\{p, q, r\}$ the E-cells are regular r-gonal bipyramids. In less regular structures one can always construct the distorted bipyramid with, as base, the Voronoi cell face threaded by the considered edge. In the spherical space S_3, the base is part of a geodesic spherical surface and its area can be calculated using Gauss-Bonnet theorem. This area can be expressed in term of δ, the deficit angle along the edge:

$$(14) \qquad \delta = 2\pi - r\theta$$

θ is the dihedral angle of the cell $\{p, q\}$ of the polytope $\{p, q, r\}$. As a consequence, the volume V_E of the E-cell is given with a good approximation by :

$$(15) \qquad V_E = l\delta R^2/3$$

where l is the edge length and R is the sphere radius. The total volume V of S_3 is then :

$$(16) \qquad V = 2\pi^2 R^3 = \sum_{edges} l\delta R^2/3.$$

This relation is similar to :

$$(17) \qquad \sum_{edges} l\delta = \frac{1}{2} \int \mathcal{R}g^{1/2}d^3x$$

which appears in Regge calculus [39]. $\mathcal{R}$ is the scalar curvature and g the metric tensor determinant. The Regge calculus is a tetrahedral discretization procedure for integrals on curved manifolds, which becomes more and more accurate with increasing number of mesh points. The space enclosed by a tetrahedron is regarded as flat and the curvature is concentrated on the edges. We can test the validity of this approximation by calculating the volume of S_3 , discretized in the regular polytopes $\{p, q, r\}$. Note that here the polyhedral discretization is not necessarily tetrahedral. The table below gives the results where s is the ratio of calculated volume over the exact unit radius S_3 volume. We take l as the geodesic distance instead of the euclidean. As expected S tends to 1 when the number N of edges increases.

p, q, r	δ	l	N	V	s
2, 2, 2	6.2831	3.1416	2	13.1594	0.667
3, 3, 3	2.5903	1.8234	10	15.7445	0.798
4, 3, 3	1.5707	1.0472	32	17.5459	0.889
3, 3, 4	1.3593	1.5707	24	17.082	0.865
3, 4, 3	0.5513	1.0472	96	18.4737	0.936
5, 3, 3	0.1798	0.2709	1 200	19.4902	0.987
3, 3, 5	0.1283	0.6283	720	19.3605	0.981

Table 5
Evaluation of the S_3 volume, discretized by regular polytopes

In the case of negative curvature, the E-cell volume can be approximated by :

$$(18) \qquad V_E = |l|\delta R^2/3$$

as long as it is not too large. In this case the calculated volume is larger than the exact one.

So we have shown how a curved space can be faithfully represented by an euclidean network (Regge skeleton). In terms of disclinations, in a 2D space, there is an exact relation :

$$(19) \qquad \sum_i n_i \delta_i = < K > = < R^{-2} >$$

where $< K >$ is the average Gaussian curvature per unit area and n_i the number of disclination points with angular deficit δ_i, per unit area. In the 3D curved space, there is an approximate relation

$$(20) \qquad \sum_i L_i \delta_i = 3 < R^{-2} >$$

since the scalar curvature $\mathcal{R}$ equals $6/R^2$. L_i is the disclination length (with angular deficit δ_i) per unit volume. It is now possible to evaluate the disclination density in euclidean close-packed structure. To such structure it corresponds an underlying corrugated geometry with regions of positive and negative curvature. In the discretized Regge-like approach the curvature is concentrated on the edges, the curvature sign depending on the number of tetrahedra sharing the given edge. The requirement of vanishing mean curvature reads on the average :

$$(21) \qquad \sum_{i,i\prime} L_i^+ \delta_i^+ + L_{i\prime}^- \delta_{i\prime}^- \simeq 0$$

where L_i^+ and $L_{i\prime}^-$ are the lengths per unit volume of positive ($\delta_i^+ > 0$) and negative ($\delta_{i\prime}^- < 0$) disclinations. The LHS is exactly zero if disclinations are infinitesimal, e-g if curvature is spread in the volume. In the present case the compensation of opposite curvature is approximate. It is not known wether it is always possible to satisfy this relation by allowing for extra dimensions for the coordinates. Let us proceed to a simple calculation in the case of the above described hierarchical polytope (in the case $\lambda = 3$) where we assume that all first-neighbour distances are equals. If $\theta = 1.2306 rd.$ is the tetrahedron dihedral angle, then, in polytope P_i, each disclination edge carry a negative weight $\delta^- = 2\pi - 6\theta = -1.102 rd.$, while the other polytope edges carry a positive weight $\delta^+ = 2\pi - 5\theta = 0.1283 rd.$. Let us introduce the ratio $\Delta = L^+/L^-$. If the above relation was exact, then $\Delta = \delta^+/|\delta^-| = 8.589$. In the iterative polytopes the Δ value is given by

$$(22) \qquad \Delta^{(i)} = \frac{6(N_{12}^{(i)} + N_{14}^{(i)} + N_{16}^{(i)})}{N_{14}^{(i)} + 2N_{16}^{(i)}}.$$

From the Perron root of the associate matrix one easily gets $\delta^{(\infty)} = 9$.

V-2 the structure factor of polytopes

V-2.1 Structure Factor in curved space

The structure factor in Euclidean space is defined by :

$$(23) \qquad S(K) = \sum_0^\infty 4\pi r^2 \rho(r) \frac{sin(K.r)}{K.r} dr$$

It is an analysis in terms of spherical wave amplitude $sin(K.r)/(K.r)$. In a space with positive curvature concentric spherical waves have an amplitude $sin(l\Psi)/(l.sin\Psi)$. As the spherical space is finite, there are only a discrete set of concentric waves : l is an integer. In order to have the same notation as in ref[65] we call this number $n + 1$. This can also be justified by comparison of the value of $sin(K.r)/K.r$ for $K \to 0$ and the value of $sin((n+1)\Psi)/((n+1)sin\Psi)$ for $n = 0$ which are both equal to unity. In order to introduce a structure factor for the polytope we make the following association of variables in spherical space and in Euclidean space:

$$(24) \qquad \begin{aligned} 4\pi sin^2\Psi &\sim 4\pi r^2 \qquad \text{(we suppose} R = 1) \\ N(\Psi) &\sim G(r) \\ n+1 &\sim K \qquad \text{(the modulus of the reciprocal space vector).} \end{aligned}$$

Note that Nelson and Widom [65] identify K to be $[n(n+2)]^{1/2}$.

We can define the atomic density

$$(25) \qquad \begin{aligned} \rho(r) &= \frac{G(r)}{4\pi r^2} \qquad \text{in euclidean space} \\ &\text{and} \\ \eta(\Psi) &= \frac{N(\Psi)}{4\pi sin^2\Psi} \qquad \text{in spherical space.} \end{aligned}$$

Now the structure factor in spherical space is defined using these analogies. The expression

$$(26) \qquad S_n = \sum_0^\pi 4\pi sin^2\Psi . \eta(\Psi) \frac{sin[(n+1)\Psi]}{(n+1)sin\Psi} d\Psi$$

is comparable to the above definition ().If we introduce a mean density η_0 it follows

$$(27) \qquad S_n = \sum_0^\pi 4\pi sin^2\Psi . [\eta(\Psi) - \eta_0] \frac{sin[(n+1)\Psi]}{(n+1)sin\Psi} d\Psi \quad + t$$

where $t = 0$ for all $n \neq 0$, and $t = N_0$ for $n = 0$ (N_0 is the total number of vertices on the polytope). Therefore S_n is related to the Fourier component of the periodic function

$$(28) \qquad sin\,\Psi \left(\frac{\eta(\Psi)}{\eta_0} - 1 \right).$$

The function $\eta(\Psi)$ is a set of delta functions and consequently :

$$(29) \qquad S_n = \frac{1}{n+1} \sum_i N_i \frac{sin[(n+1)\Psi]}{sin\Psi_i}$$

where N_i and Ψ_i are in given in table 6 for the $\{3,3,5\}$ polytope.

Ψ_i =	0	$\pi/5$	$\pi/3$	$2\pi/5$	$\pi/2$	$3\pi/5$	$2\pi/3$	$4\pi/5$	π
N_i =	1	12	20	12	30	12	20	12	1

Table 6

Values of Ψ_i and N_i in polytope $\{3,3,5\}$.

V-2.2 Structure Factor of the $\{3,3,5\}$ and hierarchical polytopes

Due to the symmetrical repartition of shells relative to the great sphere ($\Psi = \pi/2$) only S_n with even n are different from zero. For $n < 60$, $S_n \neq 0$ for $n = 0, 12, 20, 24, 30, 32, 36, 40, 42, 44, 48, 50, 52, 54, 56$. For $n > 60$, $S_n = 120 \times Int(n/60) + S_{nmod60}$ showing that $(n+1)S_n$ is the sum of a periodic function defined by $Sn = 120$ for $n = 0, 12, 20, ...56$ and a staircase function $120 \times Int(n/60)$. If we suppress the scattering due to the point at origin, as is usually done in conventional non-crystalline diffraction studies a reduced structure factor $s_n = (n+1)(S_n - 1)$ can be defined. It is associated to the usual function $i(K) = K.(I(K) - 1$. The s_n function (figure 28) oscillates around zero and can be compared to the amorphous metal structure factor.

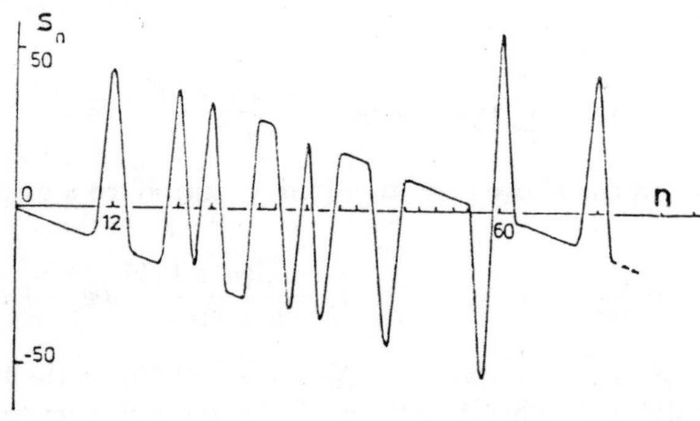

fig. 28

Reduced structure factor s_n (slightly broadened) of polytope $\{3,3,5\}$.

Consider now a hierarchical structure obtained by an iterative process which consists to replace an atom by a set of atoms with a defined local configuration. The scaling factor λ corresponds to the ratio between the atomic diameter before the iteration, and the atomic diameter of atoms of the inserted motive. The set of atoms have an interference function $I(K)$ which is, in this case, similar to a structure factor.

Suppose that the first iteration starts with the set of atoms defining the local order, and that the second iteration replace each atoms by a similar set (with a scaling factor λ) then the interference function is

$$(30) \qquad J(K) = I(K) \times I(\lambda K).$$

If the iteration is repeated n times

$$(31) \qquad J(K) = I(K) \times I(\lambda K) \times I(\lambda^2 K) \times ... \times I(\lambda^n K).$$

This expression should give an approached structure factor for the hierarchical polytopes. However we have neglected orientation correlations between two neighbouring motives. It is this effect which leads to interference spots in quasicrystal and in hierarchical structures. But in an amorphous structure it is reasonable to suppose that disorder screens this effect. Only a few terms contribute to $J(K)$ because for large $\lambda^n K$, $I(\lambda^n K)$ is close to 1. Figure 29 shows an interference function re-built in this way from the polytope $\{3,3,5\}$, with one iteration and $\lambda = 3$ in the above expression.

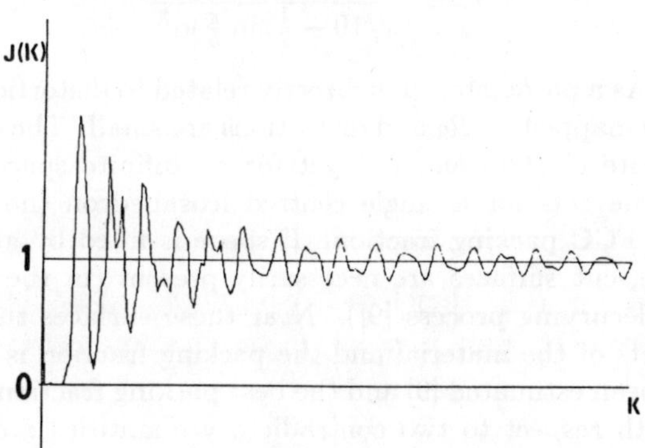

fig. 29

Interference function obtained from polytope $\{3,3,5\}$ with one iteration and $\lambda = 3$ as explained in the text.

Disorder in amorphous structures ,modeled from hierarchical structures, appears if there are different scaling factor λ from one iteration to the other or from one spatial fluctuation to another.

V-3 Specific volume variation in disordered solids : a simple model

Using the above considerations it is possible to describe in a simplified way some general trends in the density of amorphous solids:

V-3.1 The close-packed sphere filling factor[9]

The packing fraction of polytope $\{3,3,5\}$ has been previously calculated (Chapter II-3) and found to be $p_0 = 0.774$. We now try to approximate the packing fraction after mapping onto R_3. We first consider a part of S_3 (with radius R) limited by a sphere S2 of radius R_d. The enclosed (3-D) volume is

$$(32) \qquad v_d = 2\pi R^3 \left(\alpha - \frac{1}{2}\sin 2\alpha\right)$$

where $R_d = R\sin\alpha$. Upon a "geodesic" mapping S_2 is transformed into an euclidean sphere of radius $R'_d = R\alpha$ and volume

$$(33) \qquad v'_d = \frac{4}{3}\pi R'^3_d$$

So the volume expansion factor is v'_d/v_d. For the same reason there is an expansion of the sphere centered on a vertex. Upon mapping the packing fraction is changed by the expansion factor :

$$(34) \qquad C = \frac{(\alpha - \frac{1}{2}\sin 2\alpha(\frac{\pi}{10})^3}{(\pi/10 - \frac{1}{2}\sin\frac{\pi}{5})\alpha^3}$$

and $p = p_0 c$. As a parameter, α is directly related to distortions: if α is small, a small part of S_3 is mapped on R_3 and distortions are small. The calculated packing fraction is for finite clusters and not yet for an infinite space filling structure. As an example one gets for a single centred icosahedron the value $p = 0.737$, very close to the FCC packing fraction. If space is filled by gluing replicas of a mapped polytope, cut surfaces are necessarily present (in the framework of the "star-mapping" decurving process [9]). Near these surfaces the density is lower than in other parts of the material and the packing fraction is reduced. The cut surface area has been estimated [9] and the best packing fraction can be calculated by optimizing with respect to two contradictory quantities : α and the average surface area. The obtained packing fraction is $p \simeq 0.62$, very close to values measured in real space ball packing or computer simulations.

V-3.2 Specific volume variation with temperature in simple metals [66] Thermal variation of the specific volume of elements or simple compounds generally exhibit a small increase with T in the solid state, a discontinuity at the melting point and an important increase for the liquid specific volume at higher temperatures. Nevertheless if the compound is in an amorphous state the specific volume variation curve is continuous and can be divided in two regions : at low T the specific volume is a few percent higher than in the crystalline state, the two curves being approximately parallel. At higher T the glass specific volume curve is in continuity with the undercooled liquid one. Note that in many cases (as liquid copper[67] for example), the extrapolated specific volume at $0°K$ takes a value smaller than in the crystalline structure (figure 30).

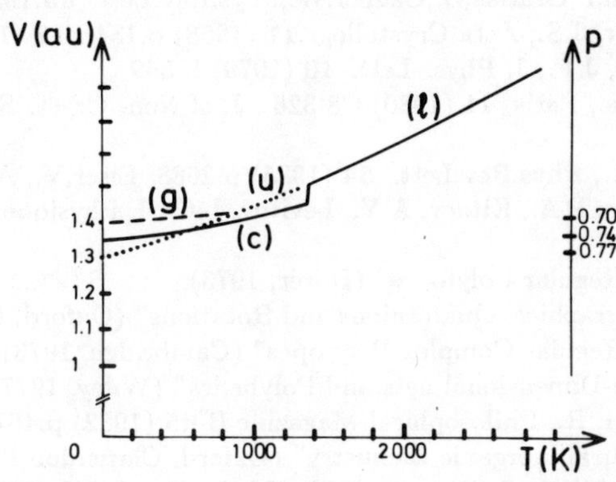

fig. 30

Variation of the copper specific volume from experimental data: (l) liquid, (c) crystal. Extrapolation to low temperature for the undercooled liquid (u). Estimated behaviour for glassy copper (g). The volume scale is that of a Copper spherical atom. p is the packing fraction.

A qualitative explanation can be given using topological arguments . The main hypothesis is that extrapolation of the liquid structure to very low T (the "undercooled" (u) curve) leads to the ideal structure defined in curved space (it is indeed found that the extrapolated value is very close to the polytope one). In the curved space we suppose that the number of defects increases with temperature, and this is the source of the volume variation. At a given temperature T_0, the defect density is large enough to achieve a complete flattening in the euclidean space. Over T_0 this hypothetical structure becomes physically observable and the usual liquid behaviours is recovered. The glass specific volume at $T = 0°K$ is that of the polytope plus a contribution due to the intrinsic defects (those necessary to decurve the polytope). This is why the glass curve should increase slowly with T (a solid state behaviour) and reach the liquid curve at T_0 .

References

[1] Ichikawa, T., Phys. Stat. Sol. (a) **29** (1975) p.293 .

[2] Sadoc, J.F., Dixmier, J. and Guinier, A., J. of Non Cryst. Sol. **12** (1973) p.46.

[3] Finney,J.L., Mat. Sci. Eng. **23** (1975) p.199 .

[4] Bennett, C.H., J. Appl. Phys. **43** (1972) p.2727 .

[5] Boudreaux, H.J., Phys. Rev. B **s3** (1981) p.1506 .

[6] Shechtman,D.,Blech,I.,Gratias,D.,Cahn,J.W.,Phys.Rev.Lett. **53**,1951(1984).

[7] Frank, F.C., Kasper, J.S., Acta Crystallogr.11 (1958) p.184 ; **12** (1959) p.483.

[8] Kleman, M., Sadoc, J.F., J. Phys. Lett. **40** (1979) L-569.

[9] Sadoc, J.F., J. Phys., Paris, **41** (1980) C8-326 ; J. of Non- Cryst. Solids, **44** (1981) p.1.

[10] Duneau,M., Katz,A., Phys.Rev.Lett. **54** (1985) p.2688; Elser,V., Acta Cryst.,**A42** (1988) p.36; Kalugin, P.A., Kitaev, A.Y., Levitov, L.C., J. Physique Lett. **46** (1985) L601.

[11] Coxeter, H.S.M., "Regular Polytopes" (Dover, 1973).

[12] Du Val, P., "Homographies, Quaternions and Rotations" (Oxford, 1964).

[13] Coxeter, H.S.M., "Regular Complex Polytopes" (Cambridge, 1973).

[14] Wells, A.F., "Three-Dimensional nets and Polyhedra" (Wiley, 1977).

[15] Sadoc, J.F., Mosseri, R., Philosophical Magazine B **45** (1982) p.467.

[16] Wells, A.F.,"Structural inorganic chemistry" (Oxford, Clarendon Press, 1962).

[17] Davidson, D.W., in "Water" (Plenum, N.Y., Ed. by Franks, F., 1973) p. 115.

[18] Connell, G.A.N., Temkin, R.J., Physical Review B **9** (1974) p.5323.

[19] Dixmier, J., Gheorgiu, A., Theye, M.L., J. Phys. C **17** (1984) p.2271.

[20] Polk, D.E., J. of Non-Cryst. Solids **5** (1971) 365.

[21] Sachdev, S., Nelson, D.R., private communication.

[22] Di Vincenzo, D.P, Physical Review B **37** (1988) p.1245.

[23] Mosseri, R.,Sadoc, J.F., Z. Phys. D **12** (1989) p.89-92.

[24] Romerio , (1987) preprint

[25] Sadoc, J.F., Mosseri, R., Journal de Physique **46** (1985) p.1809.

[26] Tilton, L., J. Res. Nat. Bur. Standard **59** (1957) 139.

[27] Coleman, M., Thomas, D., Phys. Stat. Sol. **24** (1957) K 111.

[28] Dandoloff, R., Dohler, G., Bilz, H., J. of Non-Cryst. Solids **35,36** (1980) 537.

[29] Gaskell, P., Philosophical Magazine B**32** (1975) p.211.

[30] Halperin,B., Nelson, D.R., Physical Review Lett. **41** (1978) p121. .

[31] Kosterlitz,J.M., Thouless, D., J. Phys. **C6** (1973) p.1181.

[32] Nelson, D.R., Physical review B **28** (1983) p.5515.

[33] Nicolis, S., Mosseri, R., Sadoc, J.F., Europhysics Letters **1** (1986) p.571.

[34] Mosseri, R., Di Vincenzo, D.P., Brodsky, M.H., Sadoc, J.F., Physical Review B **32** (1985) p.3974.

[35] Rivier, N., Philosophical Magazine B **40** (1979) 859.

[36] Mosseri, R., Gaspard , J.P., J. of Non-Cryst. Solids **97-98** (1987) p.415-418.

[37] Sadoc, J.F., J. Physique Lett. **44** (1983) L-707.

[38] Nelson, D.R., Phys. Rev. Lett. **50** (1983) p.982.

[39] Misner, C.W., Thorne, K.S., Wheeler, J.A., "Gravitation" (Freeman, San Francisco 1973).

[40] Mosseri, R., Sadoc, J.F., J. Physique Lett. **45** (1984) L-827 ; Sadoc, J.F., Mosseri, R., J. Physique **46** (1985) p.1809.

[41] Sadoc, J.F., Rivier, N., Philosophical Magazine B **55** (1987) p.537.

[42] Nelson, D.R., Sachdev, S., Proc. workshop on Amorphous Metals and Semiconductors, May 1985 (ed. Haasen, P. and Jaffe, R.) Acta Met.

[43] Mosseri, R., Sadoc, J.F., J. Non-Cryst. Solids **75** (1985) p. 115.

[44] Rivier, N., Sadoc, J.F., Europhysics Letters **7** (1988) p.526.

[45] Mosseri, R., Sadoc, J.F., Journal de Physique Coll. C3 **47** (1986) p.281

[46] Donnadieu, P., private communication.

[47] Mosseri, R., Sadoc, J.F., J. Non-Cryst. Solids **77-78** (1985) p. 179-190.

[48] Mosseri, R., Sadoc, J.F., in "Structure of Non Crystalline Materials 1982" (ed. Gaskell, P. H., Parker, J.H., Davis E.A. ; Taylor and Francis Ltd 1983) p. 137.

[49] Peters,J., Trebin, H.R., in "Quasicrystals", ICTP Trieste, july 1989 (Eds. Jaric,M., Lundqvist, S.) p.161.

[50] Cahn, J.W., Schechtman, D., Gratias, D., J.mat. Res. **1** (1986) p. 13-26.

[51] Cahn, J.W., Gratias, D., Mozer, B., J.de Phys.(France) **49** (1988) p.1225 .

[52] Janot, C., Dubois, J.M., Pannetier, J., de Boissieu, M., Fruchart, R., in "Quasicrystalline materials" ,ILL-CODEST Workshop (ed. Janot,C. and Dubois,J.M., World Scientific 1988) page 107.

[53] Duneau, M., Oguey, C., J. de Phys.(France) **50** (1988) p.135.

[54] Levitov S., J. Phys. (Paris) **50** (1989) p. 3181-3189 .

[55] Sadoc, J.F., Mosseri, R., in "Quasicrystalline materials" ,ILL-CODEST Workshop (ed. Janot,C. and Dubois,J.M., World Scientific 1988) page 215-223.

[56] Elser, V., Conway, J.H., J. Phys. A ,Math. Gen., **20** (1987) p.6161 .

[57] Conway, J.H., Sloane, N.J., "Sphere packings, lattices and groups (N.Y., Springer, 1988).

[58] Steinhardt, P.J. Nelson, D.R., Ronchetti, M., Phys. Rev. **B** (1983) p.784.

[59] Gaspard, J.P., Mosseri, R., Sadoc, J.F., Philosophical Magazine B **50** (1984) p.557.

[60] Mosseri, R., "Geometry of disordered systems", in "From crystal to amorphous" (ed. C. Godrèche, Ed. de Physique, 1988) p.1-81.

[61] Coxeter, H.S.M., Illinois J. Math. **2** (1958) 746 ; and in "Introduction to geometry" (Wiley, N.Y.. 1969) p. 411.

[62] Rivier, N., J. de Physique Colloque C9 **43** (1982) p.91.

[63] Sadoc, J.F., Mosseri, R., J. de Physique **45** (1984) p.1025.

[64] Regge, T., Il Nuovo Cimento **29** (1961) p.558.

[65] Nelson, D.R., Widom, M., Nucl. Phys. B **240** (1984) p.113.

[66] Sadoc, J.F., Mosseri, R., J. de Physique C9 **43** (1982) p.97-100.

[67] Borneman, K., Saverwald, Z. Metall. **14** (1922) 145. El Mehairy,A., Ward, R.G., Trans. Met. Soc. AIME **227** (1963) p.1228.